Food, Wine and China

The growth of the Chinese economy and the emergence of the Chinese middle class have fuelled the rapid expansion of China's outbound tourism market, with many destinations around the world trying to capitalise on the opportunities created by the growing number of Chinese visitors. This book specifically focuses on the demand for food and wine tourism experiences by Chinese tourists, which in recent years has become an important constituent of destination competitiveness.

Looking at the different ways in which individual destinations have responded to this increasing demand, this book provides a better understanding of the preferences, motivations and perceptions that underlie food and wine consumption by Chinese tourists. It also illustrates how food and wine tourism experiences have been used in a range of international destinations to specifically attract visitors from China. Including a range of case examples from the Asia-Pacific region and Europe, this book ultimately investigates the strategic directions adopted to guide destination development and marketing initiatives. Such a perspective provides a novel contribution to the still limited body of knowledge on China outbound tourism and will be of interest to upper level students, researchers and academics in Tourism and Hospitality.

Christof Pforr is Discipline Leader (Tourism, Hospitality and Events), School of Marketing, Curtin University, Western Australia.

Ian Phau is the Head of School of the School of Marketing at Curtin University, Western Australia.

Routledge Studies of Gastronomy, Food and Drink
Series Editor: C. Michael Hall, University of Canterbury, New Zealand

This groundbreaking series focuses on cutting edge research on key topics and contemporary issues in the area of gastronomy, food and drink to reflect the growing interest in these as academic disciplines as well as food movements as part of economic and social development. The books in the series are interdisciplinary and international in scope, considering not only culture and history but also contemporary issues facing the food industry, such as security of supply chains. By doing so the series will appeal to researchers, academics and practitioners in the fields of gastronomy and food studies, as well as related disciplines such as tourism, hospitality, leisure, hotel management, cultural studies, anthropology, geography and marketing.

Sustainable Culinary Systems
C. Michael Hall and Stefan Gössling

Wine and Identity
Edited by Matt Harvey, Warwick Frost and Leanne White

Social, Cultural and Economic Impacts of Wine in New Zealand
Peter Howland

The Consuming Geographies of Food
Hillary Shaw

Heritage Cuisines: Traditions, Identities and Tourism
Dallen J. Timothy

Food Tourism and Regional Development
Edited by C. Michael Hall and Stephan Gössling

Food, Wine and China: A Tourism Perspective
Edited by Christof Pforr and Ian Phau

For more information about this series, please visit https://www.routledge.com/Routledge-Studies-of-Gastronomy-Food-and-Drink/book-series/RSGFD

Food, Wine and China

A Tourism Perspective

**Edited by Christof Pforr
and Ian Phau**

Routledge
Taylor & Francis Group

LONDON AND NEW YORK

First published 2018 by Routledge

2 Park Square, Milton Park, Abingdon, Oxon OX14 4RN
605 Third Avenue, New York, NY 10017

Routledge is an imprint of the Taylor & Francis Group, an informa business

First issued in paperback 2022

Publisher's Note

The publisher has gone to great lengths to ensure the quality of this reprint but points out that some imperfections in the original copies may be apparent.

British Library Cataloguing-in-Publication Data
A catalogue record for this book is available from the British Library

Library of Congress Cataloging-in-Publication Data
A catalog record has been requested for this book

ISBN: 978-1-138-73225-4 (hbk)
ISBN: 978-1-03-233911-5 (pbk)
DOI: 10.4324/9781315188317

Typeset in Times New Roman
by Swales & Willis Ltd, Exeter, Devon, UK

Contents

Figures

Tables

Contributors

Wolfgang Georg Arlt, Director, China Outbound Tourism Research Institute, Germany

Richard C. Y. Chang, National Dong Hwa University, Taiwan

Isaac Cheah, Curtin University, Australia

Bin Dai, China Tourism Academy, China

Ross Dowling, Edith Cowan University, Australia

Stephen Fanning, Edith Cowan University, Australia

Graham Ferguson, Curtin University, Australia

Joanna Fountain, Lincoln University, New Zealand

Jeremy Galbreath, Curtin University, Australia

Grace Gao, Curtin University, Australia

Hailian Gao, University of South Australia, Australia

Louis Geneste, Curtin University, Australia

Kristina Georgiou, Curtin University, Australia

C. Michael Hall, University of Canterbury, New Zealand

Nazaruddin Haji Hamit, Curtin University, Malaysia

Songshan (Sam) Huang, Edith Cowan University, Australia

Yu-An Huang, National Chi Nan University, Taiwan

Niki Hynes, Curtin University, Australia

Xiujuan Jin, China Tourism Academy, China

David Lamb, Edith Cowan University, Australia

Sean Lee, Curtin University, Australia

Athena H. N. Mak, National Dong Hwa University, Taiwan

Alfred Ogle, Edith Cowan University, Australia

Harald Pechlaner, Eurac Research, Italy; Catholic University of Eichstätt-Ingolstadt, Germany & Curtin University, Australia

Christof Pforr, Curtin University, Australia

Ian Phau, Curtin University, Australia

Vanessa Ann Quintal, Curtin University, Australia

Piyush Sharma, Curtin University, Australia

Na Su, China Tourism Academy, China

Ben Thomas, Curtin University, Australia

Michael Volgger, Curtin University, Australia; Eurac Research, Italy & Catholic University of Eichstätt-Ingolstadt, Germany

Paull Weber, Curtin University, Australia

1 Food, wine and China

Opportunities and challenges for tourism

Christof Pforr and Ian Phau

Introduction

With more than 135 million international departures and US$261 billion in tourism expenditure in 2016, China continues to dominate international tourism (UNWTO 2017). This is remarkable as outbound tourism in China started only in the 1980s, but has grown rapidly since, in particular over the past two decades. During this period the China outbound travel market has changed and matured, evident in a shift from packaged travel arrangements towards more experienced, independent travellers. Wolfgang Georg Arlt (China Outbound Tourism Research Institute) refers to these developments as first, second and third wave of Chinese travellers (Arlt, 2006; Arlt, 2013; see also Chapter 3).

The growth of the Chinese economy and the emerging Chinese middle class has fuelled this rapid expansion in China outbound tourism. Huang and Gao (Chapter 7) point out that

> being in the middle class means adopting a lifestyle to differentiate themselves from others. Owning a car, speaking a foreign language, using advanced technology, eating out, consuming foreign food and drink, leisure activities such as golf, spas, teahouses, buying branded goods and particularly overseas holidays are perceived as the tags of being the middle class.

Important for the development of the Chinese outbound travel market was also the introduction of the Approved Destination Status (ADS) in the mid-1990s. The ADS is a bilateral arrangement between China and foreign countries that allows Chinese citizens to travel to those 'approved' countries. Australia was amongst one of the first Western countries (along with New Zealand) to be granted the ADS status in 1999. As of 2017, 125 countries have gained an Approved Destination Status (Arlt, 2006; Arlt, 2013; Airey & Chong, 2011; see also Chapter 7)

Considering that to date, estimated on the basis of Chinese passport holders, less than 10 per cent of Chinese citizens participate in international tourism indicates substantial future growth potential for the China outbound tourism market (see for instances Chapters 4 and 7). It is therefore no surprise that many destinations around the world try to capitalise on the opportunities created by China's outbound tourism market. A common strategy appears to be the development of new products, services and also strategic partnerships, which necessitate major adaptations in supply and service structures within destinations. Furthermore, the importance of fostering cross-cultural understanding through a greater appreciation of cultural differences in beliefs and behaviours is also recognised. Current research in the area of China outbound tourism has shown that it is crucial for destinations to develop a better understanding of the dynamic nature and complexity of the Chinese growth market with all its opportunities and challenges (see Chapters 3, 4 and 7).

One challenge is to ensure sustainable growth within destinations in light of the rapidly emerging potential of the China market. A further challenge lies in the area of strategic product development in order to respond effectively to the needs and wants of an increasingly individualised and diversified tourism source market. Although shopping and gambling, for example, currently appear to be a central focus of Chinese tourists, in the future, in the light of changing travel preferences of a maturing consumer segment, this might shift to other product categories such as nature, food and wine or cultural heritage activities (see Chapters 3, 4 and 7). Thus, flexible response strategies need to be embedded in an effective management of change.

With our book (*Food, Wine and China: A Tourism Perspective*) we will specifically focus on a growing demand for food and wine tourism experiences by Chinese visitors, which has become an important point of differentiation and a constituent of destination competitiveness in recent years.

Food and wine as key drivers of tourism

As food and wine related experiences have become an important part of tourism in recent years they have increasingly become key motivators and elements of destination choice (Hall et al., 2003; Hall et al., 2000; McKercher et al., 2008; Hall & Gössling, 2013; Getz, 2000; Getz et al., 2014; Fountain, 2017; Wexelbaum, 2017). Constituting about one-third of total tourist expenditure, food and beverages have thus become important tourism resources for destinations that contribute in various ways to economic and regional development in many countries around the world (e.g. Henderson, 2009; Getz, 2000; Everett, 2016; see also Chapters 2 and 3).

The important role food and wine play in a tourist experience does not come as a surprise as tourists need to eat and drink, 'something every visitor does' (WFTA, 2015). However, does this already constitute 'food and wine motivated travel'? Although almost all tourists dine out while on holiday this does not necessarily mean that they are all wine and food tourists as food and wine tourism is characterised by a variety of specific experiences. For some, engaging in food and wine related activities is the primary motivation for travel; for others it constitutes a luxurious experience or just reflects the interest to learn more about the regional and local cuisine and to experience local produce (Everett, 2016).

There are many descriptors for food related tourism, such as gastronomy tourism, culinary tourism or food tourism. A commonly used definition has been coined by Hall and Mitchell (2001: 308) who define food tourism as "visitation to primary and secondary food producers, food festivals, restaurants and specific locations for which food and tasting and/or experiencing the attributes of a specialist food production region are the primary motivating factors for travel". Focusing specifically on wine tourism, Charters (2017: 1441) adds that it "has traditionally been defined as visits to destinations primarily for the purpose of wine tasting and/or experiencing the attributes of a region where grapes are grown for wine". Michael Hall, in Chapter 2, has managed to succinctly capture the complex relationship between food, wine and tourism (see Figure 2.1).

Although China has a long-standing history of growing grapes and winemaking, wine consumption by Chinese people is a more recent, though growing phenomenon (Qiu et al., 2013). In particular since the 2000s we have been able to observe a substantial increase in wine consumption in China. Neirynck (2017: 1449) notes that "while the rest of the world has experienced a drop in wine production and consumption since 2006, China has bucked this trend. This is because of increasing disposable income among the Chinese elite and middle class." With the growth and globalisation of the Chinese economy and with the rise of an affluent urban upper-middle class in cities such as Beijing, Shanghai or Guangzhou, the consumer market in China has been transformed with a growing demand for different products such as imported Western food and beverages.

Considering the disproportion between the production levels of local wines and the significant increase in wine consumption, China constitutes an important export market opportunity for wine producing countries. According to recent data, China is already ranked as the fourth biggest wine import market globally and the biggest export destination for Australian wines (see Chapter 7).

Associated with these developments is a growing interest in wine tourism experiences that has emerged in recent years. Furthermore, considering also the significant culinary culture of China, which reflects the importance of food to Chinese people, it is no surprise that food has also become a major motivation of Chinese tourists to visit a destination (see Chapters 2 and 3).

Knowledge gap

Despite the remarkable growth of the China outbound travel market and the increasing quest for culinary experiences coupled with a steady growth in interest in wine tourism related activities, academic research focusing on food and wine tourism in the context of the China market is still very limited. Although in the past decade a growing body of literature on China outbound tourism as a topic of tourism research is noticeable (e.g. Law et al., 2016; Tse, 2015; Keating et al., 2015; Huang et al., 2015), this has not yet been extended into scholarly activity focusing on the engagement of Chinese tourists in food and wine tourism.

Addressing this gap in knowledge, our book not only contributes to a better understanding of Chinese tourists' food and wine preferences and behaviour but also to destination development, policies and strategies in response to the growing and diversifying China outbound tourism market and its linkages to food and wine tourism. The emphasis is placed on exploring the topic from a demand as well as a specific destination response perspective. Specifically, the aim of the book is two-fold:

1   As there is limited research on Chinese food and wine tourism, it seeks to contribute to a better understanding of the preferences, motivations and perceptions that underline food and wine consumption of Chinese tourists.
2   It also explores how food and wine tourism experiences have been used in a range of international destinations to specifically attract visitors from China. Presenting in particular case examples from Australia but also from New Zealand, Malaysia, Italy and the USA strategic directions adopted to guide destination development and marketing initiatives are explored.

Such perspectives provide a novel contribution to the still limited body of knowledge on China outbound tourism.

Australia in the spotlight

As outlined earlier, the booming Chinese outbound tourism market has become the fastest growing tourism source market and the largest spender in international tourism globally. These developments also resonate strongly in the Australian context, which has been selected as a case in point in our book to discuss opportunities and challenges associated with this important market segment (see Chapters 6–12 and 14).

Australia was one of the first Western countries to be granted the ADS status at the end of the 1990s. Since then, Australia has become one of the most preferred destinations for Chinese tourists with visitor numbers increasing rapidly over the past two decades from just over 100,000 in the year 2000 to 1.2 million visitors in June 2017 (Tourism Australia, 2017a; 2017b; see also Chapters 7 and 8). In addition to the ADS, in 2015, Western Australia has, for instance, become the world's first and the only Australian 'China Ready and Accredited' destination. Further, the launch of the 'China-Australia Year of Tourism 2017' also highlights the strategic importance of the China inbound market, which is estimated to grow to a value of over $18 billion by 2020 (Tourism Research Australia, 2017; Tourism Western Australia, 2015b). In 2016/17, China was already Australia's largest market for total visitor spending and second largest inbound market (behind New Zealand) for visitor arrivals (TRA, 2017).

According to Tourism Research Australia (2014), good food and wine was identified as one of the key motivators for Chinese visitors in their holiday destination choice. As the Australian food and wine sectors have a very positive image among Chinese visitors, Australia's unique food and wine experiences provide an important component to further grow the China inbound market and capitalise on its future potential.

It therefore comes as no surprise that food and wine feature prominently in recent tourism marketing campaigns at the national and state levels. For instance,

Tourism Australia's 'Restaurant Australia' campaign was launched in December 2013 to target a growing Asian market. Western Australia also moved to capitalise on its culinary offerings with the release of its *Taste 2020* strategy, which aims "to enhance the positioning of Western Australia as an extraordinary destination for gourmet produce, fresh seafood, premium wines and beverages (e.g. boutique beers and cider)" (Tourism Western Australia, 2015b).

Thus, in the context of food and wine tourism, visitors from China have become an increasingly important market for Australia. However, not only tourism but also wine from Australia has developed into a significant export, with China as the largest export destination for Australian wines (see Chapters 11 and 12). Australia has experienced rapid growth in wine production over the past three decades and is today one of the top ten wine-producing countries and is the fifth largest exporter of wine globally (Chapter 10).

Structure of the book

The book *Food, Wine and China: A Tourism Perspective* includes 17 chapters, which focus on different aspects of food and wine tourism in the specific context of the growing China outbound tourism market. After this introductory chapter by Pforr and Phau (Curtin University, Australia), in Chapter 2 (Food and wine tourism: challenges, issues and opportunities in the Chinese market) Michael Hall (University of Canterbury, New Zealand) highlights the increasing importance of food and wine as a new tourism niche market, in particular in a rural context. He emphasises that there are many synergies in the overlapping agribusiness and tourism sectors, for example in form of wineries that focus on cellar door sales or expand into the restaurant, events or accommodation sector as well as joint branding initiatives. Regional development can thus be fostered via strong supply chain linkages that help to reinforce destination identity as well as long-term place and product attachment. Further, Hall also points to the potential for increasing food and wine exports from a destination that might develop from the intersection of food, wine and tourism. These issues are then discussed in the chapter in more depth in the specific context of the Chinese market against the backdrop of Chinese food and wine culture and consumption practices. Hall concludes that some caution is warranted in light of, at times, overly optimistic assessments of the prospects of the Chinese market in the short term.

In Chapter 3 (Food and wine tourism in China) Na Su, Bin Dai and Xiujuan Jin (China Tourism Academy) highlight a shift in Chinese tourism in recent years away from rushed sightseeing towards slower leisure and vacation orientated travel behaviour that seeks more meaningful and authentic experiences and a stronger engagement with the local population. In this context, food and wine can be key vehicles to experience a destination and thus food and wine tourism have developed into some of the fastest growing sectors in China's tourism market. The authors briefly review the long history of food in China, which has given rise to Chinese tourists' particular appreciation of food experiences, and the role food plays in tourism, not only as an item of tourist expenditure but as a specific pull factor to draw some tourists to a destination. The authors also provide an interesting insight into why wine consumption in China, despite the country's long-standing history of wine making, has been a more recent phenomenon that is not only driven by a desire to display social status and to foster business relationships via gifting but more and more also by personal enjoyment. As Su and colleagues highlight, these developments have spurred growth in wine tourism within China. However, the authors also point to some barriers and challenges for future development of this tourism sector, such as increased international competition for the domestic market and the need to transform and upgrade existing vineyards in China into more attractive tourist destinations.

Chapter 4 (Chinese outbound tourists: food and beverages) by Wolfgang Georg Arlt (China Outbound Tourism Research Institute) also sets the scene to this volume with a detailed and insightful account on the development of Chinese outbound tourism, which in recent years has caught up with the USA and Germany as the biggest international tourism source markets. His specific focus is on the role food and beverages play for different market segments of the Chinese outbound travel market. Arlt highlights the varying demands and interests of sub-sets of Chinese tourists, as those who are on their first overseas trip as part of an organised package tour will certainly have very different attitudes and expectations towards their culinary experiences compared to a very affluent, seasoned independent Chinese traveller. Arlt argues that the generalised view that Chinese tourists prefer Chinese cuisines when overseas is too simplistic as a growing number of 'foodies' can also be found amongst China's outbound tourists, in particular those who are younger and travelling independently. According to Arlt's analysis they can develop into a lucrative future niche market in light of the prestige associated with culinary experiences and the value they hold in Chinese culture for socialising, networking and displaying social status. In his contribution, which attempts to shed more and also more differentiated light on Chinese tourists' culinary behaviour, Arlt refers to a study by Chang (see Chapter 6), which categorises Chinese tourists according to their dining behaviour in observers, browsers or participators as well as to a study which correlates their culinary behaviour to levels of neophobia. Interestingly Arlt sees culinary differences between consumption behaviour at home and overseas mainly in a dining context. With respect to beverages he highlights the particular role of wine and thus also wine tourism that might help foster further growth in Chinese outbound tourism in future years.

In her contribution "Motivations underlying tourist food consumption" (Chapter 5), Athena Mak (National Dong Hwa University, Taiwan) discusses the obligatory and in particular the symbolic role food plays as part of the tourist experience and explores these issues further in the context of Chinese tourists' food consumption. She highlights that tourists' food consumption is generally affected by the tourists themselves, the food offered in the respective destination as well as the destination environment and that their motivations are based on a complex interplay of drivers such as symbolic and obligatory as well as authenticity, contrast and novelty-seeking behaviours alongside prestige, extension and pleasure. The findings of Mak's survey of Mainland Chinese tourists confirm that enjoyment with respect to sensory and social aspects of food consumption, authenticity and prestige associated with the dining experience as well as assurance in terms of quality of service, hygiene standards and price vs value are important motivators for these tourists.

In Chapter 6 (Dining trajectories of Chinese tourists in Australia), Richard C.Y. Chang (National Dong Hwa University, Taiwan) explores drivers of gastronomy tourism in the specific context of Chinese visitors to Australia and highlights a range of practical implications for its tourism industry and destination marketeers. He starts off by highlighting curiosity in other food cultures as a main travel motivator, which has prompted many destinations to promote their local gastronomy as a specific tourist attraction. Against this backdrop Chang laments the current dearth in academic attention to this particular aspect of tourist behaviour. He addresses this deficit with his discussion of Chinese tourists' dining preferences while visiting Australia, who he classifies as being either 'observers', 'browsers' or 'participators' in the local cuisine. Chang emphasises the need for a diversity of food offerings to not only provide these visitors with familiar tastes that make them feel 'at home' but to also satisfy their desire to sample novel and exotic Australian food. In this context authenticity and local food characteristics are seen as being of particular importance to facilitate an immersion in Australian culture and thus provide Chinese visitors to Australia with an experiential travel experience. Chang's discussion also highlights another interesting point, the role a tour guide during package tours can and will play in enhancing Chinese tourists' satisfaction with their dining experience and as a broker of local dining etiquette.

Sam Huang (Edith Cowan University, Australia) and Hailian Gao (University of South Australia, Australia) discuss in Chapter 7 (Developing Australia's food and wine tourism toward the Chinese visitor market) potential strategies to develop Australia's food and wine resources toward the Chinese tourist market. After a brief review of the developments in Chinese outbound tourism since the 1980s, the authors highlight the dynamic, fast-changing characteristics of China's outbound tourism, which are quite different to most other tourism source markets. They provide interesting insights into these specifics, for example that Chinese tourists from so-called first tier cities like Beijing, Shanghai and Guangzhou are generally more experienced travellers who prefer to visit destinations like Europe, the USA and Australia, whereas those from second or third tier cities tend to be more novice tourists who travel in tour groups to neighbouring countries. Huang and Gao also highlight the rise in free and independent Chinese tourists, who increasingly seek to visit more exotic destinations, which can also provide a boost for regional development. They then move on to discuss opportunities and challenges in food and wine tourism catering specifically to the Chinese market by adopting a supply and also a demand perspective. Again, interesting insights are provided, such as Chinese tourists' preference for learning more about wine and wine making over pure wine tasting. With specific reference to the Australian market the authors conclude their chapter with some insightful recommendations for destinations interested in increasing their share of Chinese tourists, such as more investments in social media, developing a cultural identity of food and wine and promoting the country's high quality food and wine products, producing food and wine gifts for the Chinese markets as well as effectively bridging cultural and language gaps.

Chapter 8 (Food, wine and China: a tourism perspective from Western Australia) explores the interrelationship between Chinese visitors and wine tourism in the specific context of Western Australia. Ross Dowling (Edith Cowan University, Australia) illustrates the growing importance of wine tourism to the state over the past two decades and provides insights into the significance of the Chinese market to Australia. In line with other authors (e.g. Chapters 4 and 7) he points to the growing Chinese free and independent traveller market and discusses opportunities and challenges for food and wine tourism in Western Australia arising from this particular market segment. Against the backdrop of a description of Western Australia's food and wine industry as well as a brief snapshot of the state's food and wine tourists, Dowling discusses Western Australia's current food and wine tourism as well as its China strategies and reviews recent initiatives to further enhance the appeal of Western Australia as a wine producing and wine tourism destination to China.

Also focusing specifically on Western Australia, Alfred Ogle, David Lamb and Stephen Fanning (Edith Cowan University, Australia) explore in Chapter 9 (Are we China ready? A study of Western Australian hotels and Chinese tourists' appetites) how hotels in the state's capital city Perth, via their room service offerings, cater to Chinese travellers' food requirements and palates. The motivation for this study was the recent drive of Tourism Western Australia, the state's peak tourism organisation, to ensure that the tourism industry is 'China-ready' in order to be able to capitalise on growing Chinese visitor numbers and provide them with the best possible visitor experience. A critical analysis of a range of hotel menus unveiled that Perth hotels currently offer only a very limited selection of Chinese food to their in-house guests as part of their room service. In light of the desire of Chinese tourists to not only try and experience local cuisines during their travels but also to eat familiar Chinese food, which has been documented in a number of studies and has also been highlighted in other contributions to this volume (e.g. Chapters 3, 4 and 6), this appears to be a deficiency that should be addressed in order to become truly 'China ready'. The authors acknowledge that adapting room service menus to this particular market segment can be a challenge given the diversity of Chinese cuisines and its various regional distinctions but see value in a stronger focus on this aspect of visitor satisfaction considering the anticipated strong future growth in China outbound tourism to Western Australia.

In Chapter 10 (Preference for Australian premium wines in China: the effect of tourism) Graham Ferguson, Isaac Cheah and Sean Lee (Curtin University, Australia) explore the effects of country of origin, domain specific cultural capital and brand cues on brand choices in the premium Chinese wine consumer segment. Adopting an interesting novel perspective, the authors investigate potential links of tourism with brand choice and brand loyalty, focusing specifically on returning Chinese students who have studied in Australia. They can be seen as long-term tourists who thus have a much greater chance to become immersed in the Australian culture, be exposed to wine drinking on social occasions and to adopt local brands, including premium wines. This is an interesting approach as the impact of tourism on wine consumption and vice versa is commonly considered in the sense that (short-term) tourism exposes consumers from a new market to wine drinking in an established wine culture and can thus assist in building brand specific knowledge. Chinese tourists to Australia can therefore play an important role in enhancing the reputation and uptake of Australia's premium wines in China. In particular temporary residents like overseas students can act as future brand advocates and opinion leaders on their return home and might therefore be in a position to steer premium wine selection away from purely country of origin based purchasing decisions.

Following on from the discussion of the impact of tourism on premium wine selection (Chapter 10), Jeremy Galbreath, Grace Gao, Louis Geneste, Kristina Georgiou, Niki Hynes and Paull Weber (Curtin University, Australia) present Western Australia as a wine exporting region to China (Chapter 11: China as an export market: the case of Western Australian wine). They illustrate the challenges associated with the export of wine to China, like tariffs, cultural disparities, language barriers, taste differences and access to networks, using Western Australia as a case in point. Based on a number of in-depth interviews with Western Australian wine producers, the authors identified some key barriers and risks, for example establishing and maintaining effective distribution networks, issues with payment morale and building trustful business relationships, prestige and brand image creation as well as counterfeiting attempts. They also highlight the need to potentially re-think existing business models in the Western Australian wine industry, which are currently mainly based on low volume higher price point strategies, and to adapt marketing and design to meet the expectations of the Chinese export market. Galbreath et al. conclude that while the Chinese market might present unprecedented future opportunities for wine producers, successfully navigating this export market can be a difficult and challenging undertaking where risks and benefits need to be carefully considered.

In his contribution (Chapter 12, Great Wall or red carpet? Challenges and opportunities for Australian wines in China), Piyush Sharma (Curtin University, Australia) also focuses on Australian wine exports to China. In line with Galbreath et al. (Chapter 11) the author points to the importance of brand awareness, wine distribution networks, seasonality in demand and culturally appropriate marketing and labelling as some of the challenges faced by Australian wine exporters. But he also emphasises opportunities, for example China's growing demand for premium wines, a market segment where Australia has started to make some headways, while at the same time also growing its share at the mass end of the market. In his chapter Sharma provides a profile of a typical Chinese wine consumer and discusses the main motivations to purchase wine. He also emphasises the need for Australian producers to work closely with Chinese tourism authorities and tour operators to promote Australian wine regions as a positive wine tourism experience that can also be a stimulus to create awareness, generate sales and brand loyalty. The 2015 China-Australia Free Trade Agreement, which will eliminate import tariffs and tariffs on alcoholic beverages over the next few years, will provide a strong foundation for future growth in Australian wine exports to China.

In Chapter 13 (Responses to Chinese tourists' interest in wine and food: an Italian perspective) Michael Volgger (Curtin University, Australia) and Harald Pechlaner (Catholic University of Eichstätt-Ingolstadt, Germany) explore why Italy, despite being a renowned food and wine destination, to date has not been able to engage a significant share of its growing number of Chinese visitors in food and wine related activities. They build their analysis on the findings of a study into Chinese tourists' consumer behaviour when visiting Italy. Their discussion provides useful insights into current barriers for wine and food tourism, for example issues related to language, restrictive travel itineraries, the perception of an only superficial and little authentic wine experience, unfamiliarity of Italian food and competition from cultural heritage as well as fashion and shopping as significant attraction points for Chinese visitors. Volgger and Pechlaner suggest potential strategies to redress the situation. Similar to findings of other authors in this volume (e.g. Chapters 2, 3 and 4), they also highlight the prestige and status Chinese people associate with the purchasing and gifting or consumption of wine. They suggest that the Italian wine industry should pay more attention to the symbolic value of wine and address some of the identified barriers, for example with offers or brochures specifically aimed at Chinese visitors, targeting more affluent and less time restricted independent Chinese travellers and engaging tour guides more strongly in the promotion of food and wine tourism. As the identified challenges and barriers are not unique to the Italian case, Volgger and Pechlaner's recommendations have value beyond their particular case study focus.

Adopting a broader perspective, Vanessa Quintal, Ben Thomas (both Curtin University, Australia), Yu-An Huang (National Chi Nan University, Taiwan) and Ian Phau (Curtin University, Australia) in Chapter 14 (Wine tourists' perspectives of 'New World' winescapes: Australia, USA and China) look at tourists' perceptions of wineries in emerging wine producing countries and thus by extension also newer wine tourism destinations. In their study, which focuses on Australia, the USA and Taiwan, China, they adopt the terminology of 'winescapes' to capture not only the vineyard itself but also the overall setting, atmosphere, the quality of the wine as well as signage and service standards. Interestingly, Quintal et al. conclude that across all five study sites service staff, in particular their personal skills, and complementary product offerings are key predictors of wine tourists' attitudes towards the winery and in turn can be useful predictors for their intention to revisit and recommend the winery. These findings are of interest not only from an academic perspective but also for practitioners in emerging wine tourism destinations, which will be able to adapt their offerings to better meet the expectations of these wine tourists. Particular attention should be paid to recruitment, training and retention of qualified staff, to an innovative strategy in order to stimulate complementary cross-sales and to develop an appropriate positioning strategy for the individual winery as well as the overarching wine region.

Taking a different geographical focus, Joanna Fountain (Lincoln University, New Zealand) in Chapter 15 (The potential of wine tourism to enhance Chinese holidaymakers' experiences in New Zealand: insights from Chinese ITOs) explores how far the growing importance of wine as a status symbol mainly amongst younger, wealthier and more educated Chinese consumers can be used as a leverage to grow wine tourism to New Zealand. Based on in-depth interviews with a number of New Zealand-based Chinese Inbound Tour Operators (ITOs) and in line with other contributions to this volume, Fountain identifies existing barriers and suggests strategies to facilitate future growth in this market segment. As a relatively new wine tourism destination, Chinese tourists do not seem to associate New Zealand with wine, a phenomenon also seen in the (Western) Australian context (see Chapter 8), but mainly as a nature-based holiday destination. The dense travel itineraries of Chinese tour groups also act as an impediment to include a visit to a winery as does the still limited knowledge about wine amongst many Chinese tourists. Scope was seen, however, to grow wine tourism in this visitor segment if it were packaged as something 'unique' or 'special' built on the status symbol of consuming and being knowledgeable about wine. An appealing 'winescape' was also highlighted, which reinforces Quintal et al.'s findings presented in the preceding chapter. Targeting specifically premium and VIP groups, including government officials and business people, is seen as a potential strategy to enhance the reputation and popularity of New Zealand wines, so that in the longer term wine tourism might also act as a stimulus to grow wine exports into China.

Nazaruddin Haji Hamit (Curtin University, Malaysia) focuses in Chapter 16 (At a crossroad: a study of Nyonya cuisine as intangible cultural heritage) on a very specific aspect of Chinese cuisine, the Nyonya cuisine of the descendants of Chinese migrants residing in the Malaysian port city of Malacca. Based on a study grounded in social practice theory he not only provides a lot of interesting information on this distinct cuisine but describes it as a culinary heritage under threat from modern lifestyles and a lack of written lore. On the basis that food tourism is anchored in interest in local cultures and stories about their cuisines, special knowledge that represents the respective local culture and identities, Hamit argues that culinary tourism, in particular from Chinese visitors, might be a way of preserving and reviving this unique niche of Chinese cuisine into the future.

Lastly, in the concluding chapter (Moving forward: think 'dine and wine' with the Chinese tourists), the editors Ian Phau and Christof Pforr synthesise the main themes emerging from this volume. Reviewing the book's many case studies from around the world (i.e. Australia, New Zealand, USA, Italy, Malaysia and China), Phau and Pforr bring together an array of theoretical and practical insights into the development and management of the multiplicity of food and wine related activities and experiences in the context of the growing Chinese tourist market. In doing so, they highlight that the contextualisation of food and wine consumption is important in understanding Chinese travellers, and also provides directions for future research.

The immense potential of the Chinese tourist market, particularly with regards to food and wine tourism, is fully reflected throughout this volume and the many insights from the studies presented will undoubtedly benefit tourism and hospitality academics as well as practitioners in the area. Overall the book therefore makes a significant contribution to our understanding of the biggest international tourism source market in the context of experiences related to food and wine consumption as well as implications for destination development and management.

References

Airey, D. & Chong, K. (2011). National policy-makers for tourism in China. *Annals of Tourism Research*, 37: 295–314.

Arlt, W. G. (2006). *China's Outbound Tourism*. Oxfordshire: Routledge.

Arlt, W. G. (2013). The second wave of Chinese outbound tourism. *Tourism Planning & Development*, 10: 126–133.

Charters, S. (2017). Wine tourism. In L.L. Lowry (ed.) *The SAGE International Encyclopedia of Travel and Tourism*. London: Sage (pp. 1440–1443).

Everett, S. (2016). *Food and Drink Tourism: Principles and Practice*. London: Sage.

Fountain, J. (2017). Wine tourism, New World wines, Australia/New Zealand. In L.L. Lowry (ed.) *The SAGE International Encyclopedia of Travel and Tourism*. London: Sage (pp. 1444–1447).

Getz, D. (2000). *Explore Wine Tourism: Management, Development and Destinations*. New York: Cognizant Communication Corporation.

Getz, D., Andersson, R., Robinson, R. & Vujicic, S. (2014). *Foodies and Food Tourism*. Oxford: Goodfellows Publishing.

Hall, C.M. & Gössling, S. (eds.) (2013). *Sustainable Culinary Systems: Local Foods, Innovation, and Tourism & Hospitality*. London: Routledge.

Hall, C.M. & Mitchell, R. (2001). Wine and food tourism. In N.D.R. Derrett (ed.), *Special Interest Tourism*. Australia: John Wiley (pp. 307–325).

Hall, C.M., Sharples, E., Cambourne, B. & Macionis, N. (eds.) (2000). *Wine Tourism Around the World: Development, Management and Markets*. Oxford: Butterworth-Heinemann.

Hall, C.M., Sharples, E., Mitchell, R., Cambourne, B. & Macionis, N. (eds.) (2003). *Food Tourism Around the World: Development, Management and Markets*. Oxford: Butterworth-Heinemann.

Henderson, J.C. (2009). Food tourism reviewed. *British Food Journal*, 111(4): 317–326.

Huang, S., Keating, B.W., Kriz, A. & Heung, V.C.S. (2015). Chinese outbound tourism: an epilogue. *Journal of Travel & Tourism Marketing*, 32(1–2): 153–159.

Keating, B.W., Huang, S., Kriz, A. & Heung, V.C.S. (2015). A systematic review of the Chinese outbound tourism literature: 1983–2012. *Journal of Travel & Tourism Marketing*, 32(1–2): 2–17.

Law, R., Sun, S., Fong, D.K.C., Fong, L.H.N. & Fu, H. (2016). A systematic review of China's outbound tourism research. *International Journal of Contemporary Hospitality Management*, 28(12): 2654–2674.

McKercher, B., Okumus, F. & Okumus, B. (2008). Food tourism as a variable market segment: It's all how you cook the numbers! *Journal of Travel & Tourism Marketing*, 25(2), 137–148.

Neirynck, B.C.R (2017). Wine tourism, New World wines, China. In L.L. Lowry (ed.) *The SAGE International Encyclopedia of Travel and Tourism*. London: Sage (pp. 1448–1450).

Qiu, H.Z., Yuan, J., Ye, B. & Hung, K. (2013). Wine tourism phenomena in China: an emerging market. *International Journal of Contemporary Hospitality Management*, 25(7), 1115–1134.

Tourism Australia (2017a). *China Market Profile*. Sydney: Tourism Australia.

Tourism Australia (2017b). Tourism statistics. Retrieved from http://www.tourism. australia.com/en/markets-and-research/tourism-statistics.html

Tourism Research Australia (2014). Food and wine tourism in Western Australia. Retrieved from https://www.tra.gov.au/ArticleDocuments/185/Food_and_Wine_Tourism_in_Western_Australia_11062014.pdf.aspx?Embed=Y

Tourism Research Australia (2017). International visitors in Australia (year ending June 2017). Retrieved from https://www.tra.gov.au/ArticleDocuments/250/IVS_one_pager_June2017.pdf.aspx?Embed=Y

Tourism Western Australia (2015a). WA is Australia's first 'China ready' destination. Retrieved from http://www.tourism.wa.gov.au/About%20Us/News_and_media/Article/WA_is_Australia's_first_'China_Ready'_destination/214

Tourism Western Australia. (2015b). Taste 2020 – a strategy for food and wine tourism. Perth: TWA.

Tse, T.S.M. (2015). A review of Chinese outbound tourism research and the way forward. *Journal of China Tourism Research*, 11(1): 1–18.

UNWTO (2017). *Tourism Highlights*. Madrid: UNWTO.

Wexelbaum, R. (2017). Culinary tourism. In L.L. Lowry (ed.) *The SAGE International Encyclopedia of Travel and Tourism*. London: Sage (pp. 322–325).

WFTA (2015). What is food tourism? Retrieved from http://worldfoodtravel.org

2 Food and wine tourism

Challenges, issues and opportunities in the Chinese market

C. Michael Hall

To the people, food is heaven.

(Hanshu, chapter 43, compiled up to about 115
by Ban Biao, Ban Gu, and Ban Zhao in Höllmann, 2010: vii)

Introduction

The fourth largest country in terms of land mass and the largest in terms of population, China is estimated to be home to some 1.3 billion people. The growth of the Chinese economy since becoming engaged with the international trading system in the later 1970s has been nothing short of remarkable (Lotta, 2009). In 2015, China had a GDP of US$11 trillion, making it the second largest economy in the world behind the United States of America (The World Bank Group, 2017). If the current economic growth rates are maintained, it is widely forecast that China will overtake the United States as the world's leading economy some time in the 2020s (Ito, 2010).

Undoubtedly the growth of the Chinese economy and the country's importance to international trade in goods and services has seen a massive shift in perceptions towards China with respect to its significance for business. It is therefore no surprise that China has become a focal point of interest for exporters in sectors such as tourism, where China is now a major source of outbound travellers as well as being a significant destination in its own right, as well as for the wine industry, as Chinese wine exports continue to grow. However, changes in Chinese consumption patterns and the growing international significance of the Chinese consumer are not just an economic artefact. Instead they also reflect the importance of understanding Chinese culture and politics as factors that influence not only the overall relationship of China to the countries it does business with but also individual Chinese consumption and the market potential of the country for the many businesses that wish to do business either there or with its nationals.

Höllmann (2010: vii) in discussing the changing nature of Chinese food culture observed: 'A serious history of food culture should offer more than a chronicle of epicurean indulgence and should go beyond examining the social framework of nutrition.' In the same way this chapter takes the perspective that a serious account of Chinese food and wine tourism should also offer more than a chronicle of epicurean indulgence and go beyond promotional images and storytelling to examine the advantages and disadvantages of the Chinese market to stakeholders in food and wine tourism. The chapter is divided into three main sections. First, a discussion of the advantages and disadvantages of food and wine tourism. Second, a contextualisation of Chinese food and wine culture and consumption practices with some brief comments with respect to wine and food within Chinese tourism. Third, questioning as to where in the value chain does food and wine tourism add value and for whom. The chapter then concludes on a note of caution with respect to the Chinese market though highlighting its long-term nature and significance.

Advantages and disadvantages of wine and food tourism

The position of a business or individual in the food, wine and tourism system is critical to understanding how they see the value of wine and food related tourism, as well as their capacity to benefit from it. The reality is that although perfect partners from a promotional perspective, at a regional level the economic benefits of the wine and tourism relationship are not generally shared equally in direct terms at a business level, although the indirect benefits may be substantial.

Figure 2.1 illustrates an idealised food, wine and tourism system (see also Gössling & Hall, 2013; Hall & Gössling, 2016). The system is actually the overlap of two sectoral systems and their value chains. The first is the agribusiness system in which food and wine are produced and then moved through the supply chain

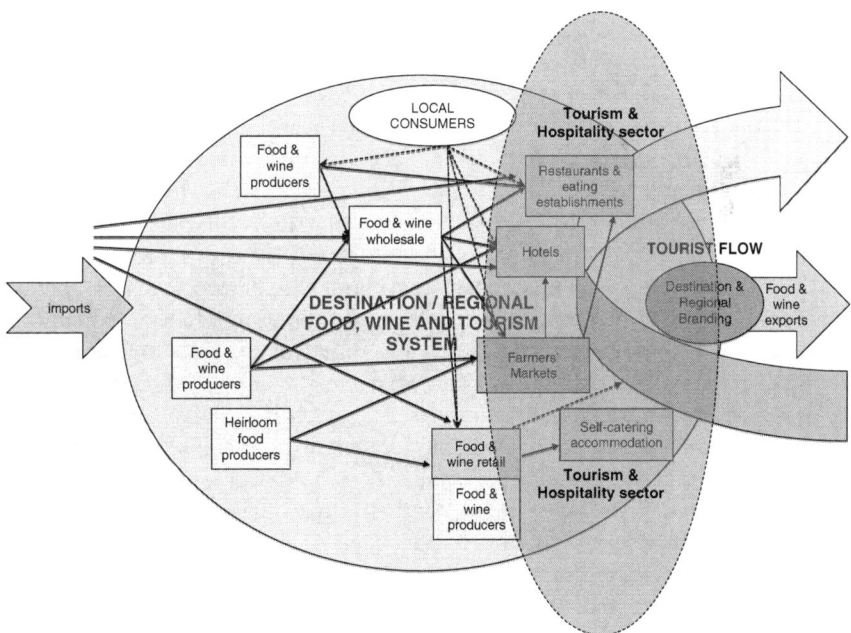

Figure 2.1 Destination/regional food, wine and tourism system

to end consumers. The second is the tourism system that is primarily focused on accommodation, attractions and restaurants. The two systems primarily intersect in restaurant and hospitality sector businesses, but linkages are also found when tourists buy food and wine direct from retailers, especially when self-catering as well as when staying with friends and relatives at the destination. However, in recent years the systems have become even more enmeshed as a result of changes within both systems that serve to blur the boundaries between them. An obvious example here are wineries that focus on cellar door sales as a means of selling wine as well as those that expand out to also cover a range of other tourism and hospitality products such as a restaurant, food services, hosting events and providing accommodation (Baird & Hall, 2014). Many farmers' markets have enlarged their business focus to serve not only their local growers and purchasers but also act as a tourist attraction and expand the promotion of local food to visitors as well as the immediate community (Hall, 2013). In the tourism sector some accommodation providers and restaurants have positioned themselves in terms of their use of local foods as a point of differentiation (Roy et al., 2016), while there is also the growth of heritage markets and heirloom food stalls, specialist local food and wine tours as well as self-guided options (Hall, 2016).

From a strategic sense a key question to ask of actors within the food, wine and tourism system is what business are you actually in (Baird & Hall, 2016)? This is not as straightforward a question as it sounds as many businesses are often unsure of who their customer base are or their perceptions may be quite different from the actual reality. Many market gardeners and small farms in periurban areas that provide direct marketing and retail, for example, invariably never regard themselves as being in the business of tourism (usually defined in terms of travel outside the home environment). However, their immediate local walk up customer base is often quite low and instead they rely on passing traffic and daytrippers for their sales, while in many cases many berry farms and orchards are often daytrip destinations in their own right. Nevertheless, their owners will often insist that they are not in the business of tourism! Yet for such businesses, and usually for the majority of food and wine producers direct sales to tourists will be mainly domestic tourists as opposed to international. Such visitors are, because of direct marketing and retail channels, usually much easier to connect to with respect to encouraging long-term consumer purchase of products, unless living in a federal system that does not enable cross-border trade. In contrast, maintaining sales to international visitors once they have returned to their own country may be much more difficult.

A further area in which the agribusiness and tourism systems intersect is with respect to branding. Destination branding for tourism and economic development purposes can have substantial implications as to how people perceive food and wine products from the same region. Similarly this situation will also occur in reverse with respect to the role of regional food and wine brands, appellations, and food of protected geographically designated origin serving to shape visitor perceptions and expectations in a broad sense as well some of the products they will wish to consume if they visit (Hall & Baird, 2014). Regional promotion, product appellations and protected geographical status of foods together with associated media stories all develop and reinforce the inferred brand of a location. However, increasingly locations are seeking to manage this relationship more so that the overall brand becomes actively promoted and reinforces the connections between wine and tourism with the belief that it will contribute to increased visitors (or at the very least increased numbers of higher spending visitors) as well as increased sales of food and wine products.

While the development of actively management place brands that are positioned to benefit both tourism and food and wine producers can be valuable the greatest returns to producers overall often depends on the extent of integration of the food, wine, hospitality and tourism systems, especially at the local level, which is where the most tangible and immediate benefits may accrue. The level of integration between the two systems is primarily based on the degree to which the accommodation, hospitality and restaurant businesses purchase their food and wine supplies locally. Importantly, this applies not only to the high-end destination flagship restaurants that often feature in food media but to the 'everyday' food suppliers. In many countries these would be the fast food (bread, burger, fish and chips, kebab, pizza, noodle, sandwich) businesses. By sheer weight of the size of their purchasing, these food businesses are invariably the most important with respect to visitor induced demand for locally grown foods and therefore for local agriculture, however they seldom feature in travel magazines. Similarly, it is important for small local wine, beer and spirit producers that their products are provided by local restaurants, bars and cafés, although in some countries or regions with state monopolies they may actually be disadvantaged (Malm et al., 2013). These issues are important as they raise fundamental questions about the value of the food, wine and tourism relationship and its potential value for producers, tourism and hospitality businesses, destinations and regions, and communities as a whole. Questions that need to be asked include 'how is success measured?', 'Are both costs and benefits being measured?' and 'At what scale is the food, wine and tourism relationship being assessed?'.

Business perspectives

The advantages and disadvantages of food and wine tourism for wine producers will be a function of their market, retail and sales strategy and their size and are noted in general in Table 2.1. A key issue for cellar door sales is the nature of the visitor and the capacity to develop a relationship with them and convert them to a regular purchaser of wine. This is usually far more achievable for domestic tourists than international ones because of the potential ease of ongoing purchase whether from retailers or mail order direct from the winery. For international tourists the immediate economic value of the visitor will clearly depend on whether the winery already has a market presence in the country the tourist is from or whether they are planning to launch there in the near future. However, both domestic and international visitors can contribute to brand awareness and positive word of mouth as well as be significant sources of market information.

Table 2.1 The advantages and disadvantages of food and wine tourism for wine
producers

Advantages	Disadvantages
Consumer exposure to product increased	Increased costs and management time
Brand awareness and loyalty developed	Capital required and return on investment (ROI)
Customer relationships created – see 'behind the scenes'. Positive customer relations may lead to both direct sales and indirect sales through 'word of mouth'	Inability to significantly increase sales or sell at an appropriate price, i.e. Because of location, accessibility and expectations arising from other cellar door experiences in region
Increased sales margins – direct sale (where the absence of distributor costs are not carried over entirely to the consumer)	Opportunity costs
New market/diversify sales base	Dealing with the wrong market
Additional sales outlet(s) – especially for smaller producers who cannot guarantee volume or constancy of supply	Seasonality issues – visitor demand may be at busiest times of year with respect to winery and vineyard activities
Product and customer marketing intelligence	Additional health and safety costs
Educational opportunities – developing the market	Biosecurity risks

Source: Hall et al., 2000; Hall 1996, 2012

Many of the disadvantages of wine producers being engaged in food and wine tourism are connected to the potential return on investment from providing for tourists. For many small producers building and maintaining cellar door facilities uses capital that may achieve a better return elsewhere, while catering to visitors also increases costs (including meeting additional health and safety regulations) and management time that may not be retrieved by increased sales. Furthermore, because of location, accessibility and expectations arising from other cellar door experiences in the region it may simply not be possible to increase sales by a sufficient extent to cover the additional costs of being open to tourists. This may also be because of the price points of cellar door sales or the nature of market demands. Furthermore, in some cases the market that is attracted may not be the most appropriate with respect to business strategy. For example, a bus tour market may provide a revenue stream but may also deter other purchasers from stopping that are seeking a more intimate cellar door experience. Similarly, international tourists, even if they make a purchase from the cellar door, may be from countries that the producer does not or does not plan to export to. A final potential disadvantage of food and wine tourism is the potential risk it poses with respect to biosecurity and the introduction of pests and diseases onto a vineyard (Hall, 2003).

The advantages and disadvantages of food and wine tourism for tourism businesses within a wine region are similarly mixed but generally tend toward the positive (Table 2.2). In terms of direct benefits winegrowers and the wine products of a region provide an attraction for tourists and therefore business

Table 2.2 Advantages and disadvantages for local tourism businesses

Advantages	Disadvantages
Spillover effects (positive)	Spillover effects (negative)
Visitors to winegrowers are a source market as well as market knowledge	Unable to cater to additional demand leading to customer dissatisfaction
Very significant for restaurants, cafés, accommodation and lodging, arts and crafts	Poor match to existing product base and the composition of the marketing mix
Provides an attraction/activity/stop within an area/itinerary (also a potential toilet stop)	Seasonality issues
Potential to extend length of stay/stop as well as return visit	
Generation of destination awareness and destination recall	
Seasonality issues	

Source: After Hall, 2012

opportunities to convey them to wineries, and provide food and accommodation while visiting. In addition, visitors may also seek to combine their travel with visits to other attractions that, by themselves, were not sufficient justification to travel to a particular location and utilise specific businesses; and in some cases may even stay longer because of the presence of food and wine tourism opportunities. With respect to possible disadvantages these mainly occur in relation to the negative spillover effects of food and wine tourism that relate to issues such as crowding, poor match to existing product base and marketing mix, and being unable to satisfy consumer demand. The critical aspect for the tourism operations that directly utilise the presence of winegrowers for their business is the relationships and networks that they establish with wineries. Without the willingness of wineries to open their cellar door to visitors the large majority of wine tourism would cease to exist, while the food tourism dimension would also be affected, although not as strongly, as local wine and food matching could still continue in restaurants, cafés and other providers.

Regional perspectives

The market bottom line for food and wine tourism at a regional level is that while every visitor must eat and drink not everyone is a food or wine tourist (Hall, 2012). The committed gastronomic tourist, that travels to a destination *only* because of the food and/or wine opportunities, is actually a very small, albeit economically significant, market segment regardless of which country they are from (Hall & Sharples, 2003). Nevertheless, there are clearly a large proportion of visitors interested in trying new foods and wines, which does potentially present market opportunities. But because everyone must eat this raises significant issues as to what is provided and available and from a regional economic development perspective, as noted above, to what extent is the selling of food to tourists linked to local food suppliers. Even if tourists will only eat the familiar it still creates opportunity for local food linkages and connections. This is one of the most important aspects to understand with respect to the benefits of wine, food and tourism linkages from a regional perspective (Table 2.3).

Outside of the immediate nexus between food and wine consumption and travel to a destination is the longer term potential for increasing export from a destination, as even if some of the immediate wines that have been tasted are unavailable to the visitor in their home location the overall positive perception of the visitor experience may create a 'halo effect' that influences subsequent purchase of other regional products (McGrath et al., 1993; Low & Davenport, 2006). Nevertheless, for a destination to be able to maximise the advantage of such effects it needs to have exported food products available to consumers, which relies on the interests and efforts of a range of producers as well as government support and, for international exports, market access to be able to maximise synergies. Indeed, such issues also become important with respect to the contribution of food and wine regional promotion and tourism destination promotion to the development and active management of place brands. Strong place brands can help create a 'feel good factor' among local stakeholders but unless it is connected to the region's export strategy and networks then its economic value may be negligible. Furthermore, while such brand synergies can be beneficial to some sectors, they may not be useful for others. For example, in the case of New Zealand the focus on the clean and green image is not regarded as helpful to the country's ICT or creative sectors. Indeed, it has helped create a stereotype of the country being a nice place to visit but lacking in business acumen (Hall, 2010, 2017). Regional brand values of food (including wine and spirits) and tourism can potentially be very good for destination and regional promotion but in order to maximise the benefits regions need to be sure of how they fit with the overall regional economic, innovation and export strategies as well as ensuring that the place brand and brand architecture are appropriate in the context of those strategies (Hall & Baird, 2014).

This section has outlined a range of issues that need to be examined in the context of the Chinese food and wine tourist market and its value for any destination.

Table 2.3 Contribution of wine and food tourism synergies at a region/destination level

Advantages	Disadvantages
Association with a quality product can build positive inferred place brand	Focusing on food and tourism connections may mean other opportunities are not explored or that a market's perception of a region is not properly understood
Can help differentiate local products	May negatively affect some local products and sectors
May help increase visitor expenditure on local product; and contribute to exports	May negatively affect non food, wine and tourism exports

Source: After Hall, 2012

However, before specifically discussing the fit of the Chinese market, it is also essential to note some of the ways within which Chinese food and wine culture is framed and its potential short- and long-term implications.

Contextualising Chinese food and wine culture and consumption

One of the most significant issues for understanding Chinese food and wine culture, as well as geographic market segmentation, is the problem of generalisation. In the case of China few countries' culinary heritage has been so misunderstood and misportrayed in the West for so many years. However, as a result of increased international trade, the Chinese diaspora and, of course, growth in tourism, the image of Chinese food and the place of wine within it is gradually changing (Camillo, 2012; Capitello et al., 2017).

As befits a country of such size, China is a land of many cuisines, although how many regional variations there are is debatable. Historically, China is the 'Land of the Five Flavours' as, for around the past 2,000 years, the five flavours (sour, bitter, sweet, pungent and salty) have been regarded as a general framework for Chinese cuisine (wood, fire, earth, metal, water) that, in turn, provides a culinary context for wine and food matching within Chinese foodways. Regionally, China has been divided into eight areas by some scholars: Sichuan, Hunan, Guangdong, Shandong, Jiangsu, Anhui, Fujian and Zhejiang; and four by others: Shanghai (east – sour), Canton (south – sweet), Sichuan (west – spicy) and Beijing (north – salty). Regardless of the number recognised the key issue is that there is substantial local and regional variety, with various dishes and products having strong place associations.

Alcohol has a long (and glorious) history in Chinese cuisine and culture, but is strongly associated with beer, distilled alcohol, and rice wine rather than grape wine (Cochrane et al., 2003). Historically, where grape wine has been drunk and figured in the culinary history of China it was invariably associated with the elite. Domestically produced Chinese grape wine was probably first recorded in the Middle Kingdom in 223 during the Wei Dynasty, in an edict from Emperor Wen extolling its virtues, although mention was made as early as 126 BC when General Zhang Qian returned from a diplomatic mission. However, it was not until the emergence of modern China from the early-mid 19th century on, that grape wine began to feature more in accounts of Chinese cuisine, with the presence of wine in the European entrepots on the Chinese coast being the major source for diffusion.

The social context of alcohol consumption is extremely important when considering Chinese approaches towards wine (Cochrane et al., 2003). Alcohol is strongly linked with traditional festivals and holidays and functions as the happiness drink at weddings or the sadness drink at funerals. Importantly, alcohol is regarded as part of the forging of social and business bonds between people, and such occasions would traditionally involve drinking games as part of the social experience by which people come to know each other better. Sharing food and alcohol as part of a meal is regarded as 'the visible manifestation of the harmony which should exist between family and friends'. Also significant is the role of alcohol in gift giving in Chinese culture that has become an important factor in the purchase of some wines by people on holiday as gifts to take back to China. In terms of gift giving, international wine should therefore be regarded as much as a symbolic product as one consumed as an alcoholic beverage of choice (Zheng & Wang, 2016).

Continuity and change in Chinese food culture: from China to Chinatown (and back)

As China's economy has internationalised its social connectivity to the world has also changed, including its sourcing of food and food tastes. As a result of inter-nationalisation and globalisation China, and especially urban Chinese, have been increasingly exposed to Western tastes, including the role of wine in cuisine. These encounters come from a variety of sources, from the Chinese diaspora and the return to home villages and family, student mobility to international institutions, international travel, and even internal migration within China itself. The mobilities of people and their accompanying foodways has meant that not only has Chinese food become a familiar part of Western culinary settings, together with a range of 'Chinese' gastronomic fusions that may be unrecognisable to many mainland Chinese (Roberts, 2004), but Western foods, along with Chinese interpretations of imagined Western foods and foodways, have also now come to China.

One of the key dimensions to understanding the emerging Chinese middle class's – as well as the elite's – attitudes towards wine is the significance of grow-ing cosmopolitanism in Chinese tastes, particularly in the urban coastal areas (Zhou et al., 2010). Much of the interest in wine, both as a symbolic product and as a hedonist purchase to drink, arises from wine consumption being related to the accrual of cultural capital and the gaining of cultural distinction, particularly from the practice of eating out (Warde et al., 1999). The acquisition of cultural capital via wine purchase and drinking arises from 'the capacity to behave properly and knowledgeably in public, to exercise discriminating taste when selecting places to go out and eat, and to facilitate conversation about and evaluation of culinary matters' (Warde & Martens, 2000: 69–70). As a result of such shifts in taste it is becoming more socially acceptable to drink lower percentage alcohol, such as wine, rather than grain-made spirits (Guo & Huang, 2015; Sun & Wong, 2015), with an associated connection to being a healthier as well as a more sophisticated drinking option (New Zealand Trade and Enterprise, 2015; Vinexpo, 2015), espe-cially among women. This has therefore created significant opportunities for the expansion of Chinese domestic winegrowing as well as for exports to the country (Zheng & Wang, 2016).

The contemporary Chinese wine market is arguably now one of the main geographic markets for wine exporters and industry growth. The Chinese wine market is the fifth largest in the world in both value and volume (International Organisation of Vine and Wine, 2016). Total wine volume reached 4.6 billion litres in 2016, a 20.4% increase from 3.86 billion litres in 2011. Total wine value reached CYN 251,467.0 million in 2015, an increase of 14.4% from CYN 219,729.3 million in 2011 (Euromonitor, 2017). Significantly, with respect to potential future growth, the per capita wine consumption in China was 1.7 litres per year in 2014, compared with the global per capita average of seven litres (New Zealand Trade and Enterprise, 2015). The domestic market is fairly competitive, with the top three domestic brands having a combined market share of over 50% (Zheng & Wang, 2016). Despite this, the Chinese wine market is still relatively immature (Zheng & Wang, 2016) with substantial room for foreign brands.

The relative youth of the Chinese wine market also supports the likelihood of potential future growth. The Chinese wine market has been defined by Liu et al. (2014) in terms of three segments based on benefit segmentation: 'extrinsic attribute-seeking customers', 'intrinsic attribute-seeking customers', and 'alcohol level attribute-seeking customers'. With regard to age, it was found that the 25–34 age category dominates all three clusters, while market analysis by Euromonitor International (2017) confirms that young consumers and middle-aged consumers are the main purchasers of wine in China. Nevertheless, even though perceived as a status symbol among the rapidly emerging middle class in China (Iskyan, 2016), consumer preferences for wine on the basis of colour highlight the continuing role of culture in framing the social practices associated with wine consumption.

China became the world's biggest market for red wine in 2013, with 1.86 billion bottles sold (Willsher, 2014). However, the substantial dominance of red wine sales in the Chinese wine marketplace suggests the continued role of cultural sensibility rather than just taste and quality considerations. The colour red is considered lucky in China and is also affiliated with the Communist government, while white is associated with death and is predominantly seen at funerals (Liu et al., 2014). Change in purchasing patterns will undoubtedly occur over time, although deep-seated cultural values will continue to influence perceptions of wine. Nevertheless, tourism will likely play a significant part in such processes, as Agriculture and Agri-Food Canada (2016) observed, 'as Chinese tourists visit more wine drinking countries wine knowledge and preferences for foreign wines also increases.'

Since 1985 Chinese outbound tourism has grown from a little over 500,000 trips to 122 million in 2016. This has meant that China has now become the leading country in terms of outbound tourism numbers and is the major international market for a number of countries. Nevertheless, the current small proportion of the country that has travelled internationally suggests that there is substantial growth still to come. As a result, China is a major target for the international marketing campaigns of a number of destinations (Hall & Page, 2017).

Food and wine has become an increasingly common element in destination positioning for the Chinese outbound market, especially given the market's perceived interest in upscale products (World Tourism Organization, 2014). Indeed, Qing et al. (2015) note that, on average, Chinese spend more on gift wines than for their own consumption. However, the purchase of some specific wine products, often associated with high status, for gift giving does not make for a viable cellar door sales experience aimed at the Chinese market at a regional scale (Lee et al., 2016), although if you are one of the few wineries whose products have achieved such a position, this clearly may be very advantageous.

In the case of Canada icewine, China has become the number one export destination valued at Can$6.7M in 2014, with icewine also being one of the top three gifts that Chinese visitors bring home from Canada. Lawrence (2015) reports that China is responsible for almost half of all icewine purchase. In Inniskillin Winery in Ontario 10% of all winery visitors are from mainland China and 90% purchase icewine following a winery tour. The average winery purchase is Can$500 per group of two to ten people and Chinese visitors are more likely to make purchases in the $1,000–$3,000 range, than any other visitor segment (Tourism Industry Association of Canada & Canadian Vintners Association, 2015). Nevertheless, despite the position of Canadian icewine as a gift product, the 'halo effect' on other Canadian wine sales to China appears quite weak, although the Canadian industry seeks to change this (Lawrence, 2015).

There is the beginnings of a domestic wine tourism market in China (Qiu et al., 2013) which may have significant long-term implications for the development of a greater culture of grape wine consumption and interest in international wine from a consumption perspective, given its potential consumer education benefits. Visitation to domestic wineries can also encourage interest in wine tourism when travelling internationally. However, perhaps more critical here is the way in which Chinese outbound travel is structured when they visit destinations, as package tours and dependence on bus and public transport services will tend to limit the number of wineries that are visited. In contrast, FITs (Free Independent Travellers) have the most potential to travel to wine regions and experience a number of wineries. With this being the case perhaps the most critical element therefore in the emergence of Chinese wine tourism that serves to benefit wineries will be time. Time for the market to shift, as it undoubtedly will, from being package based to being engaged in more independent travel, and time for wineries to be able to put the export connections in place in China to be able to benefit from exposure to the Chinese market at the cellar door.

Chinese wine tourism: adding value to who?

Wine and food are clearly becoming an increasingly important part of the Chinese international tourism experience. The opportunity to visit wineries can be a significant part of tour packages and can also assist in creating awareness of the host country's wine. However, the particular circumstances of Chinese wine purchasing and the difficulties associated with selling in China make the potential for wineries to take advantage of the Chinese market problematic. Using some of the issues raised earlier in the chapter with respect to the advantages and disadvantages of wine and food tourism we can use these issues to frame the potential of the Chinese market.

Is consumer exposure to product increased by Chinese wine tourism? Yes, undoubtedly the wine tourism experience provides opportunities for the consumer products that they will not have tried before.

Is brand awareness and loyalty developed? We do not know which brands though it appears likely that in the short term brand awareness is likely linked to high status brands rather than national wine brands. The contribution of wine tourism to Chinese consumption loyalty remains unknown

Will wine tourism create customer relationships? This will depend on the structures in place to create long-term customer relationships. For the majority of wineries, at least in the short to mid-term, the answer is probably no given the level of investment that would be required.

Will Chinese wine tourism lead to increased sales margins? Yes, but no different than other markets. However, in many cases if wineries have to employ specialist staff in order to cater to the market then they will need to pay special attention to the market returns.

Will Chinese wine tourism provide a new market and diversify the sales base? This will only be true for a limited number of wineries in the short to mid-term given the costs involved in servicing the market as well as the potential effects on existing markets. Furthermore, the capacity to export to China will also be a significant restriction in the value of the market.

Are there any benefits in additional sales outlet(s) – especially for smaller producers who cannot guarantee volume or constancy of supply? The Chinese market is probably not suitable for smaller producers unless they have extremely good ties to the Chinese specialist tour companies and associated value chain.

Is product and customer marketing intelligence available? Generally there is poor intelligence gathering unless a winery is looking to export to China or does so already.

Are there any educational opportunities? Yes, but the question remains if interest can be converted to sales in the long-term.

Conclusion: is Chinese wine tourism fad or fashion?

The Chinese wine market and Chinese wine tourism has become a significant focus for many wine regions but especially their associated tourism and hospitality sectors. However, is Chinese wine tourism a fad?

Fads are sudden changes that often spread quickly and fade away rapidly. They appear to be random and are impossible to predict. Significantly, fads are not a mark of cultural capital. In contrast, fashion is a mark of cultural capital. International wine has clearly become a fashionable product in China. With significant symbolic value attached, wine has become one of the hallmarks of cosmopolitanism and an 'aspirational' middle class lifestyle that includes international travel as well. However, the key to the Chinese wine market will be to move beyond wine as a fashion statement to where it becomes a social practice and becomes part of Chinese foodways in a deeper and more engrained way.

However, becoming a social practice takes time and a range of institutions and systems need to be in place. The growth of the domestic Chinese wine industry as well as wine tourism will undoubtedly help. But wine will need to move from its gift role, as valuable as that is, to one based on consumption. In doing so, to make Chinese wine tourism work for the wider wine industry there will be a need to encourage further densification of food and wine network relationships. These networks must also be extended transnationally in order to gain the benefits of international tourism visitation. Critical to this will be the capacity of winegrowers to sell their wines into China.

Despite the possible benefits of China as a market for winegrowers substantial questions remain. To a great extent awareness of Chinese wine tourism is a function of the tour market and approved destination status. However, it is possible to put too much emphasis on high status wine at the expense of other dimensions of the destination/regional offer. Furthermore, there is a range of other research issues that need to be urgently addressed. Most research on the Chinese wine market and wine tourism and its potential does not deal with regionalism in Chinese food culture and its implications for wine preferences. Furthermore, there is inadequate research undertaken with respect to the social context of wine consumption and purchase, while more research overall is required in terms of lifestyle and social practices. Finally, for Chinese tourists there are major issues with time, budget and accessibility to get to some wine regions given the overall amount of time available to them for travel. With that being the case there are likely to be a number of wine regions for whom investing in preparing for the Chinese market may well be money misspent, at least in the short term, until wider shifts occur in Chinese foodways and wine culture, even though the tourism industry may benefit.

References

Agriculture and Agri-Food Canada (2016). *Sector Trend Analysis: The Wine Market in China*. Ottawa: Agriculture and Agri-Food Canada.

Baird, T. & Hall, C.M. (2014). 'Between the vines: wine tourism in New Zealand', in P. Howland (ed.) *Social, Cultural and Economic Impacts of Wine in New Zealand* (pp. 191–207). Abingdon: Routledge.

Baird, T. & Hall, C.M. (2016). 'Competence based innovation in New Zealand wine tourism: partial strategies for partial industrialisation', in H. Pechlaner and E. Innerhofer (eds.) *Competence-Based Innovation in Hospitality and Tourism* (pp. 197–224). Abingdon: Routledge.

Camillo, A. (2012). 'A strategic investigation of the determinants of wine consumption in China', *International Journal of Wine Business Research*, 24(1), 68–92.

Capitello, R.C., Charters, S., Menival, D., & Yuan, J. (2017). *The Wine Value Chain in China: Consumers, Marketing and the Wider World*. Cambridge, MA: Chandos Publishing.

Cochrane, J., Chen, H., Conigrave, K.M. & Hao, W. (2003). Alcohol use in China. *Alcohol and Alcoholism*, 38(6), 537–542.

Euromonitor International (2017). *Wine in China*. London: Euromonitor.

Gössling, S. & Hall, C.M. (2013). 'Sustainable culinary systems: an introduction', in C.M. Hall & S. Gössling (eds.) *Sustainable Culinary Systems: Local Foods, Innovation, and Tourism & Hospitality* (pp. 3–44). London: Routledge.

Guo, X. & Huang, Y. (2015). 'The development of alcohol policy in contemporary China', *Journal of Food and Drug Analysis*, 23(1), 19–29.

Hall, C.M. (1996). 'Wine tourism in New Zealand', in G. Kearsley (ed.) *Tourism Down Under, Tourism Research Conference* (pp. 109–119). Dunedin: Centre for Tourism.

Hall, C.M. (2003). 'Biosecurity and wine tourism: is a vineyard a farm?', *Journal of Wine Research*, 14(2–3), 121–126.

Hall, C.M. (2010). 'Tourism destination branding and its effects on national branding strategies: Brand New Zealand, clean and green but is it smart?', *European Journal of Tourism and Hospitality Research*, 1(1), 68–89.

Hall, C.M. (2012). 'Boosting food and tourism-related regional economic development', in OECD, *Food and the Tourism Experience: The OECD-Korea Workshop* (pp. 49–62). Paris: OECD Publishing.

Hall, C.M. (2013). 'The local in farmers' markets in New Zealand', in C.M. Hall & S. Gössling (eds.) *Sustainable Culinary Systems: Local Foods, Innovation, and Tourism & Hospitality* (pp. 99–122). London: Routledge.

Hall, C.M. (2016). 'Heirloom products in heritage places: farmers' markets, local food, and food diversity', in D. Timothy (ed.) *Heritage Cuisines: Traditions, Identities and Tourism* (pp. 88–103). Abingdon: Routledge.

Hall, C.M. (2017). '100% pure neoliberalism: Brand New Zealand, new thinking, new stories, Inc.', in L. White (ed.) *Commercial Nationalism: Selling the National Story in Tourism and Events* (pp. 102–125). Bristol: Channelview.

Hall, C.M. & Baird, T. (2014). 'Brand New Zealand wine: architecture, positioning and vulnerability in the global marketplace', in P. Howland (ed.) *Social, Cultural and Economic Impacts of Wine in New Zealand* (pp. 105–119). Abingdon: Routledge.

Hall, C.M. & Gössling, S. (2016). 'From food tourism and regional development to food, tourism and regional development: themes and issues in contemporary foodscapes', in C.M. Hall & Gössling, S. (eds) *Food Tourism and Regional Development: Networks, Products and Trajectories* (pp. 3–57). Abingdon: Routledge.

Hall, C.M., Johnson, G., Cambourne, B., Macionis, N., Mitchell, R. & Sharples, L. (2000). 'Wine tourism: an introduction', in C.M. Hall, E. Sharples, B. Cambourne & N. Macionis (eds.) *Wine Tourism Around the World: Development, Management and Markets* (pp. 1–23). Oxford: Butterworth-Heinemann.

Hall, C.M. & Page, S. (eds) (2017). *The Routledge Handbook of Tourism in Asia*. Abingdon: Routledge.

Hall, C.M. & Sharples, E. (2003). 'The consumption of experiences or the experience of consumption?: An introduction to the tourism of taste', in C.M. Hall, E. Sharples, R. Mitchell, B. Cambourne, & N. Macionis (eds.) *Food Tourism Around the World: Development, Management and Markets* (pp. 1–24). Oxford: Butterworth-Heinemann.

Höllmann, T.O. (2010). *The Land of the Five Flavours: A Cultural History of Chinese Cuisine*. New York: Columbia University Press.

International Organisation of Vine and Wine (OIV) (2016). *State of the Vitiviniculture World Market*. Paris: OIV.

Iskyan, K. (2016). 'China's middle class is exploding', *Business Insider*, 27 August: http://www.businessinsider.com/chinas-middle-class-is-exploding-2016-8?IR=T

Ito, T. (2010). 'China as number one: how about the renminbi?' *Asian Economic Policy Review*, 5(2), 249–276.

Lawrence, D. (2015). 'Canada hopes Chinese consumers warm up to "cool-climate" wines', *The Globe and Mail*, 26 May.

Lee, K., Madanoglu, M. & Ko, J-Y. (2016). 'Exploring key service quality dimensions at a winery from an emerging market's perspective', *British Food Journal*, 118(12), 2981–2996.

Liu, H., McCarthy, B., Chen, T., Guo, S. & Song, X. (2014). 'The Chinese wine market: a market segmentation study', *Asia Pacific Journal of Marketing and Logistics*, 26(3), 450–471.

Lotta, R. (2009). 'China's rise in the world economy', *Economic and Political Weekly*, 44(8), 29–34.

Low, W. & Davenport, E. (2006). 'Mainstreaming fair trade: adoption, assimilation, appropriation', *Journal of Strategic Marketing*, 14, 315–327.

Malm, K., Gössling, S. & Hall, C.M. (2013). 'Regulatory and institutional barriers to new business development: the case of Swedish wine tourism', in C.M. Hall & S. Gössling (eds.) *Sustainable Culinary Systems: Local Foods, Innovation, and Tourism & Hospitality* (pp. 241–255). London: Routledge.

McGrath, M.A., Sherry, J.F., Jr. & Heisley, D. (1993). An ethnographic study of an urban periodic marketplace: lessons from the Midville farmers' market. *Journal of Retailing*, 69(3), 280–309.

New Zealand Trade and Enterprise (2015). *Wine in China*. Wellington: NZTE.

Qing, P., Xi, A. & Hu, W. (2015) 'Self-consumption, gifting, and Chinese wine consumers', *Canadian Journal of Agricultural Economics/Revue canadienne d'agroeconomie*, 63(4), 601–620.

Qiu, H. Z., Yuan, J.J., Ye, B.H. & Hung, K. (2013). 'Wine tourism phenomena in China: an emerging market', *International Journal of Contemporary Hospitality Management*, 25(7), 1115–1134.

Roberts, J.A.G. (2004). *China to Chinatown: Chinese Foods in the West*. London: Reaktion Books.

Roy, H., Hall, C.M. & Ballantine, P. (2016). 'Barriers and constraints in the use of local foods in the hospitality sector', in C.M. Hall and S. Gössling (eds). *Food Tourism and Regional Development: Networks, products and trajectories* (pp. 255–273). Abingdon: Routledge.

Sun, B. & Wong, J. (2015). 'China: navigating the lucrative Chinese wine market', *Wine and Viticulture Journal*, 30(2), 68–70.

Tourism Industry Association of Canada & Canadian Vintners Association (2015). *Wine Tourism in Canada*. Ottawa: Tourism Industry Association of Canada & Canadian Vintners Association.

Vinexpo (2015). *The Wine and Spirits Market in China*. Bordeaux: Vinexpo.

Warde, A. & Martens, L. (2000). *Eating Out: Social Differentiation, Consumption and Pleasure*. Cambridge: Cambridge University Press.

Warde, A., Martens, L. & Olsen, W. (1999). 'Consumption and the problems of variety: cultural omnivorousness, social distinction and dining out', *Sociology*, 33(1), 105–127.

Willsher, K. (2014). 'China becomes biggest market for red wine, with 1.86bn bottles sold in 2013', *The Guardian*, 29 January: https://www.theguardian.com/world/2014/jan/29/china-appetite-red-wine-market-boom

The World Bank (2017). China. http://www.worldbank.org/en/country/china

World Tourism Organization (2014). *AM Reports, Volume eight: Global Report on Shopping Tourism*. Madrid: UNWTO.

Ye, B.H., Zhang, H.Q. & Yuan, J.J. (2014) 'Intentions to participate in wine tourism in an emerging market: theorization and implications', *Journal of Hospitality & Tourism Research*, DOI: 10.1177/1096348014525637.

Zheng, Q. & Wang, H. (2016). 'Market power in the Chinese wine industry', *Agribusiness: An International Journal*, 33(1), 30–42.

Zhou, J., Arnold, M., Pereira, A. & Yu, J. (2010). 'Chinese consumer decision-making styles: a comparison between the coastal and inland regions', *Journal of Business Research*, 63(1), 45–51.

3 Food and wine tourism in China

Na Su, Bin Dai and Xiujuan Jin

Introduction

With the shift away from sightseeing to more leisure and vacation orientated travel, Chinese tourists aim for a more meaningful experience and a more in-depth understanding of a destination. In this context food and wine tourism have become two of the fastest growing sectors in China's tourism market. Food and wine not only constitute an integral part of travelling but are considered as one of the best ways to encounter a place and its customs. In many cases, food and wine have even been the trigger factor for tourists to visit a particular place, and thus many cities in China are striving to brand themselves as 'Cities of Gastronomy'. This chapter aims to provide some background information on the development of China's food and wine tourism by analysing both the supply and demand sides.

China's tourism market

With increasing disposable income, China is facing exponentially growing tourism demand. According to statistics from the China National Tourism Administration (CNTA) for 2016 domestic visitors surpassed 440 million arrivals, and 120 million

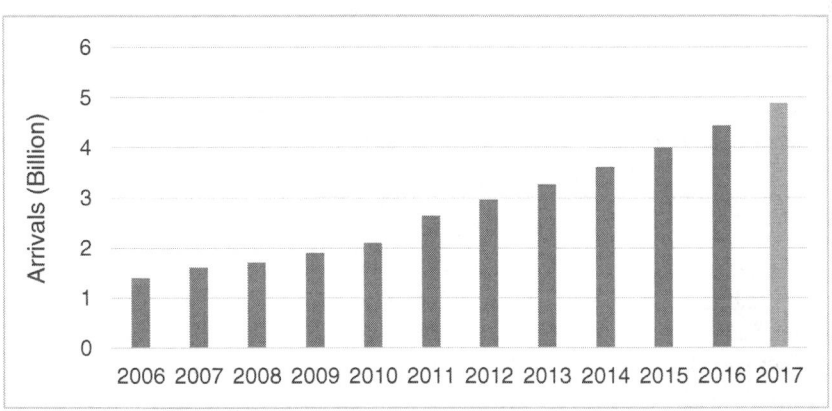

Figure 3.1 Domestic visitor arrivals, China, 2006–2017

Source: China National Tourism Administration

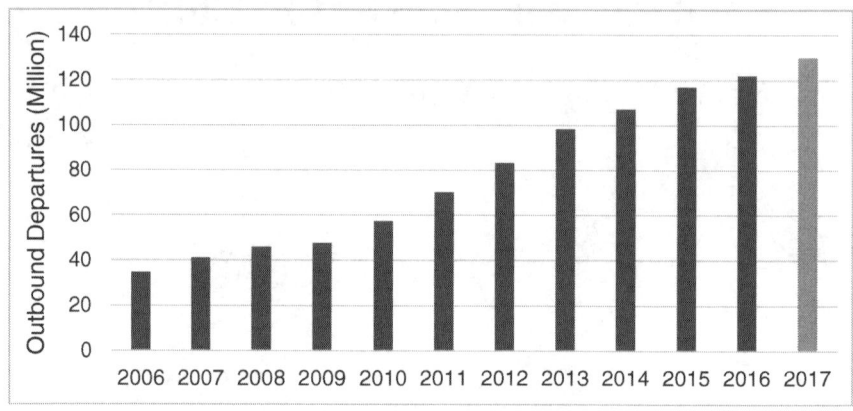

Figure 3.2 Outbound departures of China, 2006–2017
Source: China National Tourism Administration

outbound departures were recorded, which is three times more than ten years ago. This growth is predicted to continue for the next decade. It is estimated by CNTA that in 2017 the overall domestic arrivals will be close to 500 million, and there will be around 130 million tourists travelling outside of China.

Along with the growing size of China's tourism demand, China's tourism industry is also facing a shift in tourists' preferences and behaviour. According to longitudinal data obtained by the China Tourism Academy (CTA), the share of tourists who travelled for sightseeing purposes dropped to 50.8% in 2016 from 66% in 2009. At the same time the ratio of leisure tourists has grown to 52.57%, exceeding sightseeing as a key travel motivation. This implies that the traditional tourism pattern of Chinese tourists, who tended to visit many attractions within one day and glance over things hurriedly, is changing. Instead, they have started to slow down their travel pace and engage more with the local culture and residents in order to have a more authentic experience. The growth in China's middle to higher income class and the emergence of the next generation of tourists has especially led to this new trend.

In turn, these new travel patterns by Chinese tourists have also significantly influenced the choice of destination. Destinations offering vacation and leisure experiences are increasingly preferred. According to the *Global Independent Travel Report 2016* by CTA, which is based on big data analysis of an online travel agency (OTA) company in China, the top ten most desirable overseas destinations for Chinese tourists were Mauritius, Chiang Mai (Thailand), Hokkaido (Japan), Tahiti, Saipan, Paris (France), Santorini (Greece), Fiji, Venice (Italy) and the Great Barrier Reef (Australia). It can be seen from this list that the most desirable destinations are islands or cities that offer unique lifestyles, like Hokkaido or Paris.

A shift in the pattern of destination choice can also be seen in the domestic market. CTA's *Global Independent Travel Report 2016* also shows that more traditional tourism resources like natural scenery and cultural heritage were no longer

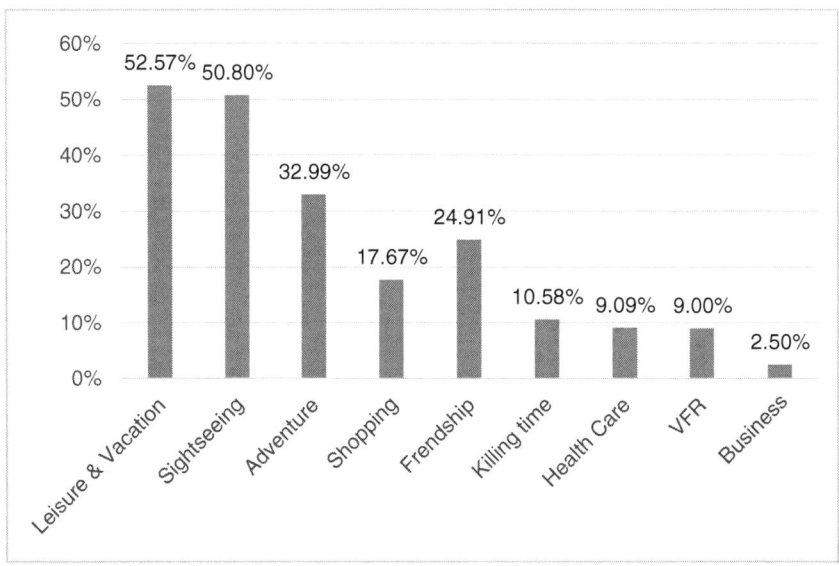

Figure 3.3 Tourism motivations of Chinese tourists, 2016
Source: Survey by China Tourism Academy (2017)

Table 3.1 Top 10 most desired destinations of Chinese tourists

Top 1	Mauritius
Top 2	Chiang Mai
Top 3	Hokkaido
Top 4	Tahiti
Top 5	Saipan
Top 6	Paris
Top 7	Santorini
Top 8	Fiji
Top 9	Venice
Top 10	Great Barrier Reef

as appealing as before to Chinese tourists. In contrast, tourists were increasingly interested in experiences that constitute an integral part of local residents' daily lives. For example, Chunxi Rd, a road in Chengdu that is used by local residents for shopping and dining, has become a must-visit destination in Chengdu. The same applies to Zhongshan Rd in Xiamen and Nanjing Rd in Shanghai.

The above changes in China's tourism demand strongly indicate that more and more tourists have started to value cultural experiences offered by destinations. In this regard, food and wine as two cultural agents of a destination will increasingly draw attention from tourists.

Food tourism in China

Food tourism refers to any tourism experience in which one learns about, appreciates and/or consumes food and drinks that reflect the local, regional or national cuisine, heritage and culture (Hall & Sharples, 2003). Food is an integral part of a travelling experience, not only because it provides energy supportive to our physical functions but is also an agent to make the local culture accessible by telling stories of a destination's history, people and customs. Food is a way for tourists to encounter a place, and to enrich and leverage their perception and experience of a destination (Chang et al., 2010).

Since more and more tourists value food experiences in destinations, food has become an important motivator for tourism. 'Foodies' can even initiate a trip for the sake of food. The 2013 *American Culinary Traveler Report* published by Mandala Research in the U.S., for example, showed that the percentage of U.S. leisure travellers who travel specifically to learn about and enjoy unique dining experiences grew from 40% to 51% between 2006 and 2013. Similarly, the report on *Chinese Catering Consumption 2015* showed that the market size of the catering industry reached 3.2 trillion YEN, with a growth of 16%, and predicted that this growth rate will continue to accelerate.

Importance of food

Food is a major contributor to a destination's economy. Tourist expenditure on food makes up approximately one third of the total of all spending in a destination (Hipwell, 2007) and it is also a major motivator for tourists to visit a destination. The China Tourism Academy (CTA), affiliated with China National Tourism

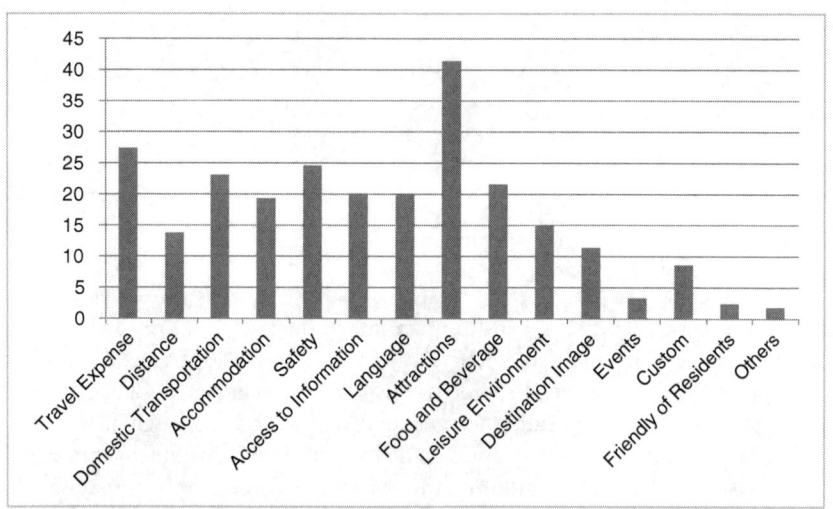

Figure 3.4 Importance of visiting factors for inbound tourists in China, 2015

Administration, has surveyed both the domestic and international Chinese tourism market over a long time period of time. Based on a quarterly conducted survey by CTA, it shows that in 2015 21.6% of inbound tourists to China declared that food was the fifth most important factor for deciding on a destination, after attractions, travel expenses, safety and domestic transportation.

It was also observed that inbound tourists in China were keen to experience natural scenery, cultural heritage and the arts, as well as tasting local food and beverages.

For domestic tourists, food and beverages were ranked as the fourth most influential factor during a travel decision, more important than accommodation and leisure atmosphere.

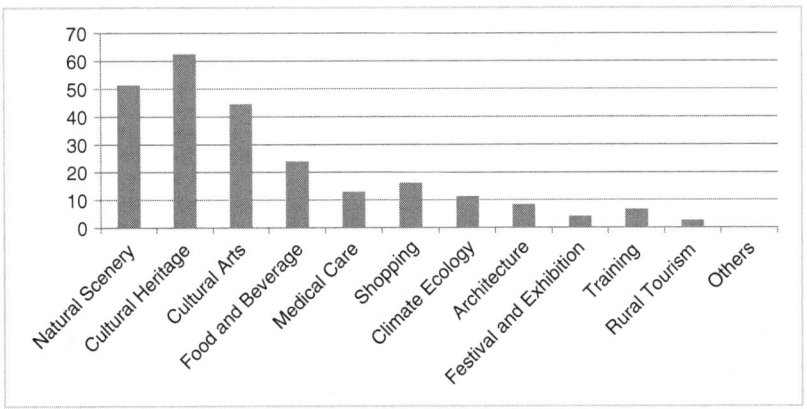

Figure 3.5 Preferred destinations and activities of inbound tourists in China, 2015

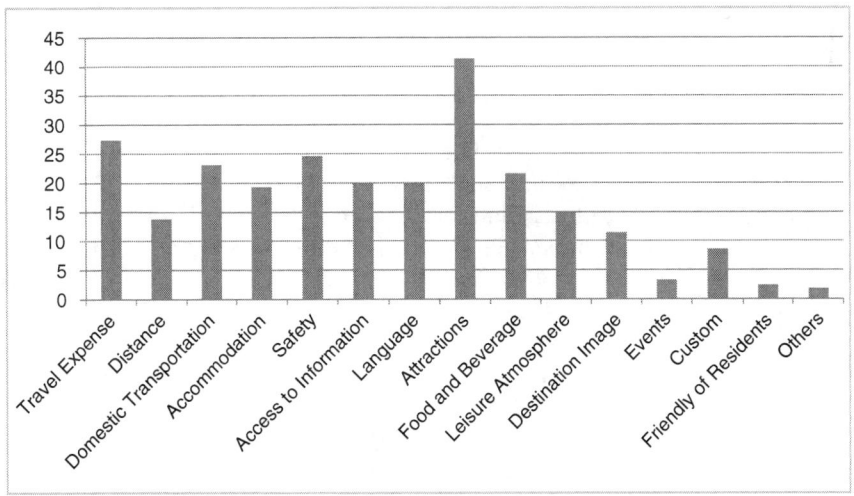

Figure 3.6 Importance of visiting factors for domestic tourists in China, 2015

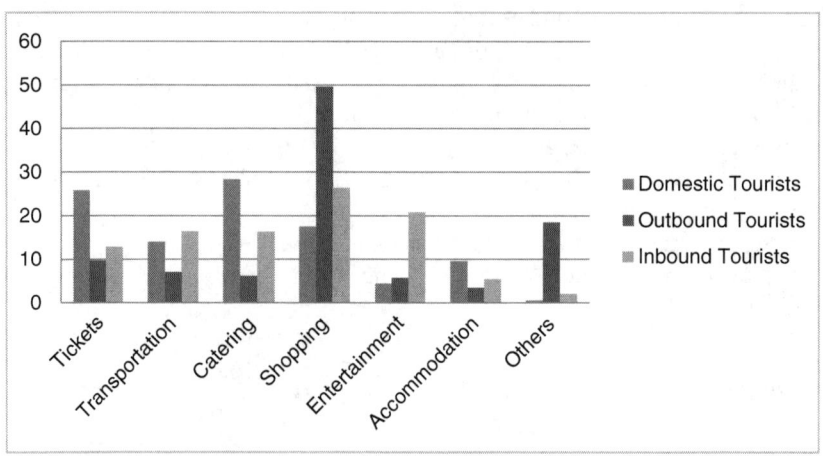

Figure 3.7 Percentage of expenditure for each tourism segment

Food can even be on the top agenda of tourists when visiting destinations that are renowned for the culinary experiences they offer. A study conducted by Chen (2015) showed that, after sightseeing, tasting local food was the second most important reason why tourists visited Chongqing, a city in the southwest of China that is famous for its spicy food and hot pot cuisine. Food is also often perceived as representing the image of a destination, as it is easy to be retrieved from memory and connected to a specific place. In this vein, Chen's (2015) study also demonstrated that tourists are more likely to perceive Chongqing as the 'Paradise of Hot Pot' rather than the 'City of Mountains' (as Chongqing is built on mountains). In this regard, food is helpful and beneficial in identifying a destination as well as enhancing its tourism identity (Boniface, 2003).

Chinese food tourism and its trends

China possesses a tremendous advantage to develop food tourism because of its well-known diversity, flexibility and adaptability of food (Chang et al., 2010). The culinary culture of China, which has developed for centuries, is overwhelming. There is an old saying in China, 'food is the first necessity of man', which reflects the importance of food to Chinese people. Learning how to plant, harvest and process food has a very long tradition in Chinese culture. Similarly, culinary skills and creations have developed over a very long time. In this sense, food is considered as an important reflection of China's national and regional culture, and a necessary way to experience and understand a region's history.

Due to this long history of food in China, Chinese tourists appreciate food experiences more than others. According to a report published by the China Tourism Academy (2016), the most frequently used keywords of Chinese independent tourists when searching destinations are 'nature', 'food' and 'shopping'.

This indicates that food is no longer a supporting factor to satisfy basic needs of tourists during the visit but has become a major motivating factor for tourists to visit a destination.

In recent years, food tourism was further boosted by TV shows and documentaries with regard to food, such as the *A Bite of China*, a series of food documentaries broadcasted on CCTV. Interestingly, the location of shooting for this gourmet TV show has been a stimulus for tourism so that the featured destinations have become top travel choices for people. In particular, rural areas in China benefit from this so called 'tongue economy' and are able to enhance their tourism resources to disseminate information about their local cultures. Triggered by the 'tongue economy', some travel agencies and travel websites have launched a series of food-themed routes, so viewers can follow the gourmet trail featured in the show in pursuit of the flavour of different foods. Ctrip, a Chinese OTA company, for example, has developed and promoted a set of themed productions, and has launched more than 20 tourist routes, mainly around the places involved in the show *A Bite of China*, such as Hainan, Yunnan, Beijing, Hunan and 20 other provinces and cities.

Chinese tourists also yearn for global food experiences. To date, some classic food tours are on offer that are strongly sought after by Chinese tourists, like Japan's Michelin restaurant tour. Hong Kong, on the other hand, is currently integrating its restaurant resources into distinctive food tours as a new stimulus to Hong Kong's tourism economy. Caesar Tourism, as a leading outbound travel service provider in China, has put hundreds of 'Snack Black and Big Stomach' food tourism products to the market. These products are related to Southeast Asian snacks, Japanese cuisine, European delicacies and fresh Australian and American tastes. They focus on more than ten high quality tourist destinations such as Canada, the United States, France, Spain, Morocco, Australia, New Zealand, Thailand, Hokkaido, Okinawa, Japan Kansai, Singapore and Taiwan.

However, the food tours offered in China's market, no matter if they incorporate domestic or global routes, are all centred on tasting experiences with little regard given to cooking and learning about local foods. It is anticipated, however, that future travellers will seek more unique and authentic-to-the-locale experiences than ever before; food tours that features cooking and learning will thus develop into an emerging consumer trend.

Wine tourism in China

Wine tourism is defined by Hall et al. (1997, p. 6) as 'visitation to vineyards, wineries, wine festivals and wine shows for which grape tasting and/or experiencing the attributes of a grape wine region are the prime motivating factors for visitors.' Wine tourism as an alternative to traditional tourism resources has increasingly developed into a way to express one's lifestyle and unique experience. This is not surprising since wine and tourism are naturally connected as both are connoted with enjoyment, relaxation and cultural experiences.

Wine tourism is conducted around wine culture, which manifests in either visible equipment or sensory experiences related to wine production, delivery, tasting and consumption. Wine tours normally include a series of activities such as experiencing wine customs, visiting often very scenic vineyards to understand grape varieties, visiting the factory to learn more about the winery and its production processes, tasting wine and engaging in leisure entertainment and shopping activities (Bruwer & Alant, 2009).

Wine tourism has been well developed in wine-producing countries such as Australia, Italy, the United States, South Africa, Moldova and others. However, although China has a 6,000-year long history of growing grapes and a 2,000-year long history of winemaking, wine consumption in China amongst the middle class only began around the 1990s, after China implemented its policy of opening up to the outside world. Chinese tourists have not engaged in wine tourism until recently, when China's wine industry experienced a rapid growth period along with China's and the global economy. In 2006, according to a report released by the International Organisation of Vine and Wine (Qiu et al., 2013), China had over one million acres of vines and produced more than 490,000 tons of wine, becoming the fifth largest country by vineyard area, and the seventh top wine producer in the world.

Today, driven by a fast growing economy, China has risen to the top six wine consuming nations in the world, with annual growth rates estimated to be seven times stronger than the global average (Qiu et al., 2013). Further, more and more Chinese people are starting to become aware of wine as an important beverage to complement a dining experience. A report released by the University of South Australia, *The China Wine Barometer* (2013), shows that Chinese consumers are most aware of red wine and beer, followed closely by champagne, white wine and whisky.

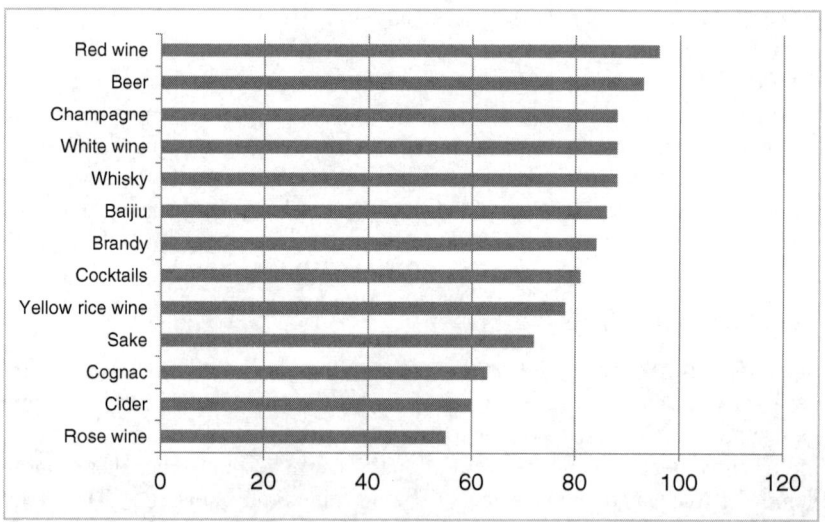

Figure 3.8 Awareness of Chinese consumers among beverages

Motivations of Chinese to consume wines

Traditionally, wine serves as an accompaniment to food, and thus its intrinsic characteristics such as flavour and aroma are strongly valued. Today wine is traded globally, even in countries where wine is not a traditional beverage. Wine often symbolises an elegant and privileged Western lifestyle. This is certainly the case in China, where people consume wine in order to display social status and to signal the lifestyle they would like to pursue. Thus, the strongest demand for wine in China comes from high-level Chinese bureaucrats and wealthy businessmen, and the most favoured wines are top ranked Bordeaux wines (Lockshin et al., 2017). Furthermore, wines are often purchased as gifts but rarely consumed for pleasure. Consumers from China and Western countries thus value very different aspects of wines. For instance, consumers in France, Italy or Germany appreciate the quality of wine, while Chinese customers purchase wines with the consideration of brand reputation, price and appearance, features that constitute a wine as a suitable gift (Qiu et al., 2013).

However, things have changed over the last few years in China, especially since the emerging middle class has triggered the mass consumption of wine. Wine is no longer connected with the image of elitism and has become a beverage that is often consumed in the home environment. Motivations for Chinese consuming wines have thus significantly changed over the past few years.

Lockshin et al.'s (2017) study is one of few that sheds light on Chinese wine consumption behaviours. The study is based on the *China Wine Barometer* (CWB), a twice-yearly survey conducted in China. According to the survey, 'wine is good for my health' is the top ranked reason why Chinese consume wine, followed by 'wine helps me to relax', 'wine helps me to create a relaxed and friendly atmosphere' and 'I like the taste of wine'.

Different drivers for drinking wine were also identified between on-premise (restaurants, bars, hotels) and off-premise (retail) situations (Lockshin et al., 2017).

Figure 3.9 Ranking of wine drinking motivations in China, March 2013

The findings of the study show that Chinese tend to choose a wine they have had before in an on-premise situation, potentially because the time for making decisions is short and they lack information on other wines to make a comparison. Recommendation from others at the same table is also an important prompt for decision-making for on-premise wine consumption. In addition, the survey also showed that consumers buy wine on-premise that matches the food they have ordered.

In a retail setting, attributes related to a wine's quality are strongly valued. Most of the top ranked drivers for off-premise wine consumption are related to quality indicators like medals and awards won by a wine, the grape variety, vintage and the country of origin. Consumers in China also tend to take recommendations from friends or family for off-premise wine consumption.

China's wine market has seen a significant shift of wine consumption across generations: Older consumers (46+ years) tend to pay more attention to their health and also tend to buy wine for personal use or for treating friends and family, while the importance of business communications and gifting ranks much lower. Consumers aged approximately 36 to 45 years, at the pinnacle of their careers, are commonly engaged in many business activities and frequently tend to choose red wine as a gift. However, they don't tend to treat their friends with wine. Consumers in the next age bracket (26 to approximately 35 years) are progressing in their careers, with more frequent social contacts, more business activities and gifting opportunities. In addition, consumers in this age group have gradually established their families and tend to prefer a healthy and fashionable lifestyle. They commonly buy red wine

Figure 3.10 Ranking of choice drivers for wine in the Chinese on-premise sector, March 2013

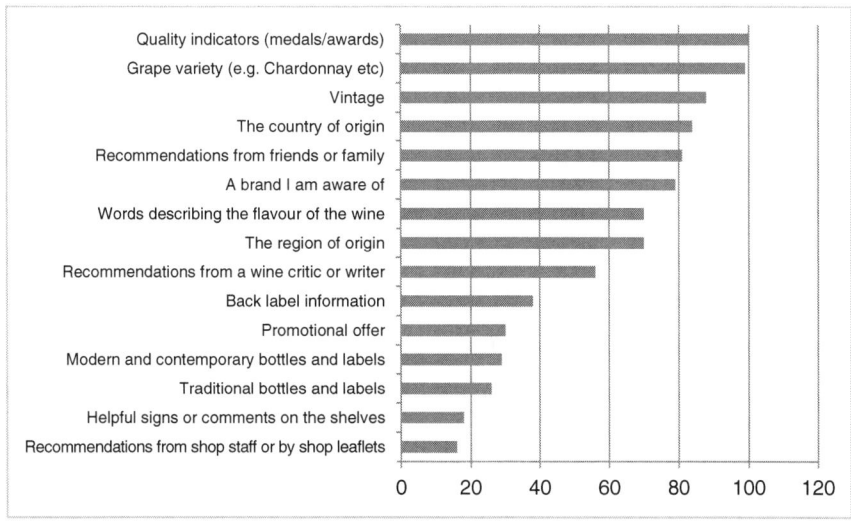

Figure 3.11 Ranking of choice drivers for wine in the Chinese off-premise sector, October 2013

mainly for personal use or for treating friends. Young Chinese wine consumers aged 18 to about 25 tend to actively pursue a romantic and also social lifestyle. They tend to purchase red wine mainly for their own enjoyment or when entertaining friends and family. Wine as a means of nurturing business relations and for gifting purposes is not very important to this consumer segment (China Wine Market, 2014).

Development of wineries in China

In light of the fast growing Chinese economy and globalisation, China has fostered the development of wineries to meet increasing demand for red wines. A report (ASKCI, 2016) in China shows there are at least 233 wineries in China, most of them located in Ningxia (36 wineries) and Xinjiang (28 wineries), the western side of the country, as well as in Shandong (53 wineries), Dongbei (23 wineries) and Hebei (42 wineries) along the east coast of the country. Wineries like Grace Vineyard, Silver Heights, Zhang Yu, Dynasty Winery and Great Wall have enjoyed growing popularity and have developed into must-visit destinations for local residents and visitors alike.

Despite competition from imported wine brands, Chinese wines have performed well in recent years. Based on a report released by a Chinese research institute (ASKCI), three Chinese wineries, Zhang Yu, Great Wall and Dynasty, held the greatest market share, accounting for almost half of all wine sales in China.

Among the Chinese wineries, Zhang Yu, which is located in Yantai, Shangdong province, has become the most popular destination for tourists to visit. As early

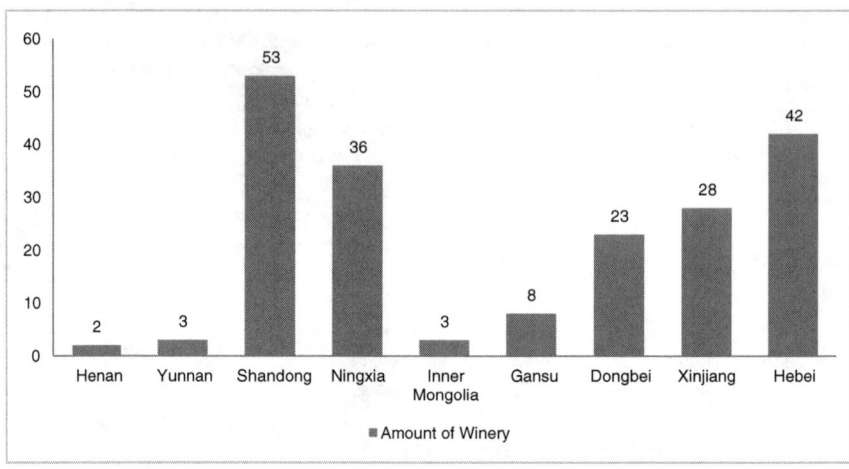

Figure 3.12 Amount of wineries in provinces of China

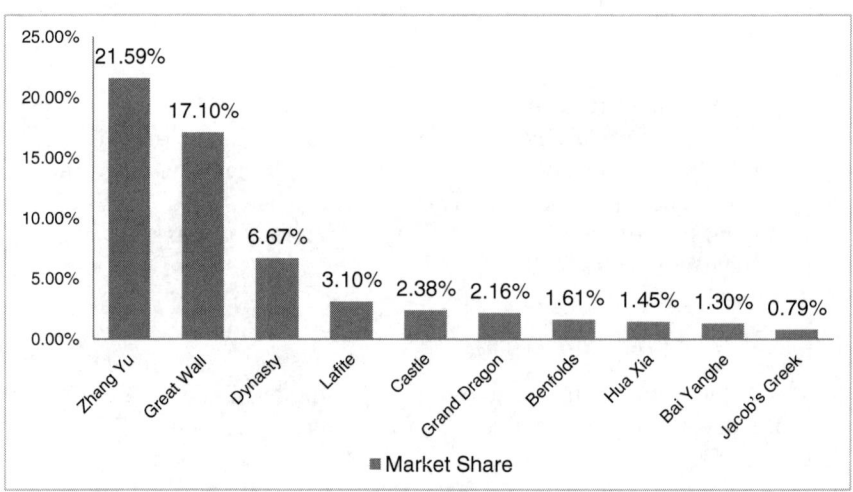

Figure 3.13 Market share of the most consumed wines in China
Source: Data cited from www.askci.com/reports/

as in 2000 the winery had already welcomed more than 56,000 tourists and generated 560 million Yuan in income. Tourists from overseas accounted for 30% of total visitation. In light of the growth of China's wine tourism market, Zhang Yu invested 42 million Yuan to expand their plaza and develop a new Wine Culture Museum Plaza as well as to renew their century-old underground wine cellars and spent 37 million Yuan to extend the areas that can be visited (Qiu et al., 2013).

Although many wineries have realised the importance of developing tourist facilities, many of them are still at the exploratory stage. They have been used to running a winery as a wine-production sector rather than a tourism spot. In order to facilitate the development of wine tourism, it appears that China's wineries have to overcome some barriers.

A study conducted by Qiu et al. (2013), for example, identified both positive and negative factors with regard to people, promotion and place:

People

As is shown in Table 3.2, wine tourists in China enjoy the ambiance of appealing natural sceneries, which constitutes an essential attribute of wineries, as well as tasting wines and grapes and learning about the history of wine and generally expanding their wine knowledge. They also appreciate the experience of customising a bottle of wine in a winery. However, wine tourists are also often reluctant to visit a winery again because they realise that there is little chance to engage in winemaking, and still lack interest in wine as it doesn't constitute a part of their culture and because they still have very limited wine knowledge. A wine's quality is of critical importance to Chinese wine tourists. It is still widely believed that wine made in China varies in quality, and thus Western-produced wines are still preferred. Against this backdrop, Chinese wineries have to win consumers' confidence in the quality of their products in order to be able to expand the country's wine tourism sector.

Promotion

China's wineries have been engaged in a variety of wine-related activities including grape picking, winemaking, wine appreciation and other leisure activities within or outside of the wineries. However, there is limited promotion of these activities and in particular visual promotional material is lacking. Another barrier to China's wine tourism is that very few wineries are run as traditional tourism facilities by connecting to online or offline travel agencies as channels of promotion. Many wineries also try to attract high-income consumers, which visit the facilities by appointment only, which makes it harder for the general public to access these wineries.

Place

China's wineries feature comprehensive functionality and fine physical conditions like modern facilities, well-designed architectures, lush vineyard landscapes and spacious tasting rooms. However, many facilities lack ingredients of local culture or other unique features that can distinguish them from others. In addition, in many instances there are no other attraction points nearby and some wineries are too focused on selling without sufficient human interaction (e.g. assisting customers in learning more about wine production and expanding their knowledge of wine).

Table 3.2 Facilitating and detrimental factors facing China's wineries

	Facilitating factors	Detrimental factors
People	• Sightseeing of the beautiful natural sceneries and architectures of the wineries. • Wine-tasting and grape-tasting. • Use the machine to make an individualised bottle of wine. • Participate in the grape picking activities in harvest season. • Learn some wine history and culture, winemaking, wine appreciation and wine drinking customs. • Generally favourable perceived service. • Understand that red wine has health benefits. • Friendly employees.	• On most visitations, little chance of engaging in winemaking activities. • Lack of knowledge of wine and assessment of wine quality. • Cultural differences between China and the West in terms of wine drinking. • Large variation in wine quality in China. • Lack of consumer confidence of local wine (i.e. wine made in China). • Mostly low in interest.
Promotion	• A proposed wine and food festival. • Demonstrating history of winemaking in China. • Knowledge of wine classification and production process. • Health implications of wine drinking. • Plan to establish promotional office and encourage collaboration among stakeholders.	• Only limited information on leisure activities within wineries. • Lack of promotion on wine-related activities that tourists can participate (e.g. winemaking). • Not enough channels for promotion. • Need genuine government support.
Place	• Impressive lushness of vineyards. • State-of-the-art wine production. • Modern and spacious tasting room. • Exotic style architectures. • Chinese and Western-style restaurants. • There are other tourism attractions nearby.	• Facilities lack ingredients of local or Chinese culture. • Nearby attractions may not be as attractive as other attractions in Shandong province. • Focus on selling, not much human interactions on wine production.

Opportunities and challenges facing China's wine tourism

Today, wine is socially more accepted and has become a more common product of choice for many Chinese. Growth in the younger generation and the middle class will further boost consumption of wine and tourism activities around it. As the *Wine Consumption Survey* conducted by the China Industry Information Network showed, China's wine consumer groups are mainly born after the 1970s as this market segment accounts for more than half of the overall wine consumption. Those born in the 1990s, more than 20% of China's population, will become a new consumption force for red wines as it is estimated that the frequency of wine consumption within this segment of the population will be even higher than that of those born in the 1980s and 1970s (China Wine Market, 2014).

China's growing middle class provides a further opportunity to increase wine related consumption because China's new wealthy class (defined as the 'three high' group, those with high spending patterns, high levels of education and high income), males, people between the age of 30 to 34 years, senior managers, professionals and technical personnel are the main consumers of wine (China Wine Market, 2014). In 2015, according to a report by CHFS, the number of China's urban middle-income class had reached 230 million. In other words, the middle class accounted for 16.42% of the total population (CHFS, 2015).

The new wealthy population subgroup also comprises a considerable number of overseas returners, who were influenced by foreign wine culture and attained a wealth of wine knowledge and habit of drinking wine. For example, in 2015, the number of people who studied overseas was nearly 523,000, and those who returned back from overseas to China more than 400,000 (National Bureau of Statistics of China R.P., 2008–2015). Their opinions on wine are often adopted by people who know very little about wine.

Pertaining to the political environment, the 'Supply-side reform' policy initiative by China's government called for the transforming and upgrading of traditional industries. It was suggested by the government that integrating tourism functions could be a way to overcome stagnations seen in many traditional industries. At the National Industrial Tourism Innovation Conference, which was held at the end of 2016, the China National Tourism Administration published the standards for industrial tourism: 'Norms and Assessment on the National Industrial Tourism Area (Site)' and the 'National Industrial

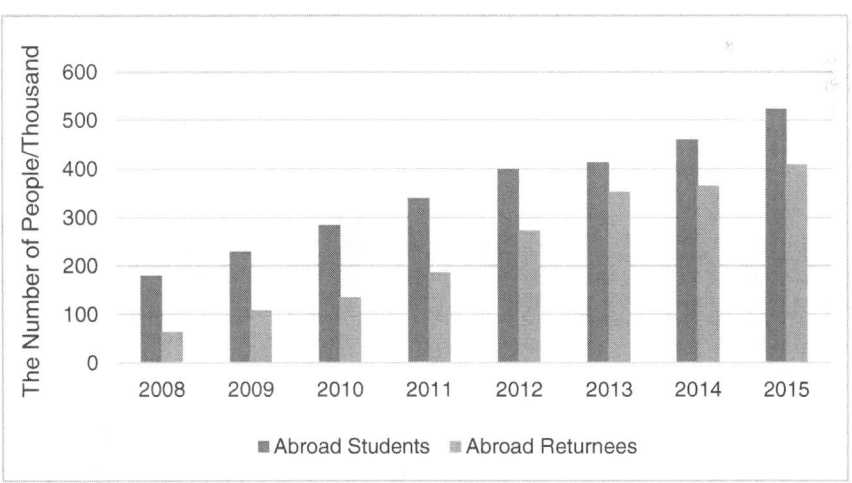

Figure 3.14 China's abroad students and abroad returnees, 2008–2015
Source: Data cited from http://www.stats.gov.cn

Tourism Development Program' (2016–2025). The China National Tourism Administration promoted 22 companies as examples of industrial tourism innovation. Ruina Castle winery owned by Zhang Yu in Shanxi Province was one of them. The China National Tourism Administration emphasised that enterprises should tap into industrial tourism resources, create a healthy environment for the development of industrial tourism and innovate more industrial tourism products. The aspiration should be to make industrial tourism a shining star in the development of tourism and create new fusion products, transforming and upgrading existing facilities to stimulate tourism investment and consumption.

Meanwhile, China's wine tourism is also facing severe challenges. First of all, consumption by officials, which used to be the main wine consuming force, has been prohibited since 2012. The Political Bureau of the CPC Central Committee introduced eight provisions to regulate the behaviour of Chinese officials. They are now, for example, prohibited from participating in any wining and dining, which necessitated a reorientation of the Chinese wine consumption to support growth in this industry.

Second, a reduction in tax for imported wines will intensify competition for domestic wineries. The Chinese government has, for example, signed free trade agreements with Chile, New Zealand and other countries that specified that import tariffs of alcoholic products would be reduced over time and eventually removed altogether. The upcoming signing of the China–Australia Free Trade Agreement might increase further Australian wine sales on the Chinese market.

In addition, Chinese investors buying wineries overseas will place domestic wineries under pressure as they can link Chinese tourists to oversea resources. Jack Ma, China's second richest person and founder of the multi-billion dollar online retailer Alibaba, also entered the growing list of Chinese investors in Bordeaux wines and became one of the highest profile château buyers in the French wine region. His company Alibaba also owns Fliggy, an online travel agency in China, which has more than 200 million registered users who access the site up to 10 million times daily. Thus, an increasing visitation of his overseas wineries by Chinese tourists can be anticipated.

Conclusion

Tourism consumption has become an integral part of people's lives all over the world. With the fast expansion of China's tourism market and a steady increase in Chinese travel experiences, traditional tourism destinations and attractions can no longer satisfy tourists' demands. An experience of uniqueness and profoundness is increasingly expected in a destination. As two cultural agents, tourists have become familiar with and have learnt about local culture through food and wine, as well as the physical scenery where they consume them. In this regard, food and wine have started to play an increasingly important role in attracting tourists and constituting memorable trips.

China, as the largest country of tourism consumption globally, will be an important market force to food and wine tourism. To understand Chinese tourists' behaviour with regard to food and wine consumption is insightful for many destinations either within or outside of China. As the dominant market force is shifting from Generation X to Generation Y, Chinese consumption of food and wine has developed many new characteristics, such as that the younger generation values a more in-depth experience of local food or wine to acquire more knowledge of the local culture; further they also value the physical environment where they dine and drink.

China's governments have launched many stimulating policies and industrial funds and venture funds to foster the development of alternative tourism. In the future, more routes and products related to food and wine tourism will emerge to satisfy the diverse needs of tourists. However, current shortcomings in food and wine tourism, insofar that tourists have little chance of engaging in food- and wine-related endeavours, need to be addressed by tour and service suppliers to meet new demand trends.

References

ASKCI (2016). Trends of Chinese red wine consumption 2012–2016. Available at http://www.chyxx.com/research/201603/398198.html (accessed 12 October, 2017).

Boniface, P. (2003). *Tasting Tourism: Travelling for Food and Drink.* Aldershot: Ashgate.

Bruwer, J. & Alant, K. (2009). 'The hedonic nature of wine tourism consumption: an experiential view'. *International Journal of Wine Business Research*, 21(3), 235–257.

Chang, R., Kivela, J., & Mak, A. (2010). Food preferences of Chinese tourists. *Annals of Tourism Research*, 37(4), 989–1011.

China Wine Market (2014). Who is the Chinese wine-drinker? Available at http://www.zhongguo-wine.com/2014/11/25/who-is-the-chinese-wine-drinker/ (accessed 16 October, 2017).

China Academy of Social Science (2015). China household finance survey. Available at http://business.sohu.com/20151117/n426657041.shtml (accessed 16 October, 2017).

Corsi, A.M., Cohen, J. & Lockshin, L. (2016). The China Wine Barometer (CWB): a look into the future. Australia Wine Australia

Hall, C.M., Cambourne, B., Macionis, N. & Johnson, G. (1997). Wine tourism and network development in Australia and New Zealand: review, establishment and prospects. *International Journal of Wine Marketing*, 9(2), 5–31.

Hall, C.M. & Sharples, L. (2003). The consumption of experiences or the experience of consumption? An introduction to the tourism of taste. In C.M. Hall, L. Sharples, R. Mitchell, N. Macionis, B. Cambourne (eds) *Food Tourism Around the World.* Sydney: B.-H (pp. 1–24).

Hall, J., Shaw, M. & Doole, I. (1997). Cross-cultural analysis of wine consumption motivations. *International Journal of Wine Marketing*, 9(2), 83–92.

Qiu, H.Z., Yuan, J., Ye, B.H. & Hung, K. (2013). Wine tourism phenomena in China: an emerging market. *International Journal of Contemporary Hospitality Management*, 25(7), 1115–1134.

Hipwell, W.T. (2007). Taiwan aboriginal ecotourism: Tanayiku Natural Ecology Park. *Annals of Tourism Research*, 34(4), 876–897.

(The) International Organisation of Vine and Wine (2006). Situation of the world viticultural sector in 2006. Available at: www.winebiz.com.au/statistics/world.asp (accessed 10 May, 2010).

Lockshin, L., Corsi, A.M., Cohen, J., Lee, R. & Williamson, P. (2017). West versus east: measuring the development of Chinese wine preferences. *Food Quality and Preference*, 56(2017), 256–265.

The China Wine Barometer (CWB). A look into the future. Available at http://research.agwa.net.au/wp-content/uploads/2014/04/GWRDC_Wave1_Final_050613_new-2.pdf

4   Chinese outbound tourists
Food and beverages

Wolfgang Georg Arlt

Introduction

> Despite the central role of dining in the holiday experience, the interface between food and tourism has received scant research attention . . . In addition, most of the existing research focus is on Western tourists; studies based on non-Western and Asian tourists are scarce.
>
> (Chang, Kivela, Mak, 2010, p. 990)

China is the largest international tourism source market with regard to both the number of international trips as well as the amount spent. Even when dismissing trips to the Special Administration Regions (SARs), Hong Kong and Macau, as being not 'real' outbound travel for mainland Chinese, China is on an equal level with the USA and Germany, which used to be the biggest sources for international travellers. Every destination has made efforts to attract more Chinese visitors, as they tend to spend a lot of money and have, in many cases, be seen as the 'saviour' of ailing destinations (Rotzinger, 2013) such as, for instance, Switzerland. One of the great discussions on how to provide good service to Chinese visitors has been the question: *Do Chinese always want to eat Chinese food when they travel?* The Hospitality Association of Switzerland and SwissTourism already published a brochure in 2004 to assist the industry to adapt to Chinese visitors. The revised version, which is still distributed to tourism service providers in several languages, provides the following advice:

> Most Chinese will not recommend to their friends or family to see specific sights such as monuments or museums, but rather, will tell them not to miss the best spring rolls in town at Wang's restaurant, or that exquisite Beijing duck. In fact, it would be unthinkable for the Chinese to enjoy something other than Chinese food when on a trip abroad.
>
> (Hotelleriesuisse, 2012, p. 22)

This chapter will take a critical look at this statement, not only by pointing out that spring rolls and Beijing duck are dishes much more favoured by foreigners than by Chinese, but by looking deeper into the role food and beverages play for the different market segments of the Chinese outbound travel market while abroad. As will be demonstrated, no meaningful answer can be given to the question what 'the Chinese' abroad are eating and drinking, simply because 'the Chinese' outbound traveller does not exist. The text will accordingly discuss to what extent among the huge number of Chinese outbound travellers the level of international experience and of previous exposure to foreign food and beverages both inside and outside of China results in different attitudes towards culinary experiences abroad.

Chinese outbound tourism market overview

The total number of border crossings from mainland China maintained a double-digit annual increase of 15% over the past 15 years, reaching approximately 145 million in 2017. For newly affluent Chinese, international travel experiences have become an important part of their consumption pattern. Nevertheless, it has to be kept in mind that less than 10% of the citizens of the People's Republic of China hold a passport, which is necessary for any travel beyond Greater China.

In 2016, for the first time, more than half of the approximately 137 million border crossings from mainland China went beyond 'Greater China', which includes the Special Administration Regions (SARs) of Hong Kong and Macau as well as Taiwan. Hong Kong SAR registered 3% fewer arrivals from mainland China in 2015 and 7% fewer in 2016. Short-term reasons were that shopping trips moved on to Seoul and Tokyo, mainlanders perceived an unwelcoming atmosphere and Shenzhen commuters were barred from daily visits to Hong Kong. In the long run, trips to the former British colony are simply not that exciting and prestigious anymore. Macau SAR stabilised its arrival numbers in 2016 after 2015 had seen a decrease, but suffered lower spending levels based on the clampdown on gambling as part of the ongoing anti-corruption campaign of the Xi Jinping government. Taiwan's tourism industry fell victim to Beijing's disapproval of the new Taiwanese government. After Chinese tour operators followed orders and almost completely stopped offering package tours to the island, the number of arrivals fell by more than a third in the second half of 2016.

Chinese tourists, however, are not travelling less than before; the focus is simply shifting from Greater China to neighbouring countries, particularly to South Korea, Japan and Australia. For all those countries China has become the most important source market. At the same time many long-haul destinations are also enjoying growing numbers of Chinese arrivals, with Europe seeing a general trend of dispersion (Arlt, 2016a) propelled by terrorist attacks in major destinations like France and Germany. As a result, many Chinese travellers are moving from main destination countries to smaller countries such as Croatia, the Czech Republic or Iceland, but also from the main cities within a major destination to smaller destinations and to second tier cities within major destinations. Chinese do not fly to Europe just for a beach holiday, so dispersion also happens in a temporal way, moving away from the summer season, supported by the Golden Weeks national holidays in February and October.

The increasing levels of air, water and indeed food pollution are additional push factors for affluent Chinese citizens to spend their holidays outside of China.

Developing market segments

As the number of Chinese tourists continues to grow, the market landscape is also changing. Many millions of Chinese are still poised to take their first overseas trips, joining group tours and visiting the world's biggest cities and most popular destinations. These package tour travellers do not only come from first tier cities (Beijing, Shanghai, Guangzhou and Shenzhen) anymore, but from second and third tier cities. Meanwhile, the number of more experienced overseas travellers continues to rise as Chinese travellers take their second, fifth or tenth trip overseas and, in the process, become more confident travelling on their own and pursuing more unique and personalised experiences. These experienced travellers are more likely to travel as Free Independent Travellers (FITs) in small groups of couples, families, friends or colleagues, eschewing typical Chinese package group tours. Those who do participate in group tours are more likely to seek out small groups that cater to their interests or travelling style ('customised tours', as they are called in China) or in 'free and easy' tours that only provide the skeleton of the trip with visa, air ticket and some hotel reservations (Arlt, 2017).

The 'second-wave' (Arlt, 2013) of Chinese travellers, who invest more time, interest and money into a service or destination than the package tourists who simply rush between various photo opportunities, shopping malls and Chinese restaurants, are accordingly getting more and more important for many destinations. They can be characterised as being younger, more independent and having a taste for more sophisticated adventures and experiences. Many of them have studied overseas and/or work in an international company in China. This trend is also supported by the fact that it has never been easier for Chinese nationals to travel to another country, whether in terms of obtaining passports, hard currency, visas or even simply accessing destination information. Most major countries and destinations have started to issue ordinary Chinese citizens with five- or ten-year multi-entry visas, providing visas on arrival, or waiving them altogether, bringing to an end the Approved Destination Status (ADS) system of group visas issued via a tour operator. OTAs (Online Travel Agencies) in China like Tuniu and Ctrip already offer simple ways to book all tourism services required on an individual basis.

Furthermore, there is hardly any country or region upon which information cannot be found on the Chinese search engine Baidu.com (the Chinese equivalent of Google) and the Chinese social media platforms WeChat and Weibo. For popular destinations, potential travellers can also be quite certain that some of their – real or virtual – friends will have already visited and can provide peer-group information and tips.

While the global tourism industry is still just starting to comprehend the 'second wave' of Chinese outbound travellers, as will be discussed below with regard to food and beverages, the 'third wave' is already on the horizon in the form of a widening age band of Chinese outbound travellers. International travel being perceived less and less as special or dangerous, more children are joining their parents or travelling in youth travel groups with their classmates or friends. As China's outbound tourism started more than ten years ago, the first cohorts of teenagers who have been travelling internationally with their parents since childhood are entering the market. They are the *linglinghous* (born after 2000), adding to the majority groups of *balinghous* and *jiulinghous* (persons born after 1980 and 1990 respectively) (Liu, 2011). From the other end of the scale, the first Chinese affluent pensioners are also joining the game. Most of them are 'best-agers' in the age group of 55–65 years and – unlike the majority of middle-aged affluent travellers – they are not 'time poor' and are able to afford to go on trips that are less frantic and last several weeks.

Chinese outbound tourists and food

As the short description of some major trends of China's outbound tourism indicates, the demands and interests of a third tier city dweller on his or her first ever trip abroad will be radically different from the demands and interests of a Shanghainese middle-aged real estate agent travelling since his/her study years in the USA. That is certainly also true when it comes to the question of food and beverage preferences while travelling abroad.

A warm-up question at the beginning of the 'Chinese Tourist Welcome' training sessions the China Outbound Tourism Research Institute (COTRI) conducts all over the world is the following: *How many Chinese restaurants exist in China?* Most participants are surprised when they are told that they know the exact figure. Fortunately there is almost always a clever Sherlock Holmes participant in the room who understands that it cannot be 854,877 or something like that, as probably every minute ten restaurants are newly opened or closed somewhere in China. So the correct answer can only be zero.

There are for example no 'European' restaurants either. There are Italian or Spanish or German and, if you look more closely, they will in fact specifically offer Milanese, Andalusian or Bavarian cuisine. China is bigger than Europe and as diverse, which results in having Hunanese, Cantonese, Shanghainese, Sichuanese and so on restaurants in China, but not 'Chinese'. Chinese travellers are in the same way as diverse as the cuisines and the question: *What does the Chinese traveller want?* can only lead to answers of insufficient complexity. 'Chinese' restaurants exist only outside of China and if they do not further specify their cuisine on the menu it is normally a warning sign (Arlt, 2014).

It should also not be forgotten that there are more than 5,000 KFC, 2,800 McDonald's, 2,300 Starbucks and almost 2,000 Pizza Hut branches covering all first and second tier and a number of third tier cities (Yum, 2016; Patnaik, 2016; Ordillas, 2016). Western fast food has become a part of the daily fare in China even for those who have never travelled abroad. There are furthermore, of course, a huge number of international restaurants offering Japanese, Italian, Lebanese etc. food to the urban Chinese.

As a young digital marketing manager from Shanghai wrote in an article about Chinese FITs:

Granted, Chinese travellers often prefer Chinese food when they travel. However, I don't think that applies as much to the younger generation of independent travellers. Many of my friends and I are foodies. We travel to eat. Exploring local cuisine is a huge part of exploring a new place to us. It's something we can't experience in China and gives us an introduction and better understanding of the local culture.

(Xu, 2013)

A China-based consultant describes the same trend, without however getting out of the mindset of talking about 'the Chinese people':

Until now, Chinese people have usually been travelling within a group and tended to eat Chinese or Asian food while travelling abroad – that could be coming to an end! Chinese people seem to be more and more conscious about local culture and food. They now want to try local specialities, especially when it comes to countries for which quality of cuisine is world famous such as France or Italy, even though Chinese and Asian restaurants are more and more numerous near famous tourist sites worldwide, aiming at attracting Chinese tourists. This may reflect a shift in Chinese tourists' attitude and mentality.

(Verot, 2014)

A study of the travel behaviour of mainland Chinese travellers on Hawaii still found, based on research conducted in 2009, that the interviewed visitors 'showed the preference of their ethnic Chinese food over local food during their tour to Hawaii' (Agrusa et.al., 2011, p. 267). A study by IPSOS in 2014, however, yielded the result that 'tourists with a high individual income and young tourists prefer Western food' (WTCF, 2014, p. 27). The wish to try 'local specialities' was mentioned by 88% of the interviewed outbound travellers (ibid.).

Chinese affluent travellers are also willing to spend considerable amounts of money for a big local dinner, even to the level of disbelief by the local tourism service providers. In the summer of 2016, eight Chinese outbound tourists spent more than US$4,000 at Abu Ghosh, a famous Israeli restaurant. The chief executive of the Israeli Incoming Tour Operators Association (IITOA) started a public debate by publishing a copy of the bill in an Israeli financial newspaper, accusing the owner of the restaurant of cheating his customers and creating a bad image of Israel as a destination. Global media and the Israeli Embassy in China got involved. Mainland Chinese official media ignored the 'incident', and discussions on Chinese social media based on international media coverage mostly agreed that, for the Chinese, there is nothing unusual about spending $500 per head on a lavish dinner. Many Chinese netizens even speculated that the group was spending government money anyway (Arlt, 2016b).

Eating food is not only about tasting new cuisines, as in the case of the Abu Ghosh restaurant; it is also about bragging and sharing the experiences with friends back home. With the help of Smartphones, WeChat pictures of the food are posted online, often leading to the demand to serve all courses of a dinner at the same time (Shaw, 2013). In line with the theory of 'co-presence' (Hannam et al., 2006), some Chinese travellers will even take a photo of the menu and upload it to share it with their friends back in China and ask, '*What should I order?*' This way they are able to bring their friends with them, even though they are not physically there.

Brand USA data shows that fine dining as a motivation for travel to the USA is the third most important reason for leisure travellers after shopping and sightseeing. Forty-eight per cent of Chinese tourists cite culinary exploration as a reason for their trip. In response to this growing trend, Brand USA developed a programme called 'Great American Food Stories', a guide that explores the bounty of regional offerings allowing tourists to 'experience the USA one dish at a time' with the help of prominent chefs around the country. The guide, available in Chinese, is seen as an opportunity to let the traditional Chinese love of food and new money meet (FlorCruz, 2014).

In Australia, food and wine have been identified as premier resources that lend themselves to attract Chinese tourists. Tourism Australia, the country's peak destination marketing organisation, launched the 'Restaurant Australia' programme and a series of collaborative marketing campaigns. A key objective has been to promote Australian food and wine tourism resources, especially those located in regional Australia, which can also serve as a dispersion tool (Mao, Huang, 2016).

An important aspect in choosing either Chinese or Western food while travelling is however not concerned with the food itself. Eating together is an important social act in Chinese culture: it enables the definition of hierarchies, gives opportunities to develop contacts and networks, and to show off knowledge and ability in ordering and commenting on food, a field which is considered an art in itself. Experiencing the atmosphere of sharing a large number of dishes around a round table with a group is of similar importance to the question of which dishes are consumed (Pearce et.al, 2013). In a groundbreaking study on Chinese outbound tourism and food (Chang et al., 2010) developed a model of Chinese tourists' food preferences (Figure 4.1).

Chinese food was seen by the participants in this Australian-focused study as offering the right kind of eating situation as well as familiar flavour and the certainty that the food would be appetising. The local Australian food however was consumed for a lot of reasons that were not connected to taste or the quality of the food but for reasons such as learning, experiencing, prestige and status. The decision on which food to choose as per tour guide recommendation was accepted on the ground of keeping harmony in the group and on 'prejudiced advocacy', based on the low expectation of the quality of Australian dishes. Generally, mainland Chinese were most keen on sampling Australian dishes that they saw mostly as a way to get prestige and status by being able to brag about it to their friends.

Based on the research, Chang et al. developed a typology of the Chinese tourists' tourism dining behaviour, distinguishing between three types of tourists: observers, browsers and participators.

The observers are generally interested in trying local food as they regard tourism dining experiences as a 'learning/education opportunity' and a means to 'explore the local culture'. However, they are sentimentally attached to Chinese food and cannot be completely withdrawn from their 'core eating behaviour' when travelling.

(Chang et al., 2010, p. 1003)

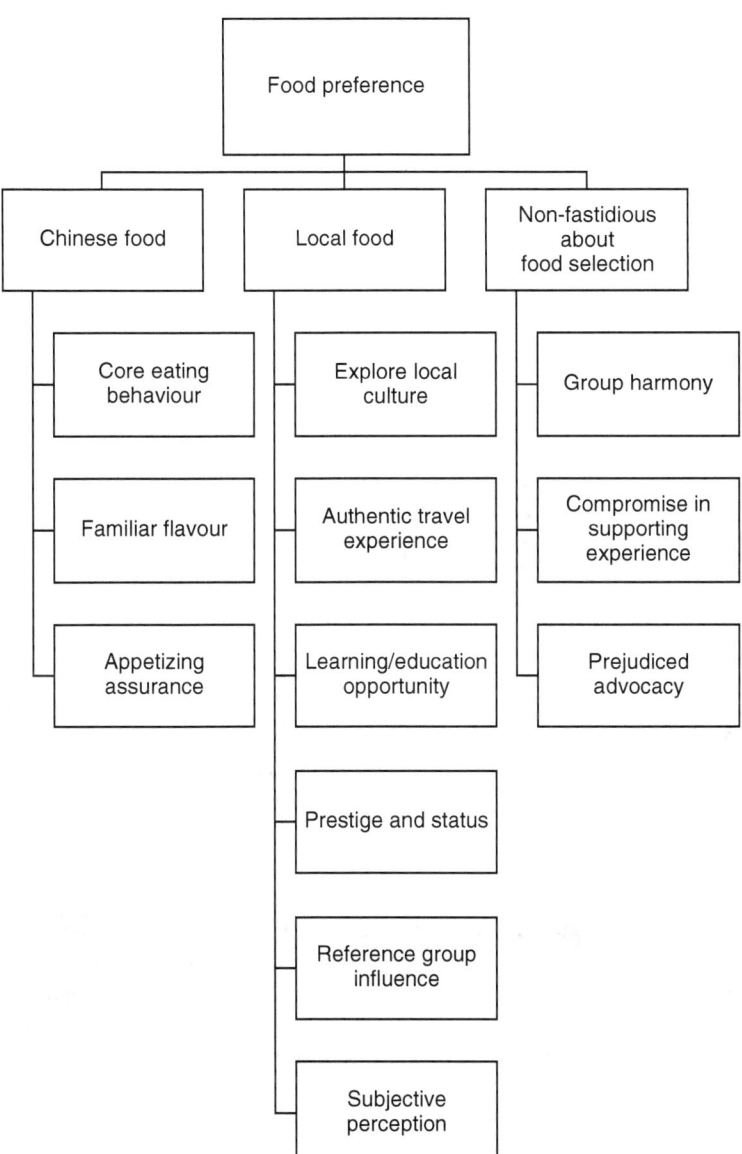

Figure 4.1 A model of Chinese tourists' food preferences

Source: Own graph based on Chang et al., 2010, p. 998

The browsers regard tourism dining experiences in a casual way, and are not fastidious about food selection when travelling. For this type of tourists, food is not a major concern in gauging the level of satisfaction for the holiday, and they are prepared to compromise their food preferences in order to preserve group harmony.

(Chang et.al., 2010, p. 1004)

Finally, the participators

are those who have great interests in local food. Similar to the observers, they also regard tourism dining experiences as a 'learning/education opportunity' and an effective way to 'explore the local culture'. Furthermore, they consider the consumption of local food as an indispensable part of an 'authentic travel experience', one that not only enriches their culinary knowledge, but also increases their cultural capital.

(Chang et al., 2010, p. 1005)

Though insightful, there are some limitations to Chang et al.'s (2010) findings. Their research was based on the analysis of only three tour groups to Australia from mainland China, Hong Kong and Taiwan respectively. Only five of the twenty-five participants in their interviews and focus groups came from mainland China. Four of them were described as experienced travellers, however the non-mainlanders had more experiences with foreign food. Furthermore only group travellers were involved, not FITs. Even the authors conclude that 'individual tourists might provide different insights' (Chang et al., 2010, p. 1008).

A new study published in 2016 tried a quantitative approach for 'understanding Chinese tourists' food consumption in the United States' (Wu et al., 2016), based on 278 questionnaires from Chinese visitors to the USA. The starting point of the study were the discrepancies identified in the literature where food was either seen as a tourist attraction or as an impediment that prevented tourists from exploring the food attractions of a travel destination. As main result Wu et al. (2016) state that 'neophobia' is the most important influencer in the acceptance or the interest in local food. Different sensory appeals do not alter tourists' attitudes, because, according to the authors, tourists already assume the food is essentially different. While higher levels of concern about food safety and differences in table manners lead to negative attitudes towards consuming unfamiliar local food in travel destinations, the main find is that 'tourists with high and low food neophobia show different psychological reactions and behavior patterns . . . Hence . . . food neophobia plays a moderating role in tourists' food-related decisions' (Wu et al., 2016, p. 4711).

A topic that cannot be ignored while discussing Chinese outbound tourists and food are the many stories about 'uncivilised behaviour' of Chinese tourist groups. Any search on Google or YouTube on 'Chinese tourists' food' results almost exclusively in links to articles and videos that show and discuss Chinese plundering buffets in seconds (Liu, 2016a), or grabbing hundreds of giant shrimps from a buffet, leaving behind most of them uneaten (Jackson, 2016), or even catching undersized sea urchins in an Italian national park and eating them on the spot (Joseph, 2016). This kind of 'feeding frenzy' (He, 2016) at buffets is however not limited to outbound tourists. The author of this chapter saw signs asking customers to behave in a civilised way and not to take more away from the buffet than they intend to eat even at several rather up-market restaurants in Chinese tourist resorts. This behaviour can therefore not simply be reduced to cultural distance (Moufakkir, 2011). A recent study of such 'uncivilised behaviour' accordingly comes to the conclusion: 'We can find that the quality of Chinese tourists is uneven, the experience and tourism knowledge are insufficient, and the fear of social morality is weakening, which will lead to a lot of uncivilised behaviours' (Wang et al., 2017).

Chinese outbound tourists and beverages

In respect to beverages, the cultural differences between consumption inside and outside China are much smaller. Coming back to the Swiss brochure of 2012, the statement 'the usual European selection – mineral water, soft drinks, beer, tea, coffee or hot water – will perfectly suit Chinese clients. In recent years, the interest in wine and coffee has also grown strongly. Green tea is the preferred kind of tea for the Chinese' (hotelleriesuisse, 2012, p. 24) is basically correct. Chinese travellers will try to avoid to be seen drunk in public in a foreign country even more than at home, so hard liquor is not extensively consumed. Green tea is indeed preferred, but many Chinese carry their own tea with them, often also for medical reasons.

Only younger urban Chinese consider spending time in a bar as attractive; in China, sitting together is in most cases connected to eating or chatting in private. Karaoke rooms would be another option, but are seldom offered outside of Eastern and Southeastern Asia.

The changes in the development of Chinese consumption patterns of beverages are reflected in the development of wine consumption. One of the learning requirements for newly rich Chinese was what to drink, and Bordeaux wine and Cognac were quickly identified as the most prestigious beverages. Red wine, not Champagne, is seen as the most 'romantic' beverage, and up to 2012, Chinese bought Bordeaux wine also as an investment, many of them losing considerable amounts of money because of lack of experience. Still China is the biggest export market for Bordeaux wine, 80 million bottles, worth 300 million Euros being shipped in 2016.

> The learning curve for Chinese wine lovers has been rapid. Consumers are no longer focused solely on only the most expensive brand names. China is now also quite vintage conscious, similar to the American market. Wine buyers in China are seeking quality and value, which is the right blend needed for long-term appreciation of fine wine.
>
> (Leve, 2016)

From almost no wine production in 1980, China is now the fifth biggest wine market in the world and the biggest for red wine. A hundred Bordeaux châteaux are now owned by Chinese companies (Samuel, 2015).

In 1996, Chinese premier Li Peng surprised his audience at the National People's Congress by toasting the Ninth Five-Year Plan with red wine: 'Drinking grape wines is helpful to our health, does not waste grain, and is good for social ethics' he announced (Cavell, 1995). In the 20 years following this statement, heavy drinking at official dinners has certainly been diminishing, as the author can confirm from his own experience. When shot glasses are filled with wine, many *ganbei* toasts can be drunk without serious effects.

Abroad the interest of trying local food also extends to drinks, especially when they are perceived as typical for the destination. Different kinds of beer are tasted in Germany (Penzialek, 2016) and travels on the whisky trail in Scotland, followed by substantial buying of vintage bottles (MacLean, 2016), have become popular for Chinese visitors.

In earlier years, Chinese airport bookstores would offer extensive encyclopedias of wine, displaying labels and discussing price levels and the prestigeousness of the winery, but saying very little about the taste of the wine. Today every first and second tier city in China offers wine shops that cater for local connoisseurs. The London-based Wine and Spirit Education Trust, the largest institution for teaching skills in the wine and spirits field in the UK, has been teaching mainland Chinese customers since 2006. In 2015, 135 accredited course providers catered to more than 13,000 participants reflecting a changing interest in wine and spirits, moving from social drinking toward connoisseurship. As Tobias Gorn, Managing Director of the London-based wine and whisky agency Campbell and Gorn, was quoted in the *China Daily*: 'wine drinking in China is no longer about Château Lafite Rothschild, Chinese consumers are now more knowledgeable about exciting niche market wines' (Liu, 2016b).

Conclusion

Referring back to the opening question whether or not '*Chinese always want to eat Chinese food when they are travelling*', a discussion thread about this on Quora, a California-based website with many Chinese readers, offers all common arguments including many myths. For example, that most Chinese are lactose intolerant and therefore cannot consume milk products, that they do not want to risk getting diarrhea during their precious outbound trip, that in many cases the members of a tour group have no choice as the tour operator uses (cheap) Chinese restaurants, but also that Chinese food is more refined and faster to eat. The younger Chinese netizens participating in the discussion however are insisting that they, as modern, educated, sophisticated persons are, of course, almost always eating local food when travelling as FITs outside of China (Quora, 2016). The 'social construction' of authentic local food (Lu, Fine, 1995), which may still lead to the search for food that is considered as authentic for a destination even though it is not (such as looking for Bavarian food in Berlin), is an important part of the self-definition of sophisticated Chinese travellers as opposed to mass market tourists.

The answer therefore not surprisingly depends on which kind of Chinese travellers one speaks to when discussing choice of food. Establishing a gliding scale, it can be said that the less experienced the travellers are internationally and the smaller the city they live in, the more likely they are to opt for a 'home away from home' option of Chinese restaurants, even if they do not offer the local Chinese cuisine they prefer and almost always offer quite inauthentic bad quality food. This does not, however, mean that hotels shouldn't offer some Chinese food, for instance at the breakfast buffet, if the customers are mostly more experienced younger Chinese travellers. Such offers not only pay necessary respect to the Chinese culture, but travel-savvy Chinese tourists might also crave for their home cuisine once in a while. As a participant of the research project of Chang et al. said: 'I won't miss the chance to try it [Australian local food] out . . . but soon, I'll miss Chinese food again. It's a long trip. How could I eat local food every day' (Chang et al., 2010, p. 997)?

References

Agrusa, J., Kim, S. & Wang, K. (2011). Mainland Chinese tourists to Hawaii: Their characteristics and preferences. *Journal of Travel & Tourism Marketing*, 28, 261–278.

Arlt, W.G. (2013). The second wave of Chinese outbound tourism. *Tourism Planning & Development,* 2013, 10(2): 126–133.

Arlt, W.G. (2014). How many Chinese restaurants exist in China? Surprisingly, zero. http://www.forbes.com/sites/profdrwolfganggarlt/2014/10/28/how-many-chinese-restaurants-exist-in-china-you-know-the-answer/#7615dd856413

Arlt, W.G. (2016a). Countries want Chinese tourists – just not all in the same place. http://www.forbes.com/sites/profdrwolfganggarlt/2016/09/22/countries-want-chinese-tourists-just-not-all-in-the-same-place/#4e91c1b45c0d

Arlt, W.G. (2016b). No, Chinese tourists weren't ripped off in Israel. http://www.forbes.com/sites/profdrwolfganggarlt/2016/09/14/no-chinese-tourists-werent-ripped-off-in-israel/#75b291401638

Arlt, W.G. (2017). Europas 'Ränder' werden für Chinesen interessanter. *ChinaContact,* 2017. (2).https://owc.de/2017/02/09/europas-raender-werden-fuer-chinesen-interessanter/

Cavell, N. (2015). How China conquered France's wine country. *New Republic*, Nov. 24, 2015. https://newrepublic.com/article/124461/china-conquered-frances-wine-country

Chang, R.C.Y., Kivela, J. & Mak, A.H. (2010). Food preferences of Chinese tourists. *Annals of Tourism Research*, 37(4), 989–1011.

FlorCruz, M. (2014). Golden Week Tourism: Chinese tourists spend lavishly on top restaurants. *Ibtimes*, Sept. 30, 2014. http://www.ibtimes.com/golden-week-tourism-chinese-tourists-spend-lavishly-top-restaurants-1696824

Hannam, K., Sheller, M. & Urry, J. (2006). Editorial: mobilities, immobilities and moorings. *Mobilities*, 1(1), 1–22.

He, H. (2016). Feeding frenzy: internet users aghast at video of Chinese tourists shovelling shrimp at buffet in Thailand. *South China Morning Post*, Mar. 20, 2016. http://www.scmp.com/news/china/society/article/1927681/feeding-frenzy-internet-users-aghast-video-chinese-tourists hotelleriesuisse, Switzerland Tourism (eds.) (2012). *Swiss Hospitality for Chinese Guests*, 1st edition 2004, revised edition 2012.

Jackson, D. (2016). Chinese tourists criticised after video of them shovelling plates with food at buffet in Thailand goes viral. *Shanghaiist*, Mar. 19, 2016. http://shanghaiist. com/2016/03/19/thai_media_slams_greedy_tourists.php

Joseph, G. (2016). Chinese tourists horrifyingly snatch sea urchins from waters and smash them with rock. *World of Buzz*, Nov. 3, 2016. http://www.worldofbuzz.com/chinese-tourists-horrifyingly-snatch-sea-urchins-waters-smash-rock/

Leve, J. (2016). China and Bordeaux wine. The complete story, current situation today. http://www.thewinecellarinsider.com/wine-topics/china-bordeaux-wine-complete-story-current-situation-today/

Liu, C. (2016a). Chinese tourists swarm Vietnamese buffet. Best to just get out of the way. *The Nanfang*, Mar. 29, 2016. https://thenanfang.com/gone-seven-seconds-chinese-tourists-overwhelm-vietnamese-buffet/

Liu, C. (2016b). UK wine school targets connoisseurs. *China Daily*, Oct. 27, 2016. http://www.chinadaily.com.cn/business/2016-10/27/content_27185704.htm

Liu, F. (2011). *Urban Youth in China*. London: Routledge.

Lu, S. & Fine, G. (1995). The presentation of ethnic authenticity: Chinese food as a social accomplishment. *The Sociological Quarterly*, 36 (3), 535–553.

MacLean, D. (2016). Chinese tourists spend up to £50k on whisky during trips to UK. *Independent*, Oct. 18, 2016. http://www.independent.co.uk/life-style/food-and-drink/chinese-tourists-whisky-buying-london-single-malt-macallan-a7367181.html

Mao, I. & Huang, S. (2016). Mainland Chinese outbound tourism to Australia: Recent progress. In: Li, X. (ed.), *China Outbound Tourism 2.0.*, New Jersey: Apple Academic Press (pp. 133–150).

Ordillas, A. (2016). Starbucks plans to double its stores in China to 5,000 by 2021, opening a new one every day. http://shanghaiist.com/2016/10/21/starbucks_expansion.php

Patnaik, S. (2016). McDonald's to add more than 1,000 outlets in China. Reuters, Mar. 31, 2016. http://www.reuters.com/article/us-mcdonalds-china-idUSKCN0WX16M

Pearce, P., Wu, M. & Osmond, A. (2013). Puzzles in understanding Chinese tourist behaviour: Towards a triple-C gaze. *Tourism Recreation Research*, 38 (2), 145–157.

Penzialek, B. (2016). Performing tourism: Chinese outbound organised mass tourists on their travels through German tourism stages. PhD paper at Catholic University of Eichstätt, Germany, 2016.

Quora (ed.) (2015). Is it true that Chinese tourists visiting Italy often prefer Chinese food over the local cuisine? https://www.quora.com/Is-it-true-that-Chinese-tourists-visiting-Italy-often-prefer-Chinese-food-over-the-local-cuisine (26 blog entries)

Rotzinger, U. (2013). Chinesen retten den Schweizer Tourismus (Chinese are saving Swiss tourism), *Blick*, Aug. 6, 2013. http://www.blick.ch/news/wirtschaft/das-ende-der-krise-chinesen-retten-den-schweizer-tourismus-id2397136.html

Samuel, H. (2015). Chinese now own 100 Bordeaux châteaux, as wine mania grows. *Telegraph*, Jan. 30, 2015. http://www.telegraph.co.uk/news/worldnews/europe/france/11380807/Chinese-now-own-100-Bordeaux-chateaux-as-wine-mania-grows.html

Shaw, D. (2013). Why Chinese flock to a Brighton chippy. *BBC*, Aug. 1, 2013. http://www.bbc.com/news/technology-23313896 (video)

Verot, O. (2014). Five new trends about Chinese tourists. *Chairman Media*, Dec. 7, 2014. http://chairmanmigo.com/trends-about-chinese-tourists/

Wang, Y., He, X. & Bi, N. (2017). An exploratory research study on uncivilised behaviour based on mass media exposure: A case study of Chinese outbound tourists. In:

Delener, E. & Schweikert, C. (eds.): *Changing Business Environment: Game Changers, Opportunities and Risks*. Global Business and Technology Association, Vienna 2017, 1021–1028.

World Tourism Cities Federation/IPSOS (eds.) (2014). Market research report on Chinese outbound tourist (city) consumption. WTCF, Beijing 2014.

Wu, K., Raab, C., Chang, W. & Krishen, A. (2016). Understanding Chinese tourists' food consumption in the United States. *Journal of Business Research*, 69: 4706–4713.

Xu, L. (2013). Independent Chinese travellers: 8 things you should know. *Jing Daily*, Jun. 6, 2013. https://jingdaily.com/independent-chinese-travelers-8-things-you-should-know/#.Vrzl0JMrLFx

Yum Annual Report (2016). http://www.yum.com/annualreport/

A note on quantitative data

All data mentioned in the text are, if not indicated otherwise, from the database of COTRI China Outbound Tourism Research Institute.

5 Motivations underlying tourist food consumption

Athena H. N. Mak

Introduction

Tourist food consumption is recognised as an important component of the tourist experience, both from the obligatory perspective (Cohen & Avieli, 2004) as well as from the symbolic perspective (Mak et al., 2017). Despite its importance, relatively little research effort has been devoted to understanding the complexity of tourist food consumption, particularly with regard to the motivational aspects. This neglect is largely due to the conventional view of its role as a 'supporting consumer experience' (Quan & Wang, 2004), one that is mainly an 'extension' or 'intensification' of the daily experience. However, with the recent escalation in utilising food and gastronomy as a key differentiator in many destinations, there is a heightened interest in understanding the key factors driving tourist food consumption.

A number of recent studies have explored the motivations underlying the consumption of local food (e.g. Kim, Eves & Scarles, 2009) and participation in specific food festivals (e.g. Chang & Yuan, 2011; Kim, Goh & Yuan, 2010). While these studies have recognised the importance of motivations in understanding tourist food consumption behaviour, their focus was restricted to a particular type of food or festival. Such a restricted focus may result in the failure to capture the complexity and heterogeneity of food consumption in tourism, thereby leaving a gap in the holistic understanding of the motivations underpinning tourist food consumption.

In an attempt to reduce this knowledge gap, this chapter first reviews the literature on the motivations driving tourist food consumption, and then presents the findings of a recent study on Chinese tourists' food consumption motivations. The Chinese tourism market was chosen as it has been recognised as the world's top outbound segment by the global tourism industry. According to UN World Tourism Organisation statistics, China continues to lead the global outbound travel market after registering double-digit growth in tourism expenditure every year since 2004. This segment has reached 135.1 million international departures in 2016, and is anticipated to continue its growth (UNWTO, 2017). However, tourist behaviour of this source market, in particular, food consumption behaviour, is relatively under-researched (Chang, Kivela & Mak, 2010). By combining the perspectives of food consumption research and tourism research, this study aims to contribute to a better understanding of the motivations underlying food consumption of this important source market.

The focus of this study includes the consumption of both local and non-local foods/cuisines, in an attempt to generate a more holistic understanding of the tourist food consumption phenomenon. In the light of the new global food culture and changing tourism dining landscape brought about by intensifying globalisation (Mak, Lumbers & Eves, 2012a), the motivations underlying the consumption of local and non-local food are both considered important (Cohen & Avieli, 2004; McKercher, Okumus & Okumus, 2008). Accordingly, in this study, food is defined as food/drink that is eaten as a meal purchased from commercial settings in a tourist destination. It includes dishes, meals and dining experiences of local and non-local cuisine/food.

Food consumption in tourism

Food consumption in the general context is recognised as a complex behaviour, with cultural, social, psychological and sensory acceptance factors all playing a role in the decision-making process (Köster, 2009; Sobal et al., 2006). Various attempts have been made to address the plethora of factors affecting food consumption in food consumption literature (e.g. Booth & Shepherd, 1988; Eertmans, Baeyens & Van den Bergh, 2001; Fotopoulos et al., 2009; Furst et al., 1996; Khan, 1981; Steptoe, Pollard & Wardle, 1995). In general, food researchers agree that these factors can be classified into three broad categories: the individual, the food, and the environment (Gains, 1994; Meiselman et al., 1999; Randall & Sanjur, 1981; Shepherd & Raats, 1996). The food itself contributes sensory attributes such as flavour, aroma, texture and appearance, whereas the environment presents cultural, social, economic and physical influences. As for the individual, socio-cultural, psychological and physiological factors are recognised to exert direct or indirect effects on food consumption behaviour. Amongst these three broad categories, factors relating to 'the individual' are widely accepted to be extremely crucial in explaining the variations in food consumption (Rozin, 2006).

Expanding on Randall & Sanjur's (1981) theoretical model, Mak et al. (2012b) propose that potential factors affecting food consumption in the context of tourism can likewise be categorised into three main categories: the tourist, the food in the destination, and the destination environment (Figure 5.1). Food in the destination presents factors such as sensory attributes, food content and cooking methods (Chang et al., 2010; Cohen & Avieli, 2004). The destination environment contributes factors such as gastronomic image/identity, marketing communications, service encounter and servicescape (i.e. physical elements in a consumption setting's built environment) (Chang, Kivela & Mak, 2011; Fox, 2007; Harrington, 2005). Arguably, these factors can be more complex than food consumption in home settings, for there is a substantial change in both the 'food' and the 'environment' components. Above all, tourists' former attitude towards food and eating

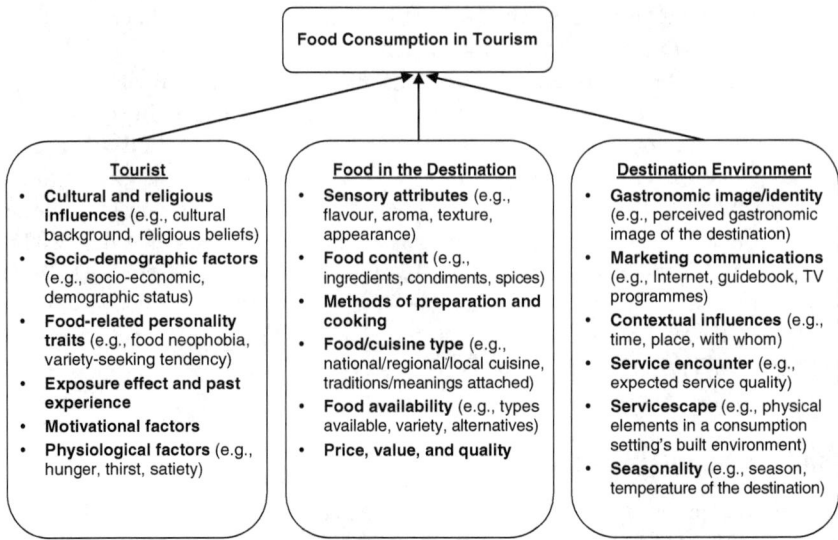

Figure 5.1 Factors affecting food consumption in tourism

Source: Adapted from 'A theoretical model for the study of food preferences', Randall & Sanjur, 1981

might change, and a different set of motivations might influence their preferences and choice of food in the new and unfamiliar environment. Limited space precludes a full discussion of the wide-ranging factors classified under these three categories. Given the focus of this study is on motivations, the motivational factors pertaining to the tourists are elaborated in the ensuing sections.

Motivations underlying food consumption in tourism

Motivation has been acknowledged as one of the most important constructs influencing different food consumption behaviour in food consumption literature. One of the most influential and well-known taxonomy is the Food Choice Questionnaire (FCQ) proposed by Steptoe et al. (1995). FCQ classifies motivations affecting food choice in general situations into nine distinctive motives: (1) health, (2) mood, (3) convenience, (4) sensory appeal, (5) natural content, (6) price, (7) weight control, (8) familiarity and (9) ethical concern. While the motives in FCQ pertain to food choice in general situations, however, considering the lack of empirical studies on tourist food consumption and the 'obligatory' nature of food consumption in the context of tourism (Mak et al., 2012b; Richards, 2002), the FCQ provides a rather comprehensive set of motivational factors in understanding the 'obligatory' aspect of tourist food consumption.

In the tourism literature, motivation has similarly been recognised as a construct exerting significant influence on tourist food consumption behaviour (Mak et al., 2012b). A number of studies have attempted to shed light on the specific motivational factors underlying tourist food consumption. For example, Fields (2002) adopted the typology of tourist motivators to elaborate on the interplay between food consumption and tourism. He proposes four motivational factors of food consumption in tourism: physical, cultural, interpersonal and status and prestige. First, food can be a physical motivator as the act of eating is predominately physical in nature involving sensory perceptions to appreciate the food or tourists' need for sustenance. Second, food can also be a cultural motivator because when tourists are experiencing new local cuisines they are simultaneously experiencing a new culture. Third, food might serve as an interpersonal motivator as meals taken on a holiday have a social function including building new social relations and strengthening social bonds. Finally, local delicacies can also be a status and prestige motivator, as tourists can build their knowledge of the local cuisine by eating as the locals do and explore new cuisines and food that they or their friends are not likely to encounter at home. Although Fields' (2002) proposition is conceptual and requires further empirical validation, it is valuable in establishing a theoretical linkage between tourist motivations and tourist food consumption motivations.

More recently, researchers have become aware of the idiosyncratic nature of food consumption in tourism and have started to adopt an interpretivist approach to investigate the motivations underlying tourist food consumption. For example, Kim et al. (2009) adopted a grounded theory approach to explore the motivational factors behind local food consumption in tourist destinations. Nine motivational factors were identified, namely, exciting experience, escape from routine, health concern, learning knowledge, authentic experience, togetherness, prestige, sensory appeal and physical environment. Kim et al.'s findings offer important insights into the motivations and factors influencing the consumption of local food. Nonetheless, since their focus was restricted to local food, they have largely adopted a general tourist motivation framework in explaining tourists' local food consumption behaviour. This may lead to overlooking of the 'obligatory' and 'extension' nature of food consumption in tourism (Mak et al., 2012a; Quan & Wang, 2004; Richards, 2002).

Another study by Chang, Kivela and Mak (2010) has adopted an ethnographic approach in exploring tourist food motivations. The participant observation technique was combined with on-site focus group interviews to examine and compare different Chinese tourists' (including mainland Chinese, Hong Kong and Taiwanese tourists) food consumption experiences while they were holidaying in Australia. Chinese tourists' food preferences were classified into three distinct categories: familiar food (Chinese food), local food (Australian food) and non-fastidiousness of food selection. The motivational factors for favouring each preference included: Chinese food – core eating behaviour, familiar flavour and appetising assurance; local food – explore local culture, authentic travel experience, learning/education opportunity, prestige and status, reference group influence and subjective perception; non-fastidious about food selection – group harmony, compromising in supporting experience and prejudiced advocacy. Chang et al.'s (2010) study was a first attempt to generate an in-depth understanding and comparison of Chinese tourists' food preferences in a Western context. Their ethnographic approach allows a revealing juxtaposition of two culturally distant food cultures, and provides a model of motivational factors underlying the Chinese tourists' food preferences from a comprehensive perspective. Their study uncovered a number of important motivations underlying preference for local food, for example, the desire to explore local culture, to seek authentic travel experience, to pursue learning/education opportunity and to acquire prestige and status. Another critical aspect of their findings is that other than the prevalent motivations for seeking local food, there was an abiding need for participants to seek familiar food. As they state, many of the participants were eager to try local food, however, they also explicitly stated local food did not match up to the criteria of a 'proper meal' in their dietary habits, and thus it was impossible for them to consume local food at every meal. This suggests that the participants perceived partaking of local food as a 'peak touristic experience', one that could satisfy their symbolic and experiential needs, but not enough to satiate their physiological needs.

Tikkanen (2007), on the other hand, has adopted Maslow's hierarchy of needs model (Maslow, 1970) to explore the relationship between different levels of motivations and various forms of food tourism in Finland. Five sectors of food tourism in Finland were identified: food tourism based on physiological needs, food tourism based on safety needs, food tourism based on esteem needs, and food tourism based on self-actualising needs. Following Tikkanen (2007), Lin and Chen (2014) identified three motivational dimensions of food and food service for Chinese group tourists, namely, physiology and safety (e.g. dining environment is clean and food hygiene is good), self-actualisation (e.g. food and food services provided reflect social status, the travel-dining experience is distinctive), and belonging and self-esteem (e.g. relationship with family and friends was enhanced by dining together on the trip). Lin and Chen's findings corroborate that Maslow's hierarchy of needs can be used as a basis for classifying tourists' food consumption motivations.

The above-reviewed studies have presented the view that motivations underpinning tourist food consumption can be highly related to tourist motivation, and yet, they also encompass more generic food choice motivations, such as seeking familiar flavour and core eating behaviour as identified in Chang et al.'s study (2010). Tourists in general welcome novelty, however they concomitantly seek a certain level of familiarity in their food consumption experiences (Mak et al., 2012a). Thus, an over-reliance of a tourism motivation framework risks the chances of ignoring the 'obligatory' and 'extension' aspects of food consumption in tourism (Mak et al., 2012b).

Following this, Mak et al. (2013) have adopted an interdisciplinary approach that incorporates both tourist motivations and food choice motivations to understand the idiosyncratic nature of tourist food consumption. Drawing on a combined repertory grid method and generalised Procrustes analysis approach, they proposed a framework with five conceptual dimensions to reflect the complex and heterogeneous nature of tourism food consumption, namely, symbolic, obligatory, contrast, extension and pleasure (Figure 5.2).

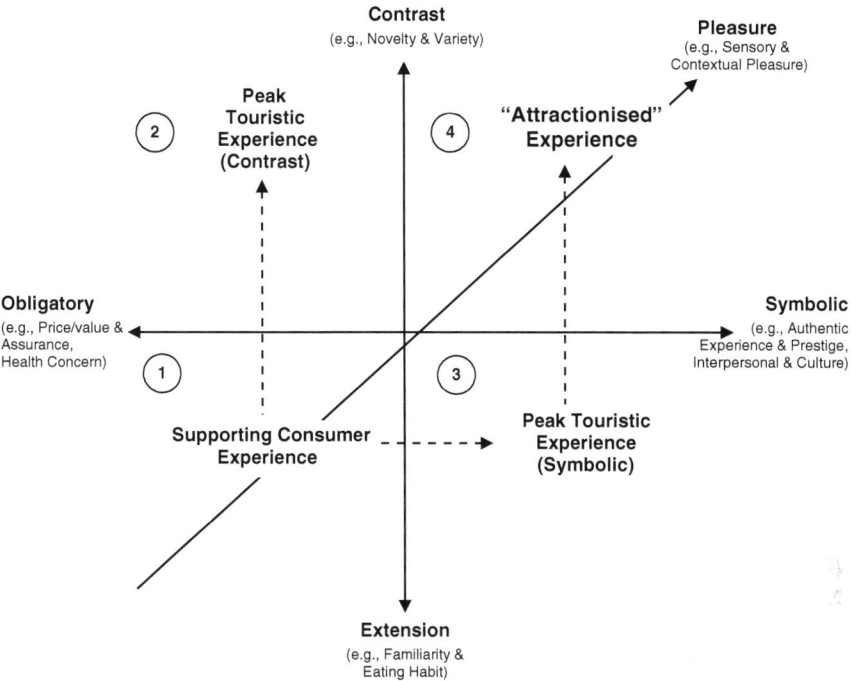

Figure 5.2 A conceptual framework for food consumption in tourism
Source: Mak et al., 2013

The 'symbolic' dimension signifies the symbolic meanings of food consumption to tourists, such as seeking authentic experience and prestige. The 'obligatory' dimension reflects the essentiality of food consumption in tourism, and includes factors such as health concern and the need to seek assurance in price and value. The 'contrast' dimension denotes the need to quest for food experiences that contrast from the tourists' daily routine, whereas the 'extension' dimension refers to the motivation to seek food experiences that extend the daily routine (Quan & Wang, 2004). Finally, the 'pleasure' dimension captures the need to seek sensory, social and contextual pleasures from the food consumption experience (Mak et al., 2013). One merit of this framework is its ability to facilitate perceptual mapping of gastronomic products into supporting consumer experience, peak contrast and symbolic touristic experience and 'attractionised' experience. Another advantage of this framework is that the five conceptual dimensions provide a useful structure for comparison of previous findings. The motivational factors of the studies reviewed are categorised according to the five dimensions (Table 5.1).

As can be seen from Table 5.1, the five conceptual dimensions capture the motivations from previous findings well except for Chang et al.'s (2010) study. That is because their study focused on group tourists and some unique factors

Table 5.1 Classification of motivational factors underlying tourist food consumption

Author(s)	Symbolic	Obligatory	Contrast	Extension	Pleasure	Others
Fields (2002) (Conceptual study)	• Cultural • Status & prestige	• Physical			• Interpersonal	
Kim et al. (2009) (Qualitative, grounded theory approach, motivations for consuming local food)	• Learning knowledge • Authentic experience • Prestige	• Health concern	• Exciting experience • Escape from routine		• Sensory appeal • Physical environment • Togetherness	
Chang et al. (2010) (Qualitative, ethnographic approach, motivations for consuming local and non-local food)	• Explore local culture • Authentic travel experience • Learning/education opportunity • Prestige & status	• Compromise in supporting experience		• Core eating behaviour • Familiar flavour • Appetising assurance	• Group harmony	• Reference group influence (by tour group members) • Subjective perception • Prejudiced advocacy (by tour leader)
Mak et al. (2013) (Qualitative & quantitative, repertory grid method & generalised Procrustes analysis approach, motivations for consuming local and non-local food)	• Authentic experience • Prestige • Cultural knowledge	• Health concern • Assurance • Convenience • Price & value	• Novelty • Variety	• Familiarity • Eating habit	• Sensory pleasure • Social pleasure • Contextual pleasure	
Lin & Chen (2014)	• Self-actualisation • Belonging & self-esteem	• Physiology & safety				
Mak et al. (2017) (Quantitative, questionnaire survey, motivations for consuming local and non-local food)	• Authentic experience & prestige • Interpersonal & culture	• Health concern • Price/value & assurance	• Novelty & variety	• Familiarity eating habit	• Sensory & contextual pleasure	

were revealed, such as reference group influence (e.g. the influence of motivation by tour group members), subjective perception (e.g. the subjective assumption that one type of food is better than the other) and prejudiced advocacy (e.g. the influence of motivation by tour leader).

More recently, Mak et al. (2017) have further identified seven tourist food consumption motivational factors based on an empirical study. The seven dimensions identified are novelty and variety, authentic experience and prestige, interpersonal and culture, price/value and assurance, health concern, familiarity and eating habit, and sensory and contextual pleasure. These dimensions provide a useful framework for understanding the key motivational factors underlying tourist food consumption in a foreign destination, as well as corroborate the five conceptual dimensions proposed by Mak et al. (2013).

Methodology

This study has employed a positivist approach to explore the motivational factors of mainland Chinese tourists. A questionnaire was developed which comprised of five sections. The first section included screening questions to identify eligible respondents. The second section contained questions about respondents' trip characteristics. The third section measured respondents' two food-related personality traits, namely, food neophobia and variety-seeking tendency. Food neophobia refers to the unwillingness to try unfamiliar foods (Pliner & Salvy, 2006). People who are more neophobic tend to expect various novel foods to taste worse and, thus, are generally less willing to taste or choose novel foods (e.g. Pliner & Hobden, 1992; Tuorila et al., 1994; Tuorila et al., 1998). Variety-seeking describes the enduring tendency of individuals to seek diversity in their food choices (Kahn, 1995). Individuals with a Higher Variety-Seeking (HVS) tendency are more inclined to seek different kind of foods across various situations. Food neophobia was measured using Pliner and Hobden's (1992) ten-item Food Neophobia Scale (FNS), and variety-seeking tendency was measured with van Trijp and Steenkamp's (1992) eight-item Variety-Seeking Tendency Scale (VARSEEK). The fourth section measured Chinese tourists' food consumption motivation based on the seven tourist food consumption motivational factors identified in Mak et al.'s study (2017). The scale contained 31 items, which were assessed with a 7-point Likert scale with response categories ranging from 1 = not important at all to 7 = extremely important. The final section enquired about respondents' socio-demographic characteristics.

A pre-test was conducted with five tourism/hospitality professors and researchers to ensure content validity and clarity of the questionnaire. Pre-test respondents were asked to complete the whole questionnaire, and also to provide any comment they had with regard to the items, wording and response options of the questionnaire. A pilot test of the questionnaire was conducted with mainland Chinese tourists in Taiwan. Pilot respondents were recruited through random interception at major tourist attractions. A total of 47 valid responses were collected and analysed using the PASW Statistics (v.18). Exploratory factor analysis results revealed that the factor loadings of two items: 'to enjoy foods that are good for my health' and 'to enquire about the ingredients in local foods before trying them' were lower than 0.40. These two items were subsequently removed (Pallant, 2007) and the final scale consisted of 29 items.

A combination of purposive and self-selection sampling approach was used to recruit respondents. The purposive sampling technique permits respondents to be chosen according to pre-selected criteria relevant to the research questions, whereas self-selection sampling method allows respondents to volunteer or self-select to participate in the study (Patton, 2002). In this way, the most appropriate respondents can be recruited and they would feel more motivated to answer truthfully.

A message was posted on a popular Chinese online forum for tourists who are interested in visiting Taiwan as FITs (http://www.weibo.com/go2tw) between October and December 2016 to invite those who had visited Taiwan over the past 12 months to participate in the survey. The following inclusion criteria were used to identify eligible respondents: (1) above 18 years of age, (2) had stayed a minimum of one night in Taiwan, and (3) had food consumption experiences in commercial settings during their stay. Criteria 2 and 3 ensured that the respondents had adequate food consumption experience to reflect upon their motivations. A small token (a postcard with Taiwan scenery) was given to those who had completed the survey. In the end, a total of 533 valid responses were obtained.

Findings and discussions

Socio-demographic and trip characteristics of the respondents

The socio-demographic characteristics of the 533 respondents are summarised in Table 5.2. The overwhelming majority of the respondents were female (82.7%). The largest age group was 25–34 (55.7%), followed by 18–24 (36.2%). Over three-quarters of the respondents were single (76.0%), followed by 12.9% who were married or in a partnership and had no children. Approximately two-thirds (67.4%) of the respondents held an undergraduate degree. The majority (28.0%) earned a personal monthly income between ¥4001–6000. Most of the respondents' occupations fell in the 'administrative/secretarial/clerical' and 'full-time student' categories (both 18.2%), followed by the 'professional/senior managerial' category (12.4%).

Nearly three-quarters of the respondents were visiting Taiwan for the first time (72.6%). The vast majority of them were visiting Taiwan for a pleasure purpose (83.5%), and most of them (71.5%) were travelling with family/friends (without children). The average length of stay was 6.5 nights (SD = 1.7). Most of the respondents were 'somewhat unfamiliar' (45.4%) or 'unfamiliar' (20.5%) with the food in Taiwan. The majority of the respondents were from Guangdong (21.4%), followed by Shanghai (12.4%), Jiangsu (10.4%), Beijing (9.4%), Fujian (8.6%), Zhejiang (7.3%) and Sichuan (6.4%).

The mean food neophobia score for the Chinese respondents was 43.77 (S.D. 4.80), which was relatively high compared with those reported in other studies,

Table 5.2 Socio-demographic and trip characteristics of the respondents

Socio-demographic characteristics	Frequency (percentage) (n = 533)	Trip characteristics	Frequency (percentage) (n = 533)
Gender		**First time/repeat visit**	
Male	92 (17.3%)	First time visitor to Taiwan	387 (72.6%)
Female	441 (82.7%)	Repeat visitor to Taiwan	146 (27.4%)
Age		**Travel companion**	
18–24	193 (36.2%)	Alone	61 (11.4%)
25–34	297 (55.7%)	With family/friends (without children)	381 (71.5%)
35–44	38 (7.1%)	With family/friends (with children)	35 (6.6%)
45–54	5 (0.9%)	Travelling in a packaged tour	46 (8.6%)
55 or above	0 (0.0%)	Others	10 (1.9%)
Marital status		**Purpose of visit**	
Single	405 (76.0%)	Pleasure	445 (83.5%)
Single, had children	4 (0.8%)	VFR	9 (1.7%)
Married/in partnership, had no children	69 (12.9%)	Business	5 (0.9%)
Married/in partnership, had children	55 (10.3%)	En route to other destination	1 (0.2%)
		Food & dining	31 (5.8%)
Educational level		Shopping	5 (0.9%)
High school	20 (3.8%)	"Bleasure" (business and pleasure)	8 (1.5%)
Vocational/college	82 (15.4%)	Others	29 (5.4%)
Undergraduate degree	359 (67.4%)		
Postgraduate degree	72 (13.5%)	**Familiarity with food in Taiwan**	
		Extremely familiar	3 (0.6%)
Personal monthly income		Familiar	29 (5.4%)
¥2000 or under	89 (16.7%)	Somewhat familiar	30 (5.6%)
¥2001–4000	117 (22.0%)	Neutral	93 (17.4%)
¥4001–6000	149 (28.0%)	Somewhat unfamiliar	242 (45.4%)
¥6001–8000	72 (13.5%)	Unfamiliar	109 (20.5%)
¥8001–10000	38 (7.1%)	Extremely unfamiliar	27 (5.1%)
¥10001–20000	52 (9.8%)		
¥20000 or above	16 (3.0%)	**Food neophobia** (mean = 43.77, SD = 4.80)	
		Neophilics (score 27–42)	193 (36.2%)

(continued)

Table 5.2 (continued)

Socio-demographic characteristics	Frequency (percentage) (n = 533)	Trip characteristics	Frequency (percentage) (n = 533)
Occupation		Average (score 43–45)	141 (26.5%)
Professional/senior managerial	66(12.4%)	Neophobics (score 46–58)	199 (37.3%)
Middle to junior managerial	11(2.1%)		
Technical/highly skilled worker	50(9.4%)	**Variety-Seeking** (mean = 30.07, SD = 3.61)	
Semi-skilled/unskilled worker	43(8.1%)	Low VS (score 18–29)	214 (40.2%)
Sales/service worker	49(9.2%)	Medium VS (score = 30)	81 (15.2%)
Administrative/ secretarial/clerical	97(18.2%)	High VS (score 31–40)	238 (44.7%)
Business owner	7(1.3%)		
Self-employed	23(4.3%)		
Homemaker	7(1.3%)		
Full-time student	97(18.2%)		
Retired	1(0.2%)		
Others	82(15.4%)		

for example, Hobden and Pliner (1995) (two separate Canadian student samples, means = 32.4 & 31.4), Koivisto and Sjödén (1996) (Swedish sample, mean = 29.25), Arvola et al. (1999) (Finnish sample, mean = 25.5), Meiselman et al. (1999) (UK undergraduate student sample, mean = 29.51), Eertmans et al. (2005) (Belgian undergraduate student sample, mean = 32.64), and Mak et al. (2017) (UK tourist sample, mean = 38.54; Taiwanese tourist sample, mean = 36.36). This indicates that the Chinese respondents in general had a relatively higher level of food neophobia and might be more inclined to seek familiar food items in their travel dining experiences. In contrast, the mean score of the variety-seeking tendency for the Chinese respondents was 30.07 (S.D. 3.61), which was relatively high compared with those reported in other studies, for example, Meiselman et al. (1999) (UK undergraduate student sample, mean = 29.45), Marshell and Bell (2004) (UK undergraduate student sample, mean = 29.33), and Mak et al. (2017) (UK tourist sample, mean = 24.43; Taiwanese tourist sample, mean = 26.52). This indicates that Chinese respondents tended to be High Variety-Seeking (HVS).

Motivational items

The means and standard deviations of the motivational items are shown in Table 5.3. The mean scores ranged from 4.20 to 6.31, suggesting all items were of more than average importance for the respondents. Furthermore, the mean scores for 25 items were above 5.0. This indicates that respondents considered 86% of all the items as

Table 5.3 Motivational items

Motivational item	Mean (n = 533)	Standard deviation
To enjoy foods that are delicious.	6.31	0.76
To sample authentic local foods.	6.30	0.82
To try the well-known foods/dishes in Taiwan.	6.28	0.70
To have an enjoyable meal with my travel companions.	6.18	0.83
To sample a wide variety of foods/cuisines in Taiwan.	6.11	0.76
To tell friends about my dining experiences in Taiwan.	6.09	0.89
To increase my knowledge about the local culture through my dining experiences.	6.08	0.90
To try out foods that are presented attractively.	6.05	0.76
To enjoy a good selection of both local and international foods in Taiwan.	6.00	0.90
To dine in restaurants with a pleasant atmosphere.	5.99	0.84
To dine in restaurants that are recommended by the media (e.g. travel guidebooks, Internet, TV).	5.96	0.91
To dine in restaurants with high hygiene standards.	5.89	1.01
To dine in restaurants with an authentic local ambience.	5.83	0.98
To dine in restaurants that provide good service.	5.83	0.94
To have foods that are prepared hygienically.	5.83	1.20
To try foods that are only available in Taiwan.	5.81	1.25
To dine in restaurants that are reasonably priced.	5.80	0.93
To be adventurous in trying out various foods in Taiwan.	5.71	1.06
To try foods that are novel to me.	5.59	1.02
To learn about local food traditions and culture.	5.56	1.08
To dine in famous restaurants in Taiwan.	5.51	1.07
To have foods that my travel companions like.	5.48	1.07
To try out foods I have never tried before.	5.47	1.06
To dine in restaurants that offer good value for money.	5.42	1.22
To enjoy foods that I am familiar with.	5.08	1.21
To have foods that match with my usual eating habit.	4.90	1.25
To dine in chain restaurants that I have been to.	4.35	1.42
To have foods that help me to maintain a healthy weight.	4.22	1.40
To dine in restaurants that are tourist-friendly.	4.20	1.47

rather important. The top five motivational items for all respondents were: (1) 'to enjoy foods that are delicious', (2) 'to sample authentic local foods', (3) 'to try the well-known foods/dishes in Taiwan', (4) 'to have an enjoyable meal with my travel companions', and (5) 'to sample a wide variety of foods/cuisines in Taiwan'.

Exploratory factor analysis results

Exploratory factor analysis (EFA) was performed to explore the underlying dimensions of the motivational items. The items were factor analysed using principal component analysis with orthogonal varimax rotation. Based on Kaiser's criterion, or the eigenvalue rule, only factors with an eigenvalue of 1.0 or more were retained for further analysis (Pallant, 2007). Seven factors were identified which were named: (1) Authenticity and prestige, (2) Novelty, (3) Familiarity and health, (4) Assurance, (5) Variety, (6) Pleasure and (7) Culture. The seven factors altogether accounted for 49.1% of the cumulative variance (Table 5.4).

Of the seven factors identified, the Pleasure factor was considered the most important motivator by the Chinese tourists (grand mean = 6.24). This factor consists of two items: 'to enjoy foods that are delicious' and 'to have an enjoyable meal with my travel companions'. It reflects tourists' desire to seek both sensory pleasure (delicious) and convivial pleasure (enjoyable meal with travel companions) in their food consumption experience. The Variety factor had the second highest grand mean (6.03). It contains three items: 'to sample a wide variety of foods/cuisines in Taiwan', 'to enjoy a good selection of both local and international foods in Taiwan', and 'to dine in restaurants with a pleasant atmosphere'. The factor represents the high propensity for Chinese tourists to seek variety in their food consumption experience.

The Authenticity and prestige factor was perceived as the third most important motivator by the Chinese respondents (grand mean = 5.90). The factor consists of eight items, including 'to sample authentic local foods', 'to try the well-known foods/dishes in Taiwan', 'to dine in restaurants with an authentic local ambience' and 'to tell friends about my dining experiences in Taiwan'. This factor depicts the desire to seek 'authenticity', in particular through experiencing famous and distinctive food in a destination to satisfy prestige-related motivations, for the subsequent sharing of the experience impresses people back home. The Assurance factor had a grand mean of 5.84, which was the fourth highest among all factors. The factor has four items: 'to dine in restaurants that provide good service', 'to dine in restaurants with high hygiene standards', 'to have foods that are prepared hygienically' and 'to dine in restaurants that are reasonably priced'. The factor indicates the importance of the need to seek a sense of assurance (in terms of service quality, hygiene standard and price) among Chinese tourists.

The Culture factor had a grand mean of 5.82. It contains two items: 'to increase my knowledge about the local culture through my dining experiences' and 'to learn about local food traditions and culture'. This factor portrays the desire to explore a local culture and to increase cultural capital through food and dining experiences. The Novelty factor had a grand mean of 5.70. The factor consists of four items: 'to try foods that are novel to me', 'to try out foods I have never tried before', 'to be adventurous in trying out various foods in Taiwan' and 'to try out foods that are presented attractively'. The factor represents the need to quest for novelty and adventure via food consumption experiences. Finally, the Familiarity and health factor contains six items, which include 'to dine in chain restaurants that I have been to', 'to enjoy foods that I am familiar with' and 'to have foods that help me to maintain a healthy weight'. Although this factor had the lowest grand mean of 4.71, it was still considered as being of more than average importance to the Chinese tourists.

Table 5.4 Results of the exploratory factor analysis

Factors and items	Factor loading	Mean	S.D.	Grand mean	Eigenvalue	Cronbach's α	Variance explained
Factor 1 – Authenticity and prestige				**5.90**	**8.42**	**0.76**	**10.50%**
To sample authentic local foods.	0.592	6.30	0.82				
To try the well-known foods/dishes in HK.	0.542	6.28	0.70				
To dine in restaurants with an authentic local ambience.	0.536	5.83	0.98				
To tell friends about my dining experiences in Taiwan.	0.530	6.09	0.89				
To dine in famous restaurants in Taiwan.	0.457	5.51	1.07				
To dine in restaurants that are recommended by the media.	0.435	5.96	0.91				
To dine in restaurants that offer good value for money.	0.421	5.42	1.22				
To try foods that are only available in Taiwan.	0.410	5.81	1.25				
Factor 2 – Novelty				**5.70**	**3.73**	**0.85**	**9.75%**
To try foods that are novel to me.	0.840	5.59	1.02				
To try out foods I have never tried before.	0.782	5.47	1.06				
To be adventurous in trying out various foods in Taiwan.	0.672	5.71	1.06				
To try out foods that are presented attractively.	0.545	6.05	0.76				
Factor 3 – Familiarity and health				**4.71**	**1.65**	**0.76**	**9.15%**
To dine in chain restaurants that I have been to.	0.675	4.35	1.42				
To enjoy foods that I am familiar with.	0.606	5.08	1.21				
To have foods that help me to maintain a healthy weight.	0.591	4.22	1.40				
To have foods that match with my usual eating habit.	0.533	4.90	1.25				
To dine in restaurants that are tourist-friendly.	0.515	4.20	1.47				
To have foods that my travel companions like.	0.487	5.48	1.07				

(continued)

Table 5.4 (continued)

Factors and items	Factor loading	Mean	S.D.	Grand mean	Eigenvalue	Cronbach's α	Variance explained
Factor 4 – Assurance				**5.84**	**1.55**	**0.83**	**8.50%**
To dine in restaurants that provide good service.	0.680	5.83	0.94				
To dine in restaurants with high hygiene standards.	0.651	5.89	1.01				
To have foods that are prepared hygienically.	0.649	5.83	1.20				
To dine in restaurants that are reasonably priced.	0.636	5.80	0.93				
Factor 5 – Variety				**6.03**	**1.28**	**0.80**	**4.35%**
To sample a wide variety of foods/cuisines in Taiwan.	0.582	6.11	0.76				
To enjoy a good selection of both local and international foods in Taiwan.	0.528	6.00	0.90				
To dine in restaurants with a pleasant atmosphere.	0.448	5.99	0.84				
Factor 6 – Pleasure				**6.24**	**1.09**	**0.72**	**3.99%**
To enjoy foods that are delicious.	0.543	6.31	0.76				
To have an enjoyable meal with my travel companions.	0.539	6.18	0.83				
Factor 7 – Culture				**5.82**	**1.02**	**0.63**	**2.87%**
To increase my knowledge about the local culture through my dining experiences.	0.509	6.08	0.90				
To learn about local food traditions and culture.	0.490	5.56	1.08				
Total variance explained							**49.09%**

Kaiser-Meyer-Olkin measure of sampling adequacy   0.905

Bartlett's test of sphericity approx. chi–square   6852.60 (df = 465, Sig. = 0.000)

Table 5.5 Inter-correlations among the motivational factors

	Factor 1	Factor 2	Factor 3	Factor 4	Factor 5	Factor 6	Factor 7
Factor 1 – Authenticity and prestige	1						
Factor 2 – Novelty	0.495**	1					
Factor 3 – Familiarity and health	0.263**	0.054	1				
Factor 4 – Assurance	0.338**	0.236**	0.525**	1			
Factor 5 – Variety	0.570**	0.581**	0.222**	0.434**	1		
Factor 6 – Pleasure	0.511**	0.429**	0.182**	0.462**	0.544**	1	
Factor 7 – Culture	0.535**	0.464**	0.237**	0.302**	0.471**	0.404**	1

* Correlation is significant at the 0.01 level (2-tailed).

Reliability and validity

The reliability of the motivation dimensions was assessed by Cronbach's alpha, one of the most common measures of scale reliability (Field, 2005). The alpha values of Factors 1–6 ranged from 0.72–0.85, indicating good internal consistency among the items of each factor (Hair et al., 2002). The alpha value of Factor 7 Culture was 0.634, which is considered an acceptable level of internal consistency, particularly for scales with five or fewer items (Field, 2005; Hair et al., 2002; Pallant, 2007).

To assess discriminant validity, correlation coefficients among the seven factors were examined by Pearson correlation analysis. The analysis results are summarised in Table 5.5. Although there were significant correlations between some factors, their correlations were only moderate and were considered acceptable according to Hair et al. (2002); for example, Factor 1 Authenticity and prestige and Factor 2 Novelty (0.495, $p < 0.01$). Other factors showed low relationship or no significant relationship with other factors (e.g., Factor 1 Authenticity and prestige and Factor 3 Familiarity and health). Thus, discriminant validity was established.

Implications and conclusion

This chapter has attempted to contribute to a more expansive understanding of the motivations underlying tourist food consumption, first by an interdisciplinary literature review that incorporates both tourist motivation literature and food consumption literature to shed light on the idiosyncratic nature of tourist food consumption, then by conducting an empirical study to investigate the motivations underlying mainland Chinese tourists' food consumption.

Despite the important findings of this study, there are several limitations that need to be considered. First, data were collected retrospectively from mainland Chinese tourists who had travelled in Taiwan. To minimise potential recall bias, the recall period was limited to the past 12 months. Second, there was an over-whelming majority of female respondents in the sample. This could be due to the popularity of the online forum among female users. The high proportion of females in the sample could result in sample bias and the results may not satisfacto-rily represent the whole population studied. Finally, the cultural distance between Chinese tourists and the host culture (i.e. Taiwan) might not be extremely large as compared with other foreign countries. Future studies may consider investigating the motivational factors in other settings, for example in Western destinations.

Notwithstanding these limitations, a number of practical implications for tour-ism and hospitality marketers can be drawn from the findings:

1   First, Pleasure was regarded as the most important motivational factor by the Chinese respondents. Most importantly, the Chinese tourists were not just motivated by the sensory pleasure of food (i.e. to enjoy foods that are deli-cious); they were also motivated by the convivial pleasure derived from the dining experience (i.e. to have an enjoyable meal with my travel companions). Convivial pleasure in the context of food consumption refers to the friendly, lively and enjoyable atmosphere of a meal (Martens & Warde, 1997). As Hayes-Conroy and Martin (2010) describe, conviviality emphasises that the enjoyment of food can and should be shared. This type of pleasure is par-ticularly important for Chinese people, as Chinese meals are fundamentally socially orientated, and there are a number of dishes to be shared around the table (Chang, 1977). Hence, both sensory and convivial pleasure of tourist food consumption should be adequately emphasised and communicated to the target market. Evidence from the food consumption literature suggests that convivial and contextual pleasure can be derived from social facilitation and physical variables (e.g. table settings) of the consumption situation (García-Segovia, Harrington & Seo, 2015; Meiselman, 1996). In this light, tourism and hospitality marketers may devote efforts to highlighting or enhancing the sensory pleasure of the food consumption experience (e.g. taste and presenta-tion), as well as the convivial pleasure (e.g. dining atmosphere, settings) of the gastronomic experience in their promotion mix. For example, the Paris Convention and Visitors Bureau (CVB) has featured meals 'at a communal table' in their official website (Paris Info, 2017a). Communal table refers to diners sharing a table, and sometimes food and drinks and conversation, with other unknown diners. Communal table dining thus offers opportunities for tourists to socialise with locals or other tourists, thereby providing them with not only sensory, but also convivial pleasure.
2   Second, Variety was considered the second most important motivational fac-tor by the Chinese respondents. This factor corroborates Chang et al.'s (2011) findings that the desire to seek variety is a significant motivator underlying Chinese tourists' dining experience in Australia (Chang et al., 2011). Their study found that Chinese tourists perceived the variety of food would con-tribute greatly to the breadth of their travel dining experience. The variety included both the variety of dishes and the diversity of meal arrangements.

In the present study, the Chinese respondents similarly valued the opportunity to sample a wide variety of food/cuisines and also to enjoy a good selection of both local and international foods in Taiwan. As Wright, Nancarrow & Kwok (2001) point out, Chinese people generally place great emphasis on the hospitality spirit and generosity to others and it is a common practice for them to order a variety of dishes when they dine out with friends or else they would feel a 'loss of face'. On the other hand, the mean score of the variety-seeking tendency for the Chinese respondents was 30.07 (S.D. 3.61), which was relatively high compared with those reported in other studies. This indicates that mainland Chinese respondents tended to be High Variety-Seeking (HVS). Accordingly, tourism and hospitality marketers may highlight the Variety factor in their gastronomic offerings to appeal to HVS Chinese tourists. For example, the Hong Kong Tourism Board (HKTB) highlights the variety of a popular local cuisine in Hong Kong – *dim sum* (literally meaning 'touch your heart') on its official website. As it describes, *dim sum* has 'as many as 150 items on a restaurant menu, and 2,000 in the entire range, it is a challenge to not find something you love' (HKTB, 2017).

3   Authenticity and prestige was regarded as the third most important motivational factor by the Chinese respondents. The desire to seek Authenticity (i.e. to sample authentic local foods; to dine in restaurants with an authentic local ambience) was found to be closely associated with Prestige motives (i.e. to tell friends about my dining experiences in Taiwan; to dine in famous restaurants in Taiwan). The desire to seek Authenticity has been recognised as an important motivator guiding Chinese tourists' food consumption (Chang et al., 2010). Fields (2002, p. 40) also points out that experiencing famous and distinctive food in a destination can satisfy Prestige motivations, for the cultural capital acquired as well as the subsequent sharing of the experience to impress friends back home. The present finding corroborates that Authenticity and Prestige are crucial in influencing Chinese tourists' food consumption. Most importantly, the results lend support to previous findings that the Authenticity motive was closely associated with Prestige motive (e.g. Mak et al., 2017). While respondents were motivated to sample authentic local food, they also desired to dine in famous restaurants and to try out well-known food/dishes in the destination. In this sense, the desire to seek authentic as well as famous/well-known dining experiences is analogous to visiting a famous 'attraction' in a destination, an 'attractionised experience' as proposed by Mak et al. (2013). As such, tourism and hospitality marketers may capitalise on this association and 'attractionise' their gastronomic offerings by combining the authentic elements of gastronomic products with prestigious or famous settings. For example, the Paris CVB has done a good job in promoting their 'attractionised' gastronomic experiences online. One example is the Jules Verne, a Michelin-starred restaurant located at the second level of the Eiffel Tower. As described on the Paris CVB website, Jules Verne reunites all the basics of authentic, classic French cuisine and presents its guests with an elegant atmosphere along with the spectacular view

of Paris. Hence, the dining experience is analogous to 'an invitation to the dream' (Paris Info, 2017b).

4   Assurance was perceived as the fourth most important motivational factor by the Chinese respondents. This indicates Chinese tourists' need to seek Assurance in their dining experience, particularly in terms of service quality (i.e. to dine in restaurants that provide good service), hygiene standards (i.e. to dine in restaurants with high hygiene standards; to have foods that are prepared hygienically) and price/value (i.e. to dine in restaurants that are reasonably priced). Tourism and hospitality marketers may need to ensure that a quality assurance mechanism is in place to address Chinese tourists' Assurance needs. One good example is the 'Taste Our Best' quality assurance scheme implemented by Visit Scotland, Scotland's national tourism organisation. The scheme has specific criteria for quality food and drink establishments (e.g. food miles kept to a minimum, friendly staff who will know all about the origins of the food on your plate, and at least 40 percent of the Scottish produce should be highlighted on the menu). The scheme provides information and locations of accredited food and drinks outlines across Scotland to ensure that tourists can enjoy quality ingredients of Scottish provenance and the freshest seasonal produce (Visit Scotland, 2017).

5   Culture was considered as the fifth most important motivational factor by the Chinese respondents. Food consumption in tourism is recognised to be influenced by Culture motivations (Chang et al., 2010; Fields, 2002). The present finding substantiates the importance of the Culture motivations for Chinese tourists (i.e. to increase knowledge about the local culture through dining experiences; to learn about local food traditions and culture). Mak et al. (2013) point out that cultural capital theory is particularly germane in explicating the Culture motivation in tourist food consumption. Cultural capital refers to a stock of knowledge and experience people acquire through the course of their lives that enables them to succeed more than someone with less cultural capital (Bourdieu, 1984). It allows individuals to interpret various cultural codes and to impart their point of views regarding certain topics within the cultural group (Bourdieu & Wacquant, 1992). Accordingly, the Chinese tourists considered acquiring cultural knowledge through their dining experiences in Taiwan was essential in allowing them to possess the cultural awareness, knowledge and skills which would be valuable in future situations. Tourism and hospitality marketers need to understand the significance of the Culture factor and come up with creative ways to showcase the cultural elements of their gastronomy. For example, in Hong Kong, guided tours through the streets of Central and Sheung Wan were offered to enable participants to experience the scenery of Hong Kong's trade and food culture (HKPolyU, 2017).

6   Novelty was regarded as the sixth most important motivational factor by the Chinese respondents. Novelty seeking has been recognised as an important motivation by a number of previous researchers (e.g. Chang et al., 2010; Fields, 2002; Mak et al., 2012a; Mak et al., 2013). For example, Mak et al.

(2013) found that over one third of their sample expressed the desire to try food that could not be easily found in their home countries. In terms of Novelty, Lee & Crompton (1992) suggest four interrelated but distinctive dimensions of novelty, namely, thrill, change from routine, boredom alleviation and surprise. Long (2004) also proposes five aspects of 'Otherness' that can be experienced via food, namely, culture, time, ethos/religion, region and socio-economic class. As a result, tourism and hospitality marketers may emphasise the unique food items or gastronomic offerings that can provide a sense of 'surprise' or 'Otherness' to appeal to novelty seeking tourists. In this light, Hong Kong Tourism Board has shone the spotlight on 'Old Meets New' experiences, with particular emphasis on culinary offerings that rejuvenate traditional Chinese cuisine, such as through molecular twist, in an attempt to provide a sense of thrill, surprise and Otherness to tourists.

7   Finally, the Familiarity and health factor was deemed as the least important by the Chinese respondents. Although this factor had the lowest grand mean of 4.71, it was still considered as of more than average importance by the Chinese respondents. This supports existing research indicating that while novelty is welcomed, a certain proportion of Chinese tourists seek a sense of familiarity in their food consumption experiences (Chang et al., 2010; Cohen & Avieli, 2004). Moreover, the mean food neophobia score for the Chinese respondents was 43.77 (S.D. 4.80), which was relatively high compared with those reported in other studies. This indicates that the Chinese respondents in general had a relatively higher level of food neophobia and might be more inclined to seek familiar food items in their travel dining experiences in Taiwan. Hence, tourism and hospitality marketers should avoid committing the 'sin of homogenisation' described by Pearce (2005, p. 2) and be aware of the food neophobia trait of the Chinese market. Efforts should be devoted to balance novelty with familiarity in order to enhance the acceptability of novel food items among food neophobics. One such effort is the 'New Asia-Singapore Cuisine', which is a new cuisine developed and promoted by the Singapore Tourism Board. The New Asia-Singapore Cuisine combines Oriental styles of ingredients, cooking process and flavours with Western presentation techniques. It has been developed as a new breed of 'gastro-attractions' that have achieved great success in terms of blending novelty and familiarity.

Food consumption in the context of international tourism can be seen as the 'tourist's paradox', an oscillation between fulfilling the 'obligatory' and the 'symbolic' facets in the encounter of food in foreign destinations (Mak et al., 2012a). The 'obligatory' facet reflects the essentiality of food consumption in tourism and the 'symbolic' facet signifies the symbolic meanings of food consumption to tourists, such as acquiring cultural capital, exploring local culture and seeking authentic experience. The seven factors confirmed the multidimensionality of motivation underlying tourist food consumption. Certain factors suggest that tourist food consumption can be highly 'symbolic' (e.g. Authenticity and prestige, and Culture), whereas other factors highlight the importance of seeking 'contrast' (e.g. Novelty and Variety) and 'pleasure' (e.g. Pleasure) through food consumption. Most importantly, the results also reveal the significance of 'obligatory' (e.g. Assurance) and 'extension' (e.g. Familiarity and health) dimensions of tourist food consumption, which are relatively overlooked in the tourism literature (Cohen & Avieli, 2004).

The findings of this study may be used to guide future research endeavours in several ways. First, future research may consider using confirmatory methods (e.g. structural equation modelling) to validate the motivational factors with other cultures or in other settings. Second, the results indicate that Pleasure was the most important motivational factor. The pleasures associated with food consumption are rarely explored in tourism and hospitality literature (Mak et al., 2013). Valentine (1999) points out that much of tourism literature has ignored the 'bodily pleasure' to be derived from eating, and the present findings provide evidence to support the significance of 'sensory pleasure' in food consumption in tourism. Most significantly, the findings suggest that 'convivial pleasure' can be equally important. This corroborates previous evidence in food consumption literature about the importance of convivial and contextual elements in food consumption. Future research efforts towards understanding the significance of sensory and convivial pleasure would be worthwhile. Last but not least, food consumption in tourism ought to be understood as a complex and heterogeneous phenomenon. To fully capture the idiosyncratic nature of tourist food consumption behaviour, it is necessary to transcend the tourist motivation framework and incorporate perspectives from other disciplines, such as food consumption literature. Future research efforts in the field may be enhanced by adopting such an interdisciplinary approach.

References

Arvola, A., Lähteenmäki, L. & Tuorila, H. (1999). Predicting the intent to purchase unfamiliar and familiar cheeses: the effects of attitudes, expected liking and food neophobia. *Appetite*, *32*(1), 113–126.

Booth, D.A. & Shepherd, R. (1988). Sensory influences on food acceptance: the neglected approach to nutrition promotion. *British Nutrition Foundation Nutrition Bulletin*, 13(39–54).

Bourdieu, P. (1984). *Distinction: A Social Critique of the Judgement of Taste*. Cambridge, MA: Harvard University Press.

Bourdieu, P. & Wacquant, L. (1992). *An Invitation to Reflexive Sociology*. Chicago: University of Chicago Press.

Chang, K.C. (1977). *Food in Chinese Culture: Anthropological and Historical Perspectives*. New Haven: Yale University Press.

Chang, R.C.Y., Kivela, J. & Mak, A.H.N. (2010). Food preferences of Chinese tourists. *Annals of Tourism Research*, 37(4), 989–1011.

Chang, R.C.Y., Kivela, J. & Mak, A.H.N. (2011). Attributes that influence the evaluation of travel dining experience: when East meets West. *Tourism Management*, 32(2), 307–316.

Chang, W. & Yuan, J.J. (2011). A taste of tourism: visitors' motivations to attend a food festival. *Event Management*, 15(1), 13–23.

Cohen, E. & Avieli, N. (2004). Food in tourism: attraction and impediment. *Annals of Tourism Research*, 31(4), 755–778.

Eertmans, A., Baeyens, F. & Van den Bergh, O. (2001). Food likes and their relative importance in human eating behaviour: review and preliminary suggestions for health promotion. *Health Education Research*, 16(4), 443–456.

Eertmans, A., Victoir, A., Vansant, G. & Van den Bergh, O. (2005). Food-related personality traits, food choice motives and food intake: mediator and moderator relationships. *Food Quality and Preference*, 16, 714–726.

Field, A. (2005). *Discovering Statistics Using SPSS* (2nd ed.). London: Sage Publications Ltd.

Fields, K. (2002). Demand for the gastronomy tourism product: motivational factors. In A.M. Hjalager & G. Richards (eds.), *Tourism and Gastronomy* (pp. 37–50). London: Routledge.

Fotopoulos, C., Krystallis, A., Vassallo, M. & Pagiaslis, A. (2009). Food Choice Questionnaire (FCQ) revisited. Suggestions for the development of an enhanced general food motivation model. *Appetite*, 52(1), 199–208.

Fox, R. (2007). Reinventing the gastronomic identity of Croatian tourist destinations. *International Journal of Hospitality Management*, 26, 546–559.

Furst, T., Connors, M., Bisogni, C.A., Sobal, J. & Falk, L.W. (1996). Food choice: a conceptual model of the process. *Appetite*, 26, 247–266.

Gains, N. (1994). The repertory grid approach. In H.J.H. MacFie & D.M.H. Thomson (eds.), *Measurement of Food Preferences* (pp. 51–76). London: Blackie Academic and Professional.

García-Segovia, P., Harrington, R.J. & Seo, H.-S. (2015). Influences of table setting and eating location on food acceptance and intake. *Food Quality and Preference*, 39(0), 1–7. doi: http://dx.doi.org/10.1016/j.foodqual.2014.06.004

Hair, J.F.Jr., Anderson, R.E., Tatham, R.L. & Black, W.C. (2002). *Multivariate Data Analysis* (6th ed.). Upper Saddle River, NJ: Prentice Hall.

Harrington, R.J. (2005). Defining gastronomic identity: the impact of environment and culture on prevailing components, texture and flavours in wine and food. *Journal of Culinary Science and Technology*, 4(2–3), 129–152.

Hayes-Conroy, A. & Martin, D.G. (2010). Mobilising bodies: visceral identification in the Slow Food movement. *Transactions of the Institute of British Geographers*, 35(2), 269–281.

HKPolyU (2017). Culture Promotion Committee. Retrieved 26 Mar, 2017, from http://www.polyu.edu.hk/cpeo/cpc/event.php?lang=en&category=highlight&id=191

HKTB (2017). Dim Sum. Retrieved 22 Sep, 2017, from http://www.discoverhongkong.com/eng/dine-drink/what-to-eat/must-eat/dim-sum.jsp

Hobden, K. & Pliner, P. (1995). Effects of a model on food neophobia in humans. *Appetite*, 25(2), 101–113.

Kahn, B.E. (1995). Consumer variety-seeking among goods and services. *Journal of Retailing and Consumer Services*, 2, 139–148.

Khan, M.A. (1981). Evaluation of food selection patterns and preferences. *CRC Critical Reviews in Food Science and Nutrition*, 15, 129–153.

Kim, Y.G., Eves, A. & Scarles, C. (2009). Building a model of local food consumption on trips and holidays: a grounded theory approach. *International Journal of Hospitality Management*, 28, 423–431.

Kim, Y.H., Goh, B.K. & Yuan, J. (2010). Development of a multidimensional scale for measuring food tourist motivations. *Journal of Quality Assurance in Hospitality and Tourism*, 11(1), 56–71.

Koivisto, U.-K., & Sjödén, P.-O. (1996). Food and general neophobia in Swedish families: parent-child comparison and relationships with serving specific foods. *Appetite*, 26, 107–118.

Köster, E.P. (2009). Diversity in the determinants of food choice: a psychological perspective. *Food Quality and Preference*, 20, 70–82.

Lee, T. & Crompton, J. (1992). Measuring novelty seeking in tourism. *Annals of Tourism Research*, 19, 732–737.

Lin, Y.-C. & Chen, C.-C. (2014). Needs assessment for food and food services and behavioural intention of Chinese group tourists who visited Taiwan. *Asia Pacific Journal of Tourism Research*, 19(1), 1–16. doi: 10.1080/10941665.2012.724017

Long, L.M. (2004). Culinary tourism: a folkloristic perspective on eating and Otherness. In L.M. Long (ed.), *Culinary Tourism* (pp. 20–50). Kentucky: The University Press of Kentucky.

Mak, A.H.N., Lumbers, M. & Eves, A. (2012a). Globalisation and food consumption in tourism. *Annals of Tourism Research*, 39(1), 171–196.

Mak, A.H.N., Lumbers, M., Eves, A. & Chang, R.C.Y. (2012b). Factors influencing tourist food consumption. *International Journal of Hospitality Management*, 31(3), 928–936.

Mak, A.H.N., Lumbers, M., Eves, A. & Chang, R.C.Y. (2013). An application of the repertory grid method and generalised Procrustes analysis to investigate the motivational factors of tourist food consumption. *International Journal of Hospitality Management*, 35, 327–338.

Mak, A.H.N., Lumbers, M., Eves, A. & Chang, R.C.Y. (2017). The effects of food-related personality traits on tourist food consumption motivations. *Asia Pacific Journal of Tourism Research*, 22(1), 1–20.

Martens, L. & Warde, A. (1997). Urban pleasure? *Food, health, and identity*, 131.

Maslow, A.H. (1970). *Motivation and Personality*. New York: Harper & Row.

McKercher, B., Okumus, F. & Okumus, B. (2008). Food Tourism as a Viable Market Segment: It's All How You Cook the Numbers! *Journal of Travel & Tourism Marketing*, 25(2), 137–148.

Meiselman, H.L., Mastroianni, G., Buller, M. & Edwards, J. (1999). Longitudinal measurement of three eating behaviour scales during a period of change. *Food Quality and Preference*, 10, 1–8.

Meiselman, H.L. (1996). The contextual basis for food acceptance, food choice and food intake: the food, the situation and the individual. In H.L. Meiselman & H.J.H. Macfie (eds.), *Food Choice, Acceptance and Consumption* (pp. 239–263). Boston: Springer.

Pallant, J. (2007). *SPSS Survival Manual: A Step-by-Step Guide to Data Analysis using SPSS for Windows* (3rd ed.). Berkshire, UK: Open University Press.

Paris Info (2017a). A meal at a communal table. Retrieved 23 Sep, 2017, from https://en.parisinfo.com/where-to-eat-in-paris/a-meal-in-an-unusual-setting/a-meal-at-a-communal-table

Paris Info (2017b). Where to eat in Paris? – Le Jules Verne. Retrieved 22 Sep, 2017, from https://en.parisinfo.com/paris-restaurant/70615/Le-Jules-Verne

Patton, M.Q. (2002). *Qualitative Research and Evaluation Methods* (3rd ed.). Thousand Oaks, CA: Sage.

Pearce, P.L. (2005). *Tourist Behaviour: Themes and Conceptual Schemes*. Clevedon: Channel View Publications.

Pliner, P. & Hobden, K. (1992). Development of a scale to measure the trait of food neophobia in humans. *Appetite*, 19(2), 105–120.

Pliner, P. & Salvy, S.J. (2006). Food neophobia in humans. In R. Shepherd & M. Raats (eds.), *The Psychology of Food Choice* (pp. 75–92). Oxfordshire: CAB International.

Quan, S. & Wang, N. (2004). Towards a structural model of the tourist experience: an illustration from food experience in tourism. *Tourism Management*, 25(3), 297–305.

Randall, E. & Sanjur, D. (1981). Food preferences: their conceptualisation and relationship to consumption. *Ecology of Food and Nutrition*, 11(3), 151–161.

Richards, G. (2002). Gastronomy: an essential ingredient in tourism production and consumption? In A.M. Hjalager & G. Richards (eds.), *Tourism and Gastronomy* (pp. 3–20). London: Routledge.

Rozin, P. (2006). The integration of biological, social, cultural and psychological influences on food choice. In R. Shepherd & M. Raats (eds.), *The Psychology of Food Choice* (pp. 19–39). Oxfordshire: CAB International.

Shepherd, R. & Raats, M. (1996). Attitudes and beliefs in food habits. In H.L. Meiselman & H.J.H. MacFie (eds.), *Food Choice, Acceptance and Consumption*. London: Chapman and Hall.

Sobal, J., Bisogni, C.A., Devine, C.M. & Jastran, M. (2006). A conceptual model of the food choice process over the life course. In R. Shepherd & M. Raats (eds.), *The Psychology of Food Choice* (pp. 1–18). Oxfordshire: CAB International.

Steptoe, A., Pollard, T.M. & Wardle, J. (1995). Development of the motives underlying the selection of food: the food choice questionnaire. *Appetite*, 25, 267–284.

Tikkanen, I. (2007). Maslow's hierarchy and food tourism in Finland: five cases. *British Food Journal*, 109(9), 721–734.

Tuorila, H., Andersson, A., Martikainen, A. & Salovaara, H. (1998). Effects of product formula, information and consumer characteristics on the acceptance of a new snack food. *Food Quality and Preference*, 9, 313–320.

Tuorila, H., Meiselman, H.L., Bell, R., Cardello, A.V. & Johnson, W. (1994). Role of sensory and cognitive information in the enhancement of certainty and liking for novel and familiar foods. *Appetite*, 23, 231–246.

UNWTO (2017). *UNWTO Tourism Highlights* (2017 Edition). Madrid: United Nations World Tourism Organisation.

Valentine, G. (1999). Consuming pleasures: food, leisure and the negotiation of sexual relations. In D. Crouch (ed.), *Leisure/Tourism Geographies: Practices and Geographical Knowledge* (pp. 164–180). London: Routledge.

Van Trijp, H.C.M. & Steenkamp, J.-B.E.M. (1992). Consumers' variety seeking tendency with respect to foods: measurement and managerial implications. *Eur Rev Agric Econ*, 19(2), 181–195. doi: 10.1093/erae/19.2.181

Visit Scotland (2017). Taste Our Best – quality assurance scheme. Retrieved 23 Sep, 2017, from http://www.visitscotland.com/about/food-drink/taste-our-best

Wright, L.T., Nancarrow, C. & Kwok, P.M.H. (2001). Food taste preferences and cultural influences on consumption. *British Food Journal*, 103(5), 348–357.

6 Dining trajectories of Chinese tourists in Australia

Richard C. Y. Chang

Introduction

Food was considered a form of sustenance rather than having any features of being a leisure or pleasure activity in ancient times. Due to industrialisation and colonisation during the 18th century, food items were extensively exchanged among different countries which facilitated the acculturation between Western and Asian cultures. Consequently, the curiosity in other food cultures has become a main trigger to the emergence of gastronomy tourism (Boniface, 2003). Gastronomy has been recognised as an expression of a local culture and dining out at a destination has become a major conduit for tourists to experience the local culture of a destination. Richards and Hjalager (2002) propose that dining out at a destination is often considered a pleasurable sensory experience to fulfil an experiential part of tourists' holiday dreams. For this reason, tourists are increasingly seeking the dining out experience to complement or as an alternative to other touristic activities. Indeed, travel dining experiences and local gastronomy have increasingly been promoted together as a combined 'attraction' by many destinations. Despite the central role of the travel dining experience as part of the holiday experience, tourist dining behaviour has been overlooked in the hospitality and tourism literature because eating and drinking are often taken for granted as a 'supporting consumer experience' which refers to an extension of tourists' daily routines (Quan & Wang, 2004).

China has become the leading contributor to the international tourism market in terms of volume and expenditure (Wang, Fong & Law, 2016). The number of outbound travelers rose by 6% to reach 135 million and the expenditure by Chinese travellers grew by 12% to reach US$261 billion in 2016 (UNWTO, 2017). As for the Australian market, China remains Australia's second largest inbound market in visitor arrivals and the largest market in terms of total expenditure (Tourism Research Australia, 2017). Both visitor numbers and receipts from the Chinese market increased 10%, reaching 1.2 million and AUD$9.8 billion respectively between June 2016 and June 2017. With the significant growth in this market, it is important for the Australian tourism industry not only to provide diversified cuisines for Chinese tourists to explore but also to understand the food preferences and food consumption motives of Chinese tourists to make them feel 'at home'. On the other hand, Australian local foods could be promoted as an 'attraction' for those who want to sample novel and exotic local food products. But at the same time, being confronted with an unknown cuisine could also be an 'impediment' which could potentially deter Chinese tourists from visiting Australia (Cohen & Avieli, 2004). Therefore, Australia's tourism sector faces challenges in understanding and learning about Chinese food culture and Chinese tourists' consumption traits, and how to cater to the needs of Chinese tourists. By synthesising the qualitative findings of Chang, Kivela and Mak's 2010 and 2011 studies, this chapter aims to delineate Chinese tourists' (including mainland Chinese, Hong Kong and Taiwanese tourists) dining trajectories in Australia, with particular focus on their food preferences, the attributes in evaluating travel dining experiences and a typology of their dining experience.

Food, culture and Chinese food culture

The relationship between food and culture has been extensively studied in the literature. It is found that there are many diverse and interesting ways in deciphering the cultural meaning of food. According to Atkins and Bowler (2001), culture is a major determinant of what we eat. In addition, Mäkelä (2000) suggests that culture defines how food is coded into 'acceptable' or 'unacceptable'. Due to cultural differences, what is considered as 'edible' food in one culture might be considered as 'inedible' in another. For example, most Western Europeans are of the opinion that internal animal organs are 'bad food', whereas Oriental people regard it as 'good food' and believe them to be very nutritional and healthy. Culture determines individual's eating and dining habits and, once such habits are established, they are likely to be continuing and persistent (Fieldhouse, 1986). However, as culture is a learned phenomenon, an individual's eating habit could still be changed during learning and socialisation processes (Beardsworth & Keil, 1997). This is particularly true in the case of tourists' adaptation of local food. As Chinese tourists are facing unfamiliar foods and cultural situations, Chinese food culture might be a basic guideline when choosing various foods in Australia. For example, Fieldhouse (1986) suggests that the sensory qualities of food are important to the acceptability and degree of preference. As such, any one of the sensory qualities, for instance food presentation, has influences on the constitution of the acceptability in any given foods or dishes. In this regard, Australian food needs to match Chinese tourists' mental construct of acceptability or it would be abandoned and considered as inedible because the food does not 'look right'. Therefore, cultural factors play a pivotal role in tourists' food consumption behaviour. It is important to understand the cultural background of Chinese tourists to cater to their palate.

Chinese food culture is known for its flexibility and adaptability because the Chinese are usually painstaking in their quest for edible food (Chang, 1977). To prepare a proper meal in Chinese culture, it must have suitable quantities of '*fan*' (rice and other starch foods) and '*ts'ai*' (vegetable and meat dishes). Typically, '*fan*' is the essential element of a meal and '*ts'ai*' is served as a complement to make '*fan*' palatable. The 'flavour principles' that distinguish Chinese cuisine include soy sauce, rice wine and ginger mixture (Rozin, 1983). In addition, Chinese people believe that food not only satisfies physical needs but also has a medicinal value. The selection of the right food at any particular time must be dependent upon one's health condition at that time. Hence, in Chinese food culture, the complementary forces of *yin* (cold foods) and *yang* (hot foods) are critical to food selection (Wu, 1995). Foods commonly classified as 'hot' are fatty meats, oily nuts, spices and ginger, whereas 'cold' foods include most fruits (except mango and lychee), bamboo shoots and crabs. In particular, the symbolic meanings of food play an auspicious role especially during special occasions and celebratory feasts. For example, long noodles signify long life at birthdays and fish and oyster dishes are often served in the Chinese New Year banquet because their names sound like desirable states – the words for 'fish' and 'surplus' sound alike, as do the words for 'oyster' and 'happy events'. The Chinese style of dining out also differs significantly from that in Western culture. Typically, Chinese restaurants are noisy; and not the least bit romantic. This is because Chinese feel more comfortable in inviting a large group of people so as to sample more dishes. Additionally, the Chinese see dining out as a social activity to build up personal connections with guests (Wright, Nancarrow & Kwok, 2001). Due to the impacts of globalisation, the increasing availability of ethnic restaurants in China and the widely accessible information about foreign cuisines on the Internet have contributed to an increased exposure to Western food culture (Cohen & Avieli, 2004). It has started to change consumer food preferences in China (Bhandari & Smith, 2000). Hence, it is interesting to try and understand whether Chinese tourist food preferences remain constant while travelling in a Western country.

Chinese tourists' food preferences

By joining tour groups originating from mainland China, Hong Kong and Taiwan and conducting on-site participant observation and focus group interviews, Chang, Kivela and Mak (2010) have explored Chinese tourists' food preferences when holidaying in Australia. Three distinct themes (see Figure 6.1) were identified to delineate their food preference, namely 'Chinese food', 'local food', and 'non-fastidious about food preference'.

Chinese food

Not every Chinese tourist was accustomed to the new culinary environment and may have inclinations towards Chinese food. Motivational factors identified under such preferences include, *core eating behaviour, appetising assurance* and *familiar flavour*. It was found that some Chinese tourists had a steadfast preference towards Chinese food when travelling in Australia. Interestingly, even for those Chinese tourists who were enthusiastic to try local food, it was impossible for them to consume local food in every meal. Although the intake of local food could be perceived as a 'peak touristic experience', it did not match up with the criteria of a 'proper meal' in their eating habit. It could be argued that appreciating local food might satisfy the experiential desires but might not be able to satiate the physiological needs of Chinese tourists. In this respect, the Chinese tourists' '*core eating behaviour*' has led to their preference of Chinese food. In particular, if the taste of Australian food could not satisfy Chinese tourists' palate, Chinese food comparatively became a more reliable way for them to obtain gustatory pleasure. This is because Chinese food was familiar to the tourists and offered them a sense of '*appetising assurance*'. Finally, some Chinese tourists were relatively intransigent with their own '*familiar flavour*' which includes familiar food items and familiar cooking methods to make both Chinese food and local food palatable.

Local food

Chang, Kivela and Mak's (2010) study indicates that many Chinese tourists were excited and enthused by having the chance to sample local food in Australia. They were eager to be involved in the local gastronomic experience. The motivational factors identified include; *explore local culture, authentic travel experience, learning/ education opportunity, prestige and status, reference group influence* and *subjective perception*. In fact, '*exploring local culture*' was explicitly demonstrated

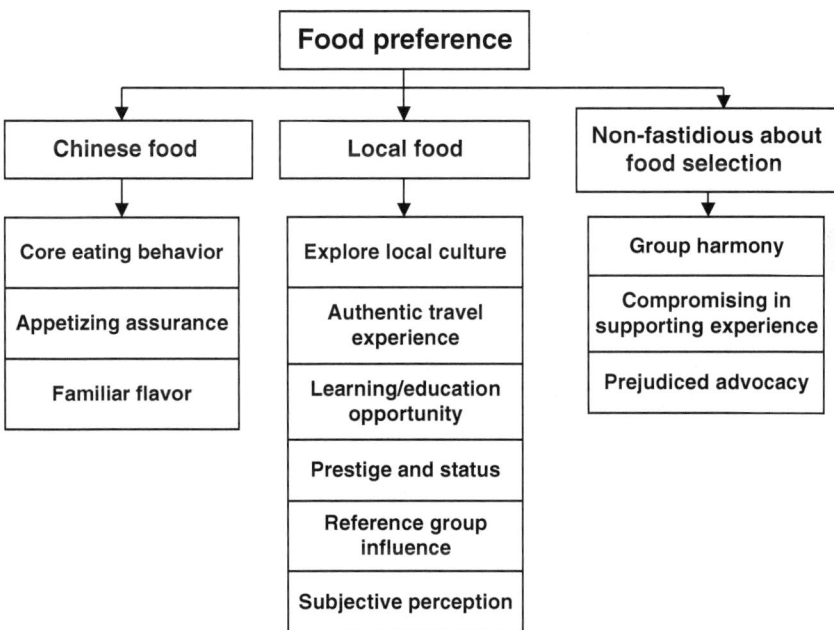

Figure 6.1 A model of Chinese tourists' food preference
Source: Chang, Kivela, & Mak (2010)

in many studies as the reason for tourists tasting local food (Chang et al., 2010; Mak et al., 2012; Mak et al., 2017). This reflects that Chinese tourists may perceive the intake of local food as a way to appreciate the local culture. Many tourists may be interested in anything that could represent the culture of the destination, and food and dining certainly provide many opportunities for tourists to encounter various facets of the local culture (Kivela & Johns, 2003). Accordingly, this explains the reason why Chinese tourists are enthusiastic in trying local food. Other than to explore local culture, another important reason for Chinese tourists to sample local food was the quest of an '*authentic' travel experience*'. There is an old saying: 'when in Rome, do as the Romans do', and Chinese people generally believe that when visiting a foreign country, one should follow the customs of those who live in it in order to show respect and blend in with the local people. As the Chinese maxim suggests: 'travelling ten thousand miles is better than reading ten thousand books'. Evidently, Chinese tourists regard sampling local food as a '*learning opportunity*' that could enhance their cultural capital and fulfill their need for self-actualisation (Maslow, 1970). In the past, outbound travel of Chinese residents was restricted due to economic constraints, tight political control and visa restrictions. And even with today's increased disposable income per capita, the ability to travel abroad still symbolises social status and privilege in China (Zhang, Pine & Lam, 2005). Thus, to a certain extent, the ability to travel abroad and taste local food are akin to conspicuous consumption which also could represented '*prestige and status*' to Chinese tourists. Due to the influence of the Chinese collectivist culture (Hofstede, 2001), it was observed that Chinese tourists are interested in trying local foods that are recommended by their '*reference group*', such as friends and travel blogs, as they are in pursuit of 'fashionability' (Finkelstein, 1998). It is worth noting that some Chinese tourists hold the '*subjective perception*' that Chinese food in Australia is not as tasty as in their home country, therefore they are inclined to try local food or other foreign foods.

Non-fastidious about food preference

It is interesting to note that some Chinese tourists were not fastidious about either Chinese food or local food when travelling in Australia. Those tourists seemed to have very little interest in food, viewing it simply as a form of sustenance rather than a source of pleasure. Because Chinese people place importance in '*group harmony*', it is plausible that Chinese tourists would accommodate their friends' food preference even when they travel abroad. In addition, some Chinese tourists have a preconceived idea that food and accommodation experiences do not match up to 'ontological comfort of home' (Quan & Wang, 2004). Accordingly, they are prepared to '*compromise with supporting experience*' because meal arrangements are not their main concern since they joined the tour. It was also found that a tour leader's '*prejudiced advocacy*' influenced Chinese tourists' preconception of travel dining which was seen as relatively unimportant compared with other 'peak touristic experiences'. The Chinese tourists who held this attitude mostly belonged to the elderly generation.

Attributes in evaluating travel dining experiences

Based on an ethnographic approach, Chang, Kivela and Mak (2011) have conducted focus group interviews and on-site participant observation in Australia to obtain in-depth context specific information about Chinese tourists' (originating from mainland China, Hong Kong and Taiwan) travel dining experiences. They further identified six attributes that affect Chinese tourists' evaluation of their travel dining experiences, namely 'Chinese tourists' own food culture', 'the contextual factor of dining experience', 'variety of food selection', 'perceptions towards Australia', 'service encounter' and 'tour guide's performance'.

Chinese tourists' own food culture

Kivela and John (2003) contend that each cultural group has their own distinctive food culture, which will subsequently determine which flavours and culinary precepts are acceptable to that cultural group. This food culture affects an individual's evaluation of food from other cultural groups, a proposition confirmed by Chinese tourists' evaluation of Australian food. For example, some Chinese tourists found kangaroo meat unpalatable because of the *unfamiliar flavour* and texture and some indicated that they disliked the *cooking methods* of Australian barbeque foods. It could be explained that Chinese tourists may lack previous experience or knowledge of Australian food that made it hard for them to ascertain the food quality. Thus, they could only evaluate local food quality based on their own food culture value. In addition, some Chinese tourists were unaccustomed to the cooking methods of Australian food and suggested ways to increase their palatability. It is obvious that the suggestions came from their own culinary precept. This implies that even when Chinese tourists are enthralled by Australian food, they still cannot withdraw from their own food cultural values when it comes to the evaluation of local food.

The contextual factor of dining experience

It was found that *authenticity* and *experiential factors* were the two significant aspects of the contextual factor for Chinese tourists when evaluating their dining experience in Australia. That is to say, Chinese tourists were concerned whether the local food represented local characteristics so that they could immerse themselves in the Australian culture. At the same time, they might place lesser value on the food quality at Australian restaurants if they were content with the experiential parts of the dining experience. It should be noted that previous literature has suggested that the tangible factor of food quality and the intangible factor of experiential service constitute the totality of dining satisfaction (Lee, Yoon & Lee, 2007). Yet, the findings by Chang et al. (2011) reveal that Chinese tourists might set the tangible food quality aside when the intangible value fulfilled their experiential needs. Wu et al. (2016) also point out that Chinese tourists tend not to alter their attitude towards local food because of the sensory appeal as they assume the food is essentially different from their own food culture. As such, many of the Chinese tourists were well aware of the differences in food culture between China and Australia and they realised that local food quality might not be able to gratify their gustatory pleasure. In effect, as long as the local food and dining ambience characterised the Australian culture, the weighting of the tangible food quality could be reduced to a minimum in Chinese tourists' evaluation process.

Variety of food selection

The variety of food selection was another important attribute that influenced Chinese tourists' dining satisfaction in Australia. Such a notion is an extension of their dining out behaviour in their home country. Finkelstein (1998) suggests that people have a desire to look for unusual or interesting features from their dining out experience so that they can broaden their culinary experience. This is also true for Chinese tourists as they conveyed their expectation of seeking a variety of food selection while travelling in Australia. From a cultural perspective, Chinese tourists may expect the local food to be representative of a culture different to their own food culture. After all, the more local food Chinese tourists have eaten, the more they understand the Australian culture. Thus, cultural appreciation could be one of the motives for Chinese tourists' inclination to sample a variety of food. In essence, variety should be supplemented with culinary differentiation for Chinese tourists to appreciate Australian food culture. Warde and Martens (2000) emphasise that the legitimacy of consumer culture means consumers can choose freely among enormous volumes of goods and services. In addition, the Chinese emphasise hospitality spirit and generosity to others. They often order many dishes when they dine out with friends or else they would feel a '*loss of face*' (Wright et al., 2001). Consequently, the Chinese generally prefer to have a variety of dishes in dining out occasions. In short, the legitimacy of consumer culture and dining out behaviour of the Chinese further support the claim of Chinese tourists' liking for the variety of dishes and meal experiences during their holidays.

Perceptions towards Australia

Another attribute that affected Chinese tourists' evaluation of travel dining experiences is the perception that they have of Australia. In particular, their impressions significantly influenced their beliefs and attitudes on what food should be eaten as well as what service should be delivered in Australia. Chinese tourists' perceptions on what food should be eaten related to Australia's gastronomic identity. As a gastronomic identity can represent a destination's character (Bessiere, 1998), it might be one of the reasons that attracted them to visit Australia. If this is so, the gastronomic identity would exert influence on their expectations on what food should be eaten during their dining experiences in Australia. On the other hand, most of the Chinese tourists considered that their ability to travel to Australia represented their social prestige. In this respect, a high level of service should also be provided so as to display the participant's superiority. This expectation of a high level of service is somehow correlated to their perception of conspicuous consumption behaviour. In light of this, it is important for marketing organisations to create a dining environment related to this perception so that tourists can realise their dreams when they are in Australia (Lin et al., 2017).

Service encounter

Andersson and Mossberg (2004) contend that server-customer interaction is critical when evaluating customer satisfaction; however, in the realm of tourists' dining experiences, the delivery of service served various purposes for Chinese tourists. For example, in the context of Chinese tourists dining out in an Australian restaurant, the Chinese could educate their children by communicating with foreign staff. In addition, by imitating others' table etiquettes, participants had more opportunities to appreciate cultural phenomena while at the table (Ji et al., 2017). In this respect, the server-customer interaction was not only a process of service delivery but also a stage for participants' learning experience. Other than the educational function of service encounter, Chinese tourists may have different expectations and attitudes towards Chinese restaurants and foreign restaurants in Australia. First, Chinese tourists assess their service encounter in Chinese restaurants in accordance with their pre-existing evaluative criteria because they share the same cultural background. In this respect, responsiveness and empathy are deemed to be imperative to the quality of service encounter (Zehrer, Crotts & Magnini, 2011). In particular, amiability is significant to the service encounter because the Chinese believe that one should show his/her hospitality when they meet friends overseas. On the other hand, Chinese tourists may tolerate shortcomings during their service encounter in foreign restaurants. Such a notion could be supported by the following deliberations. First, service encounters in foreign restaurants are deemed as possessing educational value to the participants as it allows participants contact with local people and to learn different table etiquettes. Second, most of the Chinese tourists have anticipated the potential communication barriers due to their language deficiency. Although different 'role script' may have created the miscomprehension toward service style (Broderick, 1998), Chinese tourists reproached this shortcoming on themselves and prepared for the foreseeable hindrance. Finally, in contrast to the expectation of prompt service in Chinese restaurants, the Chinese tourists tended to disregard the service speed for they preferred to enjoy the exotic ambience and romantic experience in the foreign restaurant.

Tour guide's performance

An all-inclusive package tour is still the preferred mode of outbound travel for most Asian tourists (Wang, Hsieh & Chen, 2002; Wong & Lau, 2001). Within an all-inclusive package tour, transportation, accommodation and meal arrangements are all pre-arranged, and yet there are times that unforeseeable hindrances could affect the performance of the package tour. Under this circumstance, the tour guide is often there to be prepared for any contingency. Alternatively, the tour guide plays the cultural mediator role between tourists and the host country which entails not only introducing knowledge about customs, history and arte-facts of the destination but also providing culinary information and local food culture to the tourists (Cohen & Avieli, 2004; Mak, Wong & Chang, 2011). In the realm of the travel dining experience, a tour guide should not only facilitate the organisation of the meal experience but also offer an interpretation of the symbolic meaning of local foods and Australian food culture. Importantly, the tour guide may have manipulative power in affecting tourists' dining satisfac-tion. This is because the tour guide simultaneously understands both the Chinese and Australian food cultures. In effect, the tour guide could create value-adding input to participants' dining experience by suggesting the right choice of local food to please tourists. In addition, tour guides may render significant improve-ments towards the quality of travel dining experiences. It was found that Chinese tourists displayed reliance on their tour guides when consuming local food in Australia (Chang, Kivela & Mak, 2011). In short, it is concluded that tour guides could enhance tourists' satisfaction on dining experience in three aspects, namely, *interpretation of local food, culinary broker of local dining behaviour*, and *facili-tating of the dining experience*.

A typology of Chinese tourists' dining experience

By synthesising the above findings, this chapter further proposes a typology of Chinese tourists' dining behaviour based on food consumption motives, travel dining behaviour and attributes in evaluating dining experience. The typol-ogy differentiates three types of tourists: observers, browsers and participators (see Table 6.1).

Observers

The observers represent Chinese tourists who prefer to partake Chinese food in Australia. It is suggested that observers' dining behaviours in Australia might have been predisposed by Chinese food culture. Thus, they are in quest of *appe-tising assurance* and *familiar flavour* from the travel dining experience. To some extent, the observer is favourable towards local food because exploring local cul-ture during travelling is one of the predominant motivators that are commonly shared by them. However, the observer might dread being unaccustomed to local food and might need familiar flavours to make unpalatable food (based on their own food culture value) more palatable. Such behaviour suggests that observers tend to prefer a blended dining experience with Chinese food and local food. In this respect, the fusion of Chinese and local food might be a good option for them to partake of. It is, therefore, presumed that the observers prefer to observe, or dis-cern the Australian food culture, rather than totally immerse themselves in it. This also implies that the observers are sentimentally attached to Chinese food. It can be argued that the underlying reason behind the observers' propensity of seeking familiarity from the travel dining experience is the concern regarding palatability. In effect, in their assessment of the travel dining experience they place great emphasis on food quality, sumptuousness, dining ambience and the tour guide's performance in facilitating the dining experience.

Browsers

Browsers are Chinese tourists who are not fastidious about food selection while travelling. For this type of tourist, food is not a major concern in gauging the level of satisfaction with their holiday. They are defined as '*browsers*' because they regard any dining experience in a casual way and are not particularly concerned with their travel dining experience. The browsers are prepared to compromise on their food preference in order to reach group harmony. Although Chinese food culture might not have influenced the browsers' attitudes and behaviours towards local food in Australia, their group-orientated dining behaviour manifests the influence of Chinese culture. Due to language deficiency, this type of tourist prefers to stay under the 'shelter' of the tour guide. Security and safety, thus, have become a more prominent concern for the browsers. That is to say, browsers are willing to accept the pre-arranged meals. They are likely to follow the tour guide's lead in order to sustain their own safety and security. In this respect, the tour guide's performance becomes critical since the browsers would rely on the guide to resolve any unforeseeable hindrances. Consequently, the browsers valued the feeling of being served, and evaluate their dining experience according to the service attitude as well as the tour guide's performance on facilitating the experience.

Participators

The participators are those who have great gastronomic interest in local food. They believe the best way to appreciate local culture is through the participation in a local dining experience in Australia. In contrast to the other two types of Chinese tourists who might have reservations towards Australian food, the participators are overtly willing to savour the unique gastronomy in Australia. For example, although they join a tour group, they prefer to have free dining activities so that they can freely decide on their food choices. As the intake of local food can potentially challenge their pre-existing food culture values, the participators are more likely to disregard their ingrained dining behaviour in order to pursue a genuine contact with the local food culture. As the participators actively seek chances to experience local food in Australia, the variety of food selection becomes an important indicator for their satisfaction with their dining experience. In addition, both food selection and dining ambience should characterise Australian culture so that the participators experience an exotic dining environment. As the participators can be motivated by sampling fashionable local food in Australia, the success of their travel dining experience might therefore be determined by whether the meal arrangements have corresponded with their expectations. Finally, since the participators are keen to acquire new knowledge about the local food culture,

Table 6.1 A typology of the Chinese tourists' dining behaviour

	Observer	Browser	Participator
Food consumption motives	Appetising assurance Familiar flavour	Food choice depends on group's consensus	Explore local culture Broaden culinary experience Display prestige Novelty-seeking
Dining behaviour	Look for familiar foods Remain core eating behaviour Passively accept pre-arranged meals Prefer fusion food	Safety and security is the main concern Prefer to stay under tour guide's 'shelter' Passively accept pre-arranged meals	Disregard pre-existing dining behaviour Actively pursue new dining experience Prefer free dining activity
Evaluation attributes on dining experience	Food quality Sumptuousness Dining ambience Tour guide's performance on facilitating the dining experience	Service attitude Tour guide's performance on facilitating the dining experience	Variety of food selection Dining ambience Correspondence of expectation and perceived meals Tour guide's performance on the interpretation of the local food.

they also rely on a tour guide's interpretation so as to obtain more insights into the local food of the destination. Without knowing the symbolic meanings behind local food, the participators might consider the intake of local food a meaningless exercise and accordingly, appreciation and acceptance of the local food could be significantly diminished.

Healthy eating while travelling and family influence on eating behaviour

As mentioned earlier, the complementary forces of *yin* (cold foods) and *yang* (hot foods) are imperative to the Chinese food culture. Yet, it is suggested that some Chinese tourists may disregard harmonious eating behaviour due to being tempted by sumptuous meals (Chang, 2007). Wu (1995) suggests that the concept of *yin-yang* equilibrium becomes significant particularly when the Chinese feel physically uncomfortable. Thus, it can be assumed that Chinese tourists tend to ignore the *yin-yang* balance while travelling since it is a peripheral concern to the food intake if they are in a healthy state. Particularly, the pursuit of a hedonic dining experience might induce Chinese tourists to disregard their healthy beliefs while travelling. Altogether, it can be proposed that Chinese tourists disregard their healthy eating behaviour if they decide to indulge themselves in sumptuous meals while on holiday. On the other hand, nutritional concerns significantly affect Chinese tourists' choice of food during holidays. According to the Chinese food culture, nutritional foods and sufficient food quantity are critical to sustain one's energy and strength (Lu, 1986). Thus, a sufficient food supply will prevent tourists from traveling with an 'empty stomach' which in turn would affect their tourist satisfaction.

Breakfast is another dining context when Chinese tourists may be reluctant to alter their eating behaviour. Chang (2007) contends that many Chinese tourists are unaccustomed with Australian breakfast settings and that they are unlikely to try unfamiliar food at breakfast when traveling in Australia. This could be explained by a Chinese aphorism: 'The whole day's work depends on a good start in the morning'. In this respect, a contented breakfast might bring good mood as well as provide sufficient nutrition for the tourists to sustain their travel activities.

The increasing influence children have on familial eating behaviour has been duly reflected in Chinese tourists' travel eating behaviour. As mentioned earlier, dining out when travelling is regarded as a learning opportunity that can increase children's cultural knowledge and sensitivity. For that reason, dining opportunities in Australia are perceived as learning experiences that allow children to have a better understanding of the local culture, customs and language. Therefore, Chinese parents are willing to adjust their eating behaviour in order to follow the preference of their children. This involves compromising on their ingrained eating habits and attitudes and minimising the concern for palatability and edibility. In other words, parents forgo their gustatory pleasure in exchange for dining experiences that allow their children to better learn about and experience a destination's culture.

Conclusion

This chapter depicts Chinese tourists' dining trajectories in Australia based on the key findings of Chang, Kivela and Mak's studies (2010, 2011). It can be concluded that most of the Chinese tourists in Australia pursued gustatory pleasure in their travel dining experience no matter if they partake Chinese or local foods. This is because Chinese people place emphasis on eating and regard it as the first and foremost priority in life (Chang, 1977). It was revealed that fusion food is the ideal way to satiate tourists' desire for savouring local food, while at the same time, corresponding to their pre-existing eating behaviour. In addition, Chinese tourists are inclined to avoid immutable eating behaviour during their trip to Australia. This finding corroborates with propositions in the literature that a travel dining experience is not sought to be a repetition and replication of routine dining behaviour at home (Finkelstein, 1998; Kivela & Crotts, 2006; Warde & Martens, 2000). Thus, Chinese tourists expect a diversified travel dining experience that breaks the dullness of their routine meal arrangements. As joining a tour group is the preferred mode of outbound travel for most Chinese tourists, Chinese travellers commonly prefer to have free dining activities to be integrated into their pre-arranged meal arrangements in order to allow them the chance to seek experiential dining experiences. However, free dining activities are only feasible if they take place in a designated area with the assistance and suggestions from the tour guide.

This chapter also delineates six attributes that affect Chinese tourists' evaluation of their travel dining experience. First, the effect of tourists' food culture and culinary precepts create a discrepancy between expectation and perceived food quality for some Chinese tourists. Second, many Chinese tourists are prepared to compromise on their gustatory pleasure as long as the local food and dining ambience characterise the Australian culture. Third, a variety of food selection confers great value to the breadth of the travel experience. Fourth, most Chinese tourists have initial perceptions towards their gastronomy identity and expect a certain level of service when visiting Australia, which in turn influences their travel dining satisfaction. Fifth, Chinese tourists tolerate obstacles during intercultural service encounter because they foresee communication barriers between themselves and service staff. Finally, dining satisfaction can be enhanced if a tour guide serves as the mediator to alleviate language and cultural barriers faced by tourists.

While the healthy eating concept is deeply rooted in Chinese food culture (Chang, 1977), this chapter also reveals that many of Chinese may disregard their healthy beliefs if they are overwhelmed by sumptuous foods when travelling. Conversely, Chinese tourists pay attention to sufficient nutrition and the content of breakfast because these two aspects determine whether they can travel in a healthy state. Previous literature has reiterated that one's eating behaviour is predominantly shaped and influenced by parents (Beardsworth & Keil, 1997). However, it is suggested that such a top-down influence will be altered if tourists are accompanied by their children. The reason why those tourists cater to their children's food preference while travelling is because they quest for familial contentment. In addition, parents consider the travel dining experience as an educational experience that allows their children to have a better understanding of local culture, customs and language.

A number of practical implications for the tourism industry and destination marketers can be drawn from the findings. First, the chapter suggests that most Chinese tourists are keen to be involved in the dining experience. Travel agencies may therefore want to incorporate some free dining activities into meal arrangements made for the tour group. However, free dining activities should be undertaken in a designated area with guidance and suggestions provided in advance from the tour guide. Otherwise, Chinese tourists may feel abandoned and vulnerable due to language barriers. In addition, previous studies have suggested that destination marketers should engage tourists in the food-production process (Hall & Sharples, 2003; Kim, Eves & Scarles, 2009). This is also applicable to Chinese tourists as participation in dining activities can be the most memorable part of their travel experience. In this respect, destination marketers should develop dining activities that involve tourists taking part in cooking and tasting lessons, for them to see and experience how food is grown and prepared, and to explain to tourists the cultural significance of local food. For example, chefs can influence dining satisfaction by engaging in conversation with customers and interpreting menu items in *teppenyaki* restaurants (Chen, Peng & Hung, 2016). This chapter has revealed that Chinese tourists may lack knowledge about Australian food culture, which makes it difficult for them to savour the authenticity of Australian food. Therefore, it becomes indispensable for destination marketers to provide comprehensive information and to educate Chinese tourists how to enjoy the food they try to promote. The strategy to educate tourists means that destination marketers need to provide user-friendly, attractive, informative and guided information about the destination's gastronomy. Such information may include the history of the destination's food culture, local table etiquette and the manner of food preparation. Hence, with this tourists will be able to better appreciate local food. Finally, this chapter has also provided useful insights into Chinese tourists' travel eating behaviour and their food culture values. Destination marketers could capitalise on these findings to better meet Chinese tourists' service expectations and to make their travel dining experience more enjoyable and memorable.

References

Andersson, T.D. & Mossberg, L. (2004). The dining experience: do restaurants satisfy customer needs? *Food Service Technology*, 4(4), 171–177. doi: 10.1111/j.1471-5740. 2004.00105.x

Atkins, P. & Bowler, I. (2001). *Food in Society*. London: Arnold.

Beardsworth, A. & Keil, T. (1997). *Sociology on the Menu*. London: Routledge.

Bessiere, J. (1998). Local development and heritage: traditional food and cuisine as tourist attractions in rural areas. *Sociologia Ruralis*, 38(1), 21–34.

Bhandari, R. & Smith, F.J. (2000). Education and food consumption patterns in China: household analysis and policy implications. *Journal of Nutrition Education*, 32(4), 214–224.

Boniface, P. (2003). *Tasting Tourism: Travelling for Food and Drink*. Burlington: Ashgate Publishing.

Broderick, A.J. (1998). Role theory, role management and service performance. *Journal of Services Marketing*, 12(5), 348–361.

Chang, K.C. (1977). *Food in Chinese Culture: Anthropological and Historical Perspectives*. New Haven: Yale University Press.

Chang, R.C.Y. (2007). An analysis of the Chinese leisure travellers' dining-out experiences while holidaying in Australia and its contribution to their visit satisfaction. (PhD), The Hong Kong Polytechnic University, Hong Kong.

Chang, R.C.Y., Kivela, J. & Mak, A.H.N. (2010). Food preferences of Chinese tourists. *Annals of Tourism Research*, 37(4), 989–1011.

Chang, R.C.Y., Kivela, J. & Mak, A.H.N. (2011). Attributes that influence the evaluation of travel dining experience: when East meets West. *Tourism Management*, 32(2), 307–316.

Chen, A., Peng, N. & Hung, K.P. (2016). Chef image's influence on tourists' dining experiences. *Annals of Tourism Research*, 56, 154–158. doi: https://doi.org/10.1016/j.annals.2015.11.005

Cohen, E. & Avieli, N. (2004). Food in tourism: attraction and impediment. *Annals of Tourism Research*, 31(4), 755–778.

Fieldhouse, P. (1986). *Food and Nutrition: Customs and Culture*. New Hampshire: Croom Helm.

Finkelstein, J. (1998). Dining out: the hyperreality of appetite. In R. Scapp & B. Seitz (eds.), *Eating Culture* (pp. 201–215). Albany: State University of New York Press.

Hall, C.M. & Sharples, L. (2003). The consumption of experiences or the experiences of consumption? An introduction to the tourism of taste. In C.M. Hall, E. Sharples, R. Mitchell, N. Macionis & B. Cambourne (eds.), *Food Tourism Around the World: Development, Management and Markets* (pp. 1–24). Oxford: Butterworth-Heinemann.

Hofstede, G. (2001). *Cultures Consequences: Comparing Values, Behaviours, Institutions and Organisations Across Nations* (2nd ed.). Thousand Oaks: Sage.

Ji, M., Wong, I.A., Eves, A. & Leong, A.M.W. (2017). A multilevel investigation of China's regional economic conditions on co-creation of dining experience and outcomes. *International Journal of Contemporary Hospitality Management*.

Kim, Y.G., Eves, A. & Scarles, C. (2009). Building a model of local food consumption on trips and holidays: a grounded theory approach. *International Journal of Hospitality Management*, 28, 423–431.

Kivela, J. & Crotts, J.C. (2006). Tourism and gastronomy: gastronomy's influence on how tourists experience a destination. *Journal of Hospitality and Tourism Research*, 30(3), 354–377.

Kivela, J. & Johns, N. (2003). Restaurants, gastronomy and tourists: a novel method for investigating tourists' dining out experiences. *Journal of Tourism*, 51(1), 3–19.

Lee, C., Yoon, Y.-S. & Lee, S.-K. (2007). Investigating the relationships among perceived value, satisfaction and recommendations: the case of the Korean DMZ. *Tourism Management*, 28, 204–214.

Lin, P.M.C., Qiu Zhang, H., Gu, Q. & Peng, K.L. (2017). To go or not to go: travel constraints and attractiveness of travel affecting outbound Chinese tourists to Japan. *Journal of Travel & Tourism Marketing*, 34(9), 1184–1197. doi: 10.1080/10548408.2017.1327392

Lu, H.C. (1986). *Chinese System of Food Cures: Prevention and Remedies*. New York: Sterling Publishing Co.

Mak, A.H.N., Lumbers, M., Eves, A. & Chang, R.C.Y. (2017). The effects of food-related personality traits on tourist food consumption motivations. *Asia Pacific Journal of Tourism Research*, 22(1), 1–20. doi: 10.1080/10941665.2016.1175488

Mak, A.H.N., Lumbers, M., Eves, A. & Chang, R.C.Y. (2012). Factors influencing tourist food consumption. *International Journal of Hospitality Management*, 31(3), 928–936.

Mak, A.H.N., Wong, K.K.F. & Chang, R.C.Y. (2011). Critical issues affecting the service quality and professionalism of the tour guides in Hong Kong and Macau. *Tourism Management*, 32(6), 1442–1452.

Mäkelä, J. (2000). Cultural definitions of the meal. In H. L. Meiselman (ed.), *Dimensions of the Meal: The Science, Culture, Business and Art of Eating* (pp. 7–18). Gaithersburg, ML: Aspen Publications.

Maslow, A.H. (1970). *Motivation and Personality*. New York: Harper & Row.

Quan, S. & Wang, N. (2004). Towards a structural model of the tourist experience: an illustration from food experience in tourism. *Tourism Management*, 25(3), 297–305.

Rozin, E. (1983). *Ethnic Cuisine: The Flavour Principle Cookbook*. Lexington, MA: The Stephen Greene Press.

Tourism Research Australia (2017). International visitors in Australia year ending June 2017. Retrieved Oct 6, 2017, from https://www.tra.gov.au/ArticleDocuments/250/IVS_one_pager_June2017.pdf.aspx?Embed=Y

UNWTO (2017). UNWTO tourism highlights 2017 edition. Retrieved Oct 16, 2017, from http://www.e-unwto.org/doi/pdf/10.18111/9789284419029

Wang, C.H., Hsieh, A.T. & Chen, W.Y. (2002). Is the tour leader an effective endorser for group package tour brochures? *Tourism Management*, 23, 489–498.

Wang, L., Fong, D. & Law, R. (2016). Travel behaviours of mainland Chinese visitors to Macau. *Journal of Travel & Tourism Marketing*, 33(6), 854–866. doi: 10.1080/10548408.2015.1068266

Warde, A. & Martens, L. (2000). *Eating Out*. Cambridge: Cambridge University Press.

Wong, C.K.S. & Lau, E. (2001). Understanding the behaviour of Hong Kong Chinese tourists on group tour packages. *Journal of Travel Research*, 40(1), 57–67.

Wright, L.T., Nancarrow, C. & Kwok, P.M.H. (2001). Food taste preferences and cultural influences on consumption. *British Food Journal*, 103(5), 348–357.

Wu, K., Raab, C., Chang, W. & Krishen, A. (2016). Understanding Chinese tourists' food consumption in the United States. *Journal of Business Research*, 69(10), 4706–4713. doi: https://doi.org/10.1016/j.jbusres.2016.04.018

Wu, Y.M. (1995). Food and health: the impact of the Chinese traditional philosophy of food on the young generation in the modern world. *Nutrition and Food Science*, 1, 23–27.

Zehrer, A., Crotts, J.C. & Magnini, V.P. (2011). The perceived usefulness of blog postings: an extension of the expectancy-disconfirmation paradigm. *Tourism Management*, 32(1), 106–113. doi: http://dx.doi.org/10.1016/j.tourman.2010.06.013

Zhang, Q.H., Pine, R. & Lam, T. (2005). *Tourism and Hotel Development in China: From Political to Economic Success*. New York: Haworth Hospitality Press/International Business Press.

7 Developing Australia's food and wine tourism towards the Chinese visitor market

Songshan (Sam) Huang and Hailian Gao

Introduction

Chinese outbound tourism is playing an increasingly important role in the global travel market. The expenditure of Chinese outbound tourists is viewed as a means to revive a destination's economy. Australia is one of the most preferred destinations for Chinese tourists. The food and wine sector has a very positive image among Chinese visitors. To further promote this sector, Australia's service providers need to understand the food culture of the Chinese and their preference of Australian food and wine. This chapter provides an overview of the Chinese outbound tourism market with a focus on the characteristics of the tourists. Food and wine have been conceptualised towards the Chinese market by analysing Chinese food and wine preferences and expectation in travelling. By briefly revealing the Chinese food culture with the characteristics of food consumption by Chinese people, a number of strategies are proposed to help develop Australia's food and wine resources towards the Chinese tourist market.

Market overview of China outbound tourism

Brief review of China outbound tourism

Outbound tourism in China started in the 1980s and has gone through several stages: travellers first went to Hong Kong and Macau, then China's border regions in Asia, and finally foreign countries beyond Asia (Arlt, 2006). From 1949 to 1978, outbound travel was confined to groups organised by the Chinese government for officially approved purposes, such as exchange and communication with other socialist nations, or when supporting African countries (Airey & Chong, 2011).

In 1984, the State Council approved the organisation of tours for mainland residents to visit relatives in Hong Kong and Macau (Arlt, 2006), and this is believed to be the beginning of Chinese outbound tourism. Later, in 1987, border region travel for Chinese citizens began (Er, 2002). Travelling to surrounding countries was an expansion and further development of outbound tourism. Following that, travel to foreign countries started to emerge with the first approved destination being Thailand in 1988 (Er, 2002). Since then, China's outbound tourism has rapidly developed.

The Approved Destination Status (ADS) scheme was introduced in China in 1995. The ADS scheme is a bilateral tourism arrangement between the Chinese government and destination authorities, whereby Chinese tourists are permitted to travel in groups to those destinations (Airey & Chong, 2011). By 2016, mainland Chinese were permitted to visit 124 ADS countries or regions for tourism purposes, including countries in the Asia-Pacific region, Europe, South and North America, the Middle East and Africa (China National Tourism Administration, 2014). Boosted by rising disposable incomes, the relaxation of restrictions on foreign travel and an appreciating currency, the volume of international trips by Chinese travellers has increased dramatically (Figure 7.1). In 2000, only 10 million Chinese tourists had travelled across the border (World Tourism Organisation, 2013). In 2015, the number of Chinese outbound tourists increased to 120 million with a record spending of US$194 billion overseas (Ma, 2016).

With continuous growth over the past two decades, the number of Chinese tourists has reached a historic record. However, there is still great potential in the future. In 2016, only about 6% of China's citizens owned a passport, compared with 46% of Americans (Reed, 2016). With young people, especially Millennials graduating over the next decade, the number of outbound tourists will likely double to 220 million in 2025 (Kawano et al., 2015).

Australia was one of the first Western countries (along with New Zealand) to be granted the ADS status early in 1999. Since then, the number of Chinese tourists visiting Australia has increased dramatically from 115,000 in 2000 to 1,024,000 in 2015 (Tourism Australia, 2016). Australia has benefited substantially from Chinese tourism in terms of total inbound economic value. In 2015, the total trip expenditure of Chinese visitors reached AUD$8.3 billion and the Chinese remained the top tourism spenders among all international tourists, which made China Australia's most valuable inbound tourism market (Tourism Australia, 2016).

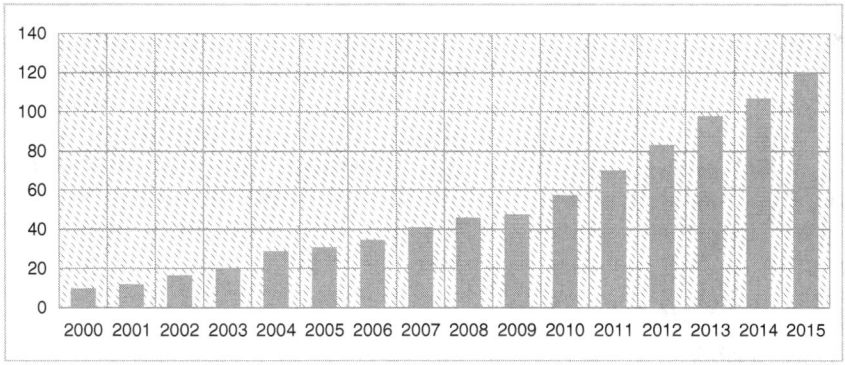

Figure 7.1 Volume of Chinese outbound tourism, 2000–2015

Source: Adapted from Annual Report of China Outbound Tourism Development 2016 (China Tourism Academy, 2016)

Tourism Australia's *China 2020 Strategic Plan* forecasted that by 2020, China outbound tourist arrivals to Australia would reach 880,000 with a total spending of AUD$7–9 billion (Tourism Australia, 2011). However, this goal was already achieved in 2015. The positive market signal demonstrates that there is still great potential for the number to grow. In 2015, the Chinese and Australian governments expanded their Air Service Agreement to allow more flights between the two countries (Tourism Australia, 2016). This has provided another opportunity to expand current air services to second tier cities in China. The Australian government has also implemented a few new visa policies to simplify visa application. For instance, the Work and Holiday Arrangement, launched in 2015, allows up to 5,000 young Chinese people with tertiary education and English skills to travel and work in Australia. Since 12 December 2016 tourist visitor visa applicants in China have been able to lodge an application in Chinese. A trial of the new 10-year 'Frequent Traveller' visitor visa also commenced in 2016 for Chinese passport holders residing in mainland China (Australian Embassy China, 2016). These new rules have greatly facilitated visa application for Chinese tourists. The year 2017 was designated as the China-Australia Year of Tourism. Both Australia and China planned events over the year. The number of Chinese tourists to Australia again reached a new level in 2017.

China outbound tourists characteristics

China's outbound tourism has unique development features compared with other countries. China's opening up and reforms provided the context for the beginning of outbound travel (Guo, Kim & Timothy, 2007). Driven by the Chinese economy and the emerging Chinese middle class, the number of outbound tourists has been growing dramatically over the years. With more disposable income, the Chinese middle class has increased expenditures in education, cultural and recreational activities, thereby boosting the tourism demand (Guo, Kim & Timothy, 2007). China's outbound tourism market also shows a dynamic, fast-changing pattern shaped by the fast-changing Chinese society. Package tours are still popular, as required by Approved Destination Status (ADS) policy. However, an increasing number of young, affluent and experienced travellers now prefer to self-organise their travel plans. Over the next decade, a large number of Millennials is set to graduate in China. With well-informed knowledge about global travel markets and less language barriers, they will play an increasingly important role in the outbound travel market (Kawano et al., 2015).

Where are they from?

Chinese outbound tourists are usually generated from more affluent areas such as Beijing, Shanghai and Guangzhou, followed by coastal provinces such as Jiangsu, Zhejiang and Shandong. Liaoning in Northeast China, Hubei in Central China and Sichuan in Southwest China are new generating provinces with a large tourist flow overseas (WTCF & Ipsos, 2014).

Tourists from the first tier cities (Beijing, Shanghai and Guangzhou) are usually more experienced and love to travel to Europe, America and other diversified destinations (Xiang, 2013). Those newly raised upper middle-class consumers from second or third tier cities are mostly just starting to travel abroad, often in tour groups (Remy & Kim, 2014). Neighbouring countries are more favoured by them.

What are their demographics?

With the increasing number of free and independent travellers, over half of Chinese outbound tourists are now the post-1980s, followed by the post-1970s and the post-1990s generations (Figure 7.2). These tourists mostly belong to the higher income groups and generally have high levels of education (at least a bachelor's degree) (WTCF & Ipsos, 2014; Xiang, 2013).

These relatively young tourists were mostly born after China's economic reform and opening up to the world, thus they are more confident and independent. Influenced by Western culture, they display different consumption patterns to their parents. They are tech-equipped, like to try new things and prefer to buy expensive products for better quality and enjoyment (Barton, Chen & Jin, 2013). Emotional satisfaction is more important and they are eager to show their identity and taste through niche brands and consumption (Barton, Chen & Jin, 2013).

What are their travel behaviours?

Travelling has formed an important part of middle and upper class family life in China (Figure 7.3). Being in the middle class means adopting a lifestyle to differentiate themselves from others. Owning a car, speaking a foreign language, using advanced technology, eating out, consuming foreign food and drink, leisure activities such as golf, spas, teahouses, buying branded goods and particularly overseas holidays are perceived as the tags of being middle class (Callick, 2016).

International travel is treated as an occasion to splurge. More than 75% of Chinese tourists believe that tourism is an important means to improve quality of life and to increase happiness (WTCF & Ipsos, 2016). Middle class Chinese travel

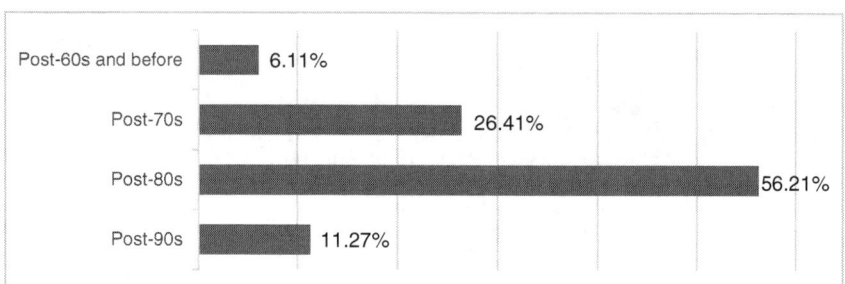

Figure 7.2 Age distribution of Chinese outbound tourists
Source: Adapted from WTCF & Ipsos (2014)

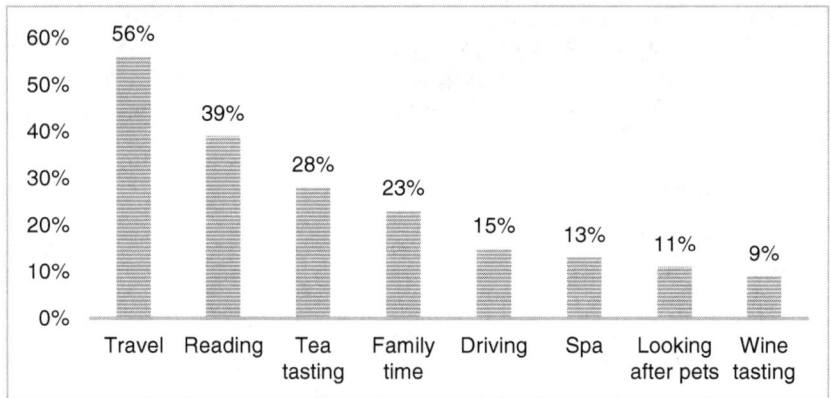

Figure 7.3 Preferred leisure pursuits by Chinese middle/upper class
Source: Adapted from Hurun Report (2014)

frequently and outbound travel has become a habit for them. They are willing to spend more for a higher level of service in an unfamiliar culture especially with language barriers (Lui et al., 2011). With more travel knowledge and experience, Chinese tourists are becoming more rational in overseas consumption and paying more attention to tourism quality. Shopping continues to be the highest spending category for Chinese outbound tourists (WTCF & Ipsos, 2014) (Figure 7.4). Souvenirs are the most purchased commodities, followed by daily necessities and luxuries, which have taken on similar percentages. China has a strong gift-giving culture, which has also stimulated the shopping spending.

Trip planning is getting more and more important for Chinese tourists, especially free and independent travellers. Planning helps increase understanding of destinations so as to reduce anxiety caused by strangeness and enhance travellers'

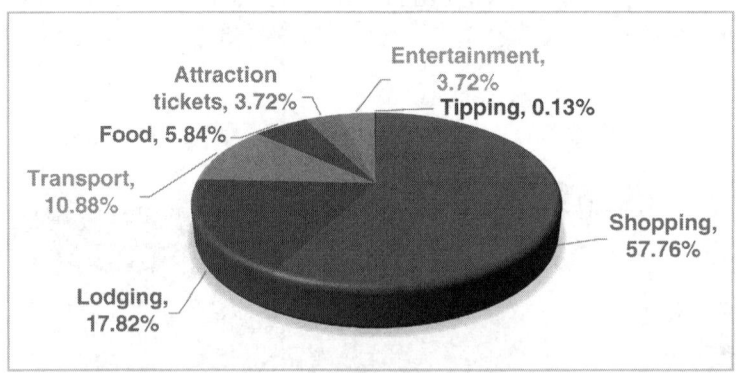

Figure 7.4 Consumptions of Chinese outbound tourists
Source: Adapted from WTCF& Ipsos (2014)

self-efficacy over the trip. Getting more information before the trip also allows tourists to learn about the local culture and customs, which satisfies their learning motivation (Xiang, 2013). Therefore, before travelling, Chinese tourists spend a few days or even several months researching travel information (Xiang, 2013).

The Internet has become a key source for Chinese travellers to acquire travel information. Experienced travellers rely on the Internet to identify the 'must see or shop' places in destinations (Remy & Kim, 2014). Tourists access the official tourism websites through search engines to acquire relevant tourism information. They tend to book tours online, search dining information and reviews online, book hotels, air and attraction tickets online and also use travel Apps to facilitate the journey (WTCF & Ipsos, 2016). WeChat, QQ groups or forums and online travelogues are the most frequently used channels as they are perceived to be more realistic, closer to life, up-to-date and interesting (Xiang, 2013). TV, newspapers, magazines and books are also used by tourists. Word-of-mouth also plays an important role to get travel information as 35% of tourists turn to friends or family for advice (Hurun Report, 2014) (Figure 7.5).

Chinese tourists like to stay connected even when travelling. Therefore, the Internet is not only critical for information searching before travelling but also important for information sharing after the trip (Figure 7.6). With the prevalence of the Internet, especially mobile Internet, forms of communication become even more diverse. WeChat and microblog are frequently used by Chinese tourists to share travel information (WTCF & Ipsos, 2014).

One of the most notable phenomena of Chinese outbound tourism is the rise of free and independent Chinese tourists. They have attracted attention as they behave differently from group tourists. They are usually younger than 45 years, travel savvy, globally connected and prefer to travel to 'exotic' locations (Arlt, 2013). They like to go off the beaten track to avoid other tourists in order to immerse

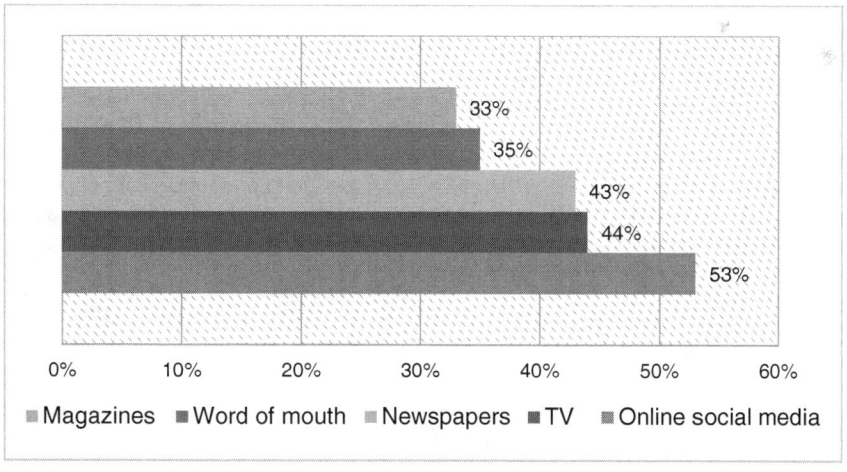

Figure 7.5 Preferred source of information
Source: Adapted from Hurun Report (2014)

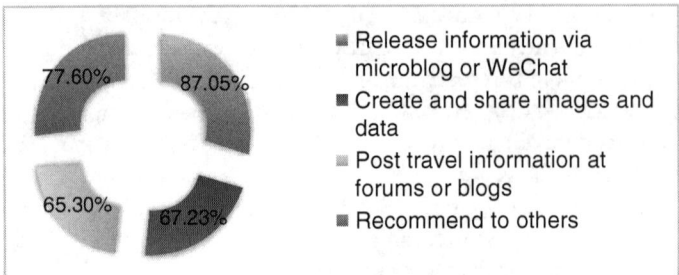

Figure 7.6 Communication behaviour of Chinese outbound tourists after the trip
Source: Adapted from WTCF & Ipsos (2014)

themselves in local life and gain the superiority of being independent (Xiang, 2013). With improved air access and tourists' ability to drive, regional destinations are becoming more attractive to Chinese free and independent tourists.

Conceptualising food and wine in tourism towards Chinese tourists

A multi-perspective framework is proposed to analyse the meanings of food and wine towards Chinese tourists. Based on the literature, two perspectives are provided: 1) the tourist experience perspective and 2) the destination perspective.

Chinese perspectives of food

Food has been especially important to the Chinese and it is a central feature of Chinese culture. Food consumption has a social and psychological value beyond its functional value (Cohen & Avieli, 2004). Historically, food has been used not only to meet people's survival needs but also to closely integrate with social and cultural activities. Chinese food culture contains a wide array of philosophy, extending from the vast variety of ingredients used, regional differences, table manners and dining etiquette to the meaning of food in experiencing life, exploring nature and interpreting society (Chen, 1994). Food in China is influenced greatly by the natural landscape. The landscape and climate from south to north, west to east determine the ingredients available and regional differences in flavours. The existence of imperial, noble and lower social classes also play a role in the preference of seasoning and cooking techniques. Therefore, food is not simply prepared to meet physical needs, but also to reflect the social life of Chinese people. Food acts as a social cement to strengthen social communication. Holiday meals permit interaction and bonding with family members, friends and strangers (Henderson, 2004). The sense of personal satisfaction and wholeness that comes from sharing food in a relaxing holiday setting is important to the well-being of Chinese (Sims, 2009). Food is also closely associated with its symbolic functions. Due to the diversity of food in different regions, people tend to associate food with ethnicity, place of birth and social status. Most calendar and family events are also marked with specific food, such as mooncakes for Moon Festival and rice dumplings for Lantern Festival. Food is also central to the staging of rituals in China as the ancestors and the gods are expected to eat the best food (Oakes, 1999).

Food experience during a trip is regarded as a critical component of a satisfactory vacation experience to Chinese tourists (Fu, Lehto & Cai, 2012). The importance of food to tourists is more prominent for the on-site and post-travel stages (Chen & Huang, 2016). Food experience will enhance tourists' experience and has great potential in contributing to tourists' overall satisfaction with the destination.

Local food products provide a great opportunity for tourists to learn about local culture (Getz, 2000). The way local people eat, the food preparation process, and the taste of local food reflects many aspects of local life (Chang, Kivela & Mak, 2010). By tasting local food and being immersed in an authentic local dining experience, tourists also feel that they are closer to the destination they visit. In addition, trying local cuisine has become proof of a traveller's unique experiences abroad, as it is perceived to be a 'fashionable and desirable' activity for tourists to indulge in food that differs from Chinese (China Travel Marketing Ltd, 2016). For the post-travel stages, demonstrating a familiarity with exotic foods on returning home is perceived to enhance the traveller's self-esteem and prestige (Henderson, 2004). Such prestige and status will make them stand out in their social circles.

Chinese perspectives of wine

Baijiu (liquor or strong distilled spirit) and beer are the dominant alcoholic beverages in China. However, with the strong economic growth and the rise of the middle class, the consumption of wine is increasing with striking speed. In 2016, China ranked as the fourth biggest wine import market behind Germany, the UK and the USA (Vinexpo Newsroom, 2017). China is also now the number one destination by value for Australian wine, accounting for 23 per cent of total export value (Wine Australia, 2016).

With the introduction and import of foreign wine, Chinese wine culture is also changing. Chinese people start to recognise the health benefits of wine (Qiu et al., 2013). For example, the consumption of red wine in China is associated with its health benefit to the heart and skin, such as prevention of cardiovascular disease, anti-fatigue and anti-ageing. Beside the health awareness of wine, the symbolic meanings have also gradually attracted more attention. Chinese people often connect elitism with drinking wine (Duan, Arcodia & Ma, 2015). With the import of Western branded wine, many Chinese associate wine drinking as a means of understanding the Western way of life and Western civilisation (Zhang et al., 2013). Chinese people consider wine drinking as social and cultural indication to help them gain social prestige, demonstrated their self-identity, lifestyle, cultural values and their social standings with particular social groups (Lee, 2009). Foreign wines are also perceived as a suitable gift option for social purposes (Ma et al., 2017).

Given the connection between wine and Western lifestyle, the products in wine tourism should include not only physical objects, but also the service, places, organisations and people. To Chinese tourists, the experience of getting to know different types of wine and learn the stories from local people is more fulfilling than simply wine tasting. Wine tourists are seeking an experience that will engage them physically, mentally and intellectually with the interaction of natural environment, food and wine, cultural events, education, as well as service (Sparks, 2007). Wine tourism experience needs to combine many factors: wine tasting, wine knowledge, wine culture as well as the environment in which the wine experience takes place (Brown & Getz, 2006). The core product of a winery visit is the wine itself (Byrd et al., 2016), but winemaking activity, wine education and customer service will also satisfy tourists' curiosity to see world-famous wineries, to taste different kinds of wine, to learn about wine culture and to experience how wine is produced. Other tourism services, such as lodging, transport, entertainment and events will provide relaxation and recreation opportunities and create a supporting environment to the wine experience.

Australian food and wine to Chinese tourists

Food, wine and destination development

Food and wine form an important part of tourist attractions in the destination. Food consumption by tourists is an economic exchange and can generate substantial revenues to strengthen the local economy (Henderson, 2004). Tourists are likely to extend their length of stay and increase expenditure on local products to further immerse themselves in the authentic experience shaped by food and wine tourism.

Wine tourism also plays a vital role in regional development as it is one of the few national industries that are concentrated outside metropolitan areas (Hall, Johnson & Mitchell, 2000). The ability to symbolise place identity and local culture contributes to the authenticity of the destination, enhances local pride, helps in job creation and sustainable tourism development (du Rand, Heath & Alberts, 2003). The food of a destination can represent the image and distinctiveness of the destination (Karim & Chi, 2010). Regions that use food and wine to position themselves will benefit highly as these resources represent the cultural experience and identity and communicate with tourists (Frochot, 2003). The connection between the tourists and the destination has the potential to enhance the visitor experience both during and after their visit (Sims, 2009). For instance, food is blended in the Italian culture and connected to the lifestyle of its people (Corigliano, 2002). Food therefore plays a key role in attracting tourists to Italy because of its reflection of Italian culture and lifestyle (Karim & Chi, 2010). Therefore, there is great potential to utilise food and wine as a core tourism resource in destination development.

Food and wine is especially important to Australia. Food and beverages are a major industry sector for the Australian economy due to their contribution to finance and employment (WFA, 2011). Australia is recognised for supplying clean and natural products with low chemical residues. This positive image has set Australia to be a competitive agricultural exporter around the world. With such a strong basis of food and wine resources, the focus now is on how to utilise tourism for the health and growth of the wine and food industry. Early in 1998, the concept of wine tourism was brought forward in the first National Wine Tourism Strategy in Australia. Since then, the number of cellar doors has more than doubled with more diverse experiences on offer (WFA, 2011). As the society is moving into experience economy, people are now seeking enriching and memorable experiences beyond merely products. It is time to further develop Australia's food and wine resources in tourism to set Australia on the path as a globally recognised destination for wine and food.

Although food and wine tourism can create huge benefit to the local destination, there are also problems for the development of this kind of tourism. Most wine producers do not have a clear perception of the tourism industry and how the two industries can cooperate (Correia & Brito, 2016). This makes it difficult for wineries to adjust their product offering and operation strategies. Without the cooperation between tourism and the wine industry, it is hard to create a satisfying tourism experience. There are also problems with the lack of investment funds and the lack of available data, information and research on wine tourism (Beames, 2003). Therefore, the cooperation of different stakeholders and the integration of key products, local lifestyle, partnerships, community support and involvement and tourism are critical for the success of food and wine tourism (Jones, Singh & Hsiung, 2015).

Australian food and wine

Australia's food and beverage is adopted and adapted since colonization (Bannerman, 2011). It is now a key asset formulating Australia as a multicultural society. A study conducted by Tourism Australia (2017b) showed that 'good food and wine' ranked third by Chinese travellers when selecting a holiday destination. Good and quality food is also one of the key factors determining Chinese tourist satisfaction visiting Australia. Chinese tourists tag Australian food and wine as 'fresh food in pristine environments', 'fresh seafood', 'fish and chips on the beach', 'high-grade meat and livestock', 'a heritage of food and wine culture', 'wineries with great food', 'spectacular outdoor dining', 'stunning food and wine trails', 'multicultural food' and 'food and wine events and local festivals'. When probed further, it was revealed that 'interesting street food', 'fresh local produce', 'a heritage of food and wine culture', 'a national style of cooking', 'award-winning restaurants' and 'vineyards/wineries' are the key elements that constitute good food and wine (Tourism Australia, 2017b).

Although Australia has premier food and wine resources that are rated high among Chinese tourists, satisfaction with food and wine is quite low (Tourism Research Australia, 2014). Chinese tourists are curious and keen to try the local flavours, but they also need the familiarity of home food. Research shows that 80% of Chinese tourists in Australia ate Chinese food every day or most days, while only 33% ate Western food every day or most days (Tourism Research Australia, 2014). Therefore, the offering of Chinese food and the opportunity to try local food both play important roles in constructing the trip experience for Chinese visitors.

Marketing strategies for developing food and wine tourism in Australia

The ever-growing China market offers increasing opportunities for destinations to get a share of it. However, destinations have to ensure that their products are adapted to the needs and expectations of Chinese tourists. Therefore, an understanding of the Chinese food culture and tourists' perception and behaviours towards food and wine products is necessary.

Realising the importance of food and wine in attracting tourists and creating satisfying experiences, Australia has endeavoured to strengthen and promote food and wine resources in its tourism planning. For instance, Tourism Australia launched the *Restaurant Australia* programme (Tourism Australia, 2017a) and a series of collaborative marketing campaigns to promote Australian food and wine. Tourism Western Australia launched '*Taste 2020 – A strategy for food and wine tourism in Western Australia for the next five years & beyond*', which is a multi-agency and industry food and wine strategy aimed at establishing WA as one of the world's foremost culinary tourism destinations (Tourism WA, 2017). Destination New South Wales's (2012) '*China Tourism Strategy 2012–20*' strives to improve the quality and range of visitor experiences by focusing on food and wine as the potential tourism attraction to develop. In '*Activating China – 2020 tourism strategy*', the South Australian Tourism Commission has exclusively positioned wine, naturalness and food as the identified strengths and sole focus to offer Chinese tourists the distinctive experiences in Australia's clean environment (South Australian Tourism Commission, 2013). These strategies will further differentiate Australia from other destinations.

The above plans and strategies provide a great context and opportunity for service providers to reshape their offering to Chinese tourists. Based on the behavioural characteristics of Chinese outbound tourists, various marketing strategies are developed. Key strategies include investing in social media for online promotion, developing cultural identity of food and wine, promoting high status image of Australia's food and wine, producing food and wine gifts and bridging the cultural gap.

Develop effective channels through Chinese social media

As mentioned earlier, Chinese tourists frequently use the Internet to seek travel-related information. They search online for travel-related content, opinions from travel review sites, travel-related Apps and to make mobile payments (WTCF & Ipsos, 2016). Therefore, digital channels will be an effective tool for destination marketing. Social media, especially WeChat (similar to Facebook) and Weibo (the Chinese version of Twitter), are popular online channels used by young Chinese consumers. Destination food and wine businesses need to develop effective and compelling content through these Chinese social media channels.

The New Zealand Tourism Board has demonstrated a successful case of using social media for tourism promotion. In 2012, a famous Chinese film and TV star Ms Chen Yao was invited by the Tourism Board to experience the scenery and she also had her wedding in New Zealand. She published her journey on her Sina Weibo, which has 35 million fans. During her journey, New Zealand Tourism's Weibo witnessed an increasing number of fans. Based on the comments on her Weibo and other travelogues, many fans subsequently visited New Zealand or expressed in interest in visiting the country in the future (Xiang, 2013).

Provide various choices in food product supply, balancing Chinese and local food elements to form satisfying experience

Quan and Wang (2004) found that food consumption can be support or peak experiences to tourists. The desire to try local food is certainly prominent amongst young Chinese travellers. Trying novel foods occasionally can give tourists a sense of joyful and memorable food experiences. Experiencing novel ingredients of foods or the way food is delivered or consumed will help tourists form peak experiences during travelling.

While tasting local food could satisfy the experiential needs of the visitors, it might not be enough to satiate their physiological needs as Chinese food offers a more reliable way for Chinese tourists to obtain gustatory pleasure and to mitigate the unfamiliar taste of local food (Chang, Kivela & Mak, 2010). Home food gives tourists a sense of the ontological comfort of home, which supports the overall experience. Tourists who stay abroad longer often miss the familiarity of Chinese food; and older travellers are more likely to stick to traditional Chinese dishes (China Travel Marketing Ltd, 2016).

A study by Tourism Research Australia (2014) shows that the share of visitors expecting to eat mainly Chinese food is 30%, which was nearly twice that for mainly Western food (15%). More than half expected to eat a combination of Chinese and Western (55%) food. Since Chinese tourists prefer both a 'taste of home' and a 'sip of the destination', it is necessary to provide Chinese tourists local dishes as well as a range of Asian-style foods to choose from. Having a menu available that is translated into Mandarin facilitates the ordering process. Otherwise, selecting a few dishes presented in Chinese as the 'dishes most popular with our Chinese guests' allows guests to have a better idea about the food that they will enjoy (China Travel Marketing Ltd, 2016). Whitsundays has set a good example in catering to Chinese tourists. During the peak season (Chinese New Year), Cruise Whitsundays prepared congee rice porridge on the boat and presented boarding tickets in traditional red envelopes, which pleased Chinese tourists and raised their satisfaction level (Freed, 2016).

Develop the cultural identity of Australian food and wine by assigning cultural meanings and telling good stories

Woodside, Sood and Miller (2008) point out that any successful business must be backed up with a good 'story' to help it attract attention and thus increase its popularity, fame or reputation. The connection between a particular destination and a certain cuisine will form a long-lasting impression in the consumers' mind. The 'story' of the product will link the destinations' natural environments, demographics, traditions and their food culture to co-promote each other (Horng & Tsai, 2010).

For centuries, Europe has been considered as the centre of good wine, especially French, Italian and Spanish wines (Banks & Overton, 2010). Consumers from Asian countries have formed a strong perception about wines from Europe. The romantic and cultural image of European countries empowered the symbolic meanings of wine, which is as appealing to many wine buyers as the wine itself (Banks & Overton, 2010; Thorpe, 2009). This has formed a strong selling point for European wines. The history of winemaking in Australia cannot compare with Europe. However, Australian wine has its own unique characteristics. The climate of Australia creates a great condition for producing high quality wines and a vast variety of grapes. Modern technology is twisted into traditional winemaking (Thorpe, 2009). The wine products have a reasonable price and genuinely good taste to attract both young and old markets. The innovation, the high quality, the huge variety as well as the favourable price form a strong identity for Australian wines compared with European wines. Marketing promotion should highlight this unique identity of Australian wines to cultivate a national brand image among consumers.

A study showed that wine tourists complained about the lack of entertainment and other activities, the accessibility of many wineries and the limitation of food choice for local dining in Australia (Ma et al., 2017). Therefore, wine tourism needs to leverage the wine experience and connect it with other service providers, with landscape and local culture to tell a story and provide a compelling reason to convert interest into bookings (WFA, 2011). For instance, Adelaide Hills in South Australia is known for its wine and German heritage. Many tours in this area have connected the German heritage in Hahndorf, wineries in Adelaide Hills as well as local produce, such as Woodside Cheese, Melba's Chocolate factory, Beerenberg's Strawberry Farm and Mount Lofty Lookout, to create a unique experience for tourists. Events also provide a great opportunity to showcase the diverse offering in a destination. Hunter Valley Wine and Food Festival, Margaret River Wine Festival, Barossa Vintage Festival and Barossa Gourmet Weekend all proved to be successful in attracting tourists and marketing the local regions.

In addition, a tour to a winery can also be arranged in a way that includes vineyard sightseeing, grape picking and visitation to the wine production base, wine cellar and wine museum together with cooking classes and winemaking. Such tours will allow visitors to learn about the wine production process and its history, to taste different wines and to purchase wine or grape-related products. Guiding a tourist to blend and create a personalised bottle of wine with his/her name on the self-designed label can enhance the travel experience and leave a vivid memory in the tourist's mind.

Assign high status and highlight the health benefits associated with food and wine

Chinese consumers perceive Western products as civilised, modern, expensive and high quality (Zhou & Hui, 2003). To attract the Chinese market, food and wine products need to be presented as unique and with a high-status image. For example, in China, the term 'aristocrats' (贵族) is frequently used in wine advertisements to show the high social status of wine in the European tradition (Zhang, 2015). Marketing a product as 'the choice of aristocrats' (贵族之选) will create an experience that is 'brag-worthy' to the consumers. Therefore, it is important to highlight the history of wine, limit editions for certain ranges and most importantly, ensure the products are presented in a luxury package to meet the Chinese demand of seeking luxury and status in purchase (Young, 2016). Penfolds wine has done well in its promotion in China. Penfolds has combined its history in wine production in Australia and the high quality of its wine especially with the high status and limited edition of the famous Grange range. Chinese people now perceive Penfolds as the symbol and the aristocrats of Australian wines.

In addition, food and wine could be promoted for its functions of 'nourishing life' (养生). Traditional Chinese medical knowledge views different foods as having different intrinsic properties to nourishing life. Products that are associated with health will be popular, especially in a time that China is facing environmental pollution and various public health problems. On-site tutorials or seminars can be organised by wineries to illustrate in detail the grapes used, the recommended food and wine match and the nutrition and health value of different types of wine. Tailored promotion always works better than standard marketing.

Develop food products as gifts

Lui et al. (2011) found that inexperienced travellers tend to focus on sightseeing when travelling, while experienced travellers are more willing to spend time and money on activities such as entertainment and shopping. In addition to food and wine tasting experience, food and wine products can also be designed as gifts to generate commercial benefits.

Food is one of the most common and favoured souvenir choices in the Chinese community (Lin, 2016). In China, gift giving is perceived to be an important vehicle for relationship building (*guanxi*). Chinese people like to buy gifts not only for their relatives, friends and colleagues, but even their neighbours

(Chan, Denton & Tsang, 2003; Zhou & Guang, 2007). Gifts in Chinese society have become a symbol of courtesy, respect, appreciation and friendship (Mok & DeFranco, 2000). Local food is a popular gift souvenir, especially among tourists from mainland China, Taiwan, Singapore and Hong Kong (du Cros & Jingya, 2013; Kong & Chang, 2012). Food is a great form to express local culture and showcase a destination's landscape and unique way of life (Harrington & Ottenbacher, 2010). Many tourist destinations endeavoured to design and produce food souvenirs and gifts to represent their unique identities (Lin, 2016).

To prepare gastronomic souvenirs and gourmet gifts, businesses can develop and design products from sensory, utility and symbolic perspectives (Lin & Mao, 2015). Sensory refers to the aesthetic property of products, such as the colour and design. Vivid colour and unique design are likely to catch the attention of tourists and raise their purchase intentions. Utility means the quality, the price and convenience of carrying. Symbolic features refer to the ability to represent a destination and its culture. Tourists prefer to buy products that are authentic and culturally embedded. Such products allow tourists to take back memories from the destination and treasure them.

Overall, the design of gifts should highlight the flavour, innovativeness, appearance, healthy features, high quality, environmentally friendly features and authentic, traditional and fun elements (Lin & Mao, 2015). Photographs and illustrations or even mascots can also be utilised in product packaging to enhance the authenticity and highlight the product's history and culture (Lin, 2016).

Break language barriers and prepare businesses to be China-ready

Most Chinese outbound tourists expressed that they face two major problems in an overseas destination: 1) the lack of Chinese signs; and 2) the lack of Chinese-speaking staff (WTCF & Ipsos, 2014). They are expecting that destination tourism websites offer Chinese language services or have Microblog/WeChat in the Chinese language for them to get sufficient information before reaching the destination. Chinese wine tourists in Australia also complained about the service at wineries. Most Chinese tourists are not familiar with wine. They would appreciate a Chinese-speaking staff who can introduce all different types of wine and give suggestions fitting their needs. However, most staff focused mainly on sales and promotion with no information on winemaking and the differences between various types of wine (Ma et al., 2017). They also feel that frontline staff are not well educated about the proper manner of interacting with Chinese tourists which leave the tourists feeling unrespected (Ma et al., 2017).

Although it might be impossible to have a Chinese-speaking staff for every service provider, providing information in Chinese can help solve the problem. Many wineries in Australia have websites in Chinese, such as Penfolds wine and McGuigan wines. Hahndorf Inn restaurant in South Australia not only has a Chinese website, but also joined Sina Weibo and WeChat (both are very popular Chinese social media) to promote to the China market. Another example is Australia's premier food, travel and lifestyle magazine *Gourmet Traveller*. The magazine started to provide a Chinese-language edition in 2016 (*Gourmet Traveller*, 2016). The Chinese edition is available at five-star hotels, casinos, duty-free shops and airport lounges all around Australia.

Beyond language, cultural training is also essential to attract and service Chinese tourists. Chinese have different expectations and practices for interpersonal interaction. They desire genuine respect and appreciate a high level of enthusiasm, a positive attitude and a good understanding of Chinese culture from service providers. Many businesses have realised the need to acquire more knowledge about Chinese tourists and have adopted strategies to cope with this challenge. For instance, Hahndorf Inn has Union Pay credit card facilities, employs a Mandarin-speaking staff member, trains staff in the etiquette behind seating visitors and is a corporate partner with China Southern Airlines (Willis, 2017). The Australian government has also offered cultural training and practical guides to help prepare businesses to be China-ready. For instance, Tourism and Events Queensland developed the '*Become China Ready* (http://teq.queensland.com/china)' programme to help prepare the tourism industry to better serve the growing China market. This programme includes various resources, such as Destination Q China Masterclass, China factsheets (including Chinese culture, preferred travelling needs and experiences, basic Mandarin learning, etc.), Digital Marketing for the China Market workshop and Learn Mandarin online. Food and wine service providers could take the opportunity to further develop their offerings to suit the needs of the China market.

Concluding remarks

Australia is known for its natural beauty as well as splendid food and wine products to Chinese visitors. Research has demonstrated that Chinese consumers' perceptions of Australia as a tour destination positively influenced their perceived image of Australian wine (Lee & Lockshin, 2011). To promote the food and wine sector, tourism and hospitality service providers need to acquire sufficient knowledge about the food culture and food preferences of Chinese tourists. Marketing strategies could focus on online marketing, especially through social media, developing Australian food and wine cultural identities, designing food and wine gifts and promoting the status and health benefits associated with food and wine. It is recommended that in future more collaborative research should be conducted to monitor the changing need and new market developments in altering and providing suitable food and wine products to Chinese tourists.

References

Airey, D.W & Chong, K. (2011). *Tourism in China Policy and Development Since 1949*. Abingdon, Oxon: Routledge.

Arlt, W.G. (2006). *China's Outbound Tourism*. Abingdon, Oxon: Routledge.

Arlt, W.G. (2013). The second wave of Chinese outbound tourism, *Tourism Planning & Development*, *10*(2), 126–133. DOI: 10.1080/21568316.2013.800350

Australian Embassy China (2016). *New visitor visa options for Chinese applicants.* Retrieved from http://china.embassy.gov.au/bjng/trialof10yearfrequenttravellervisa. html

Banks G. & Overton, J. (2010). Old World, New World, Third World? Reconceptualising the worlds of wine. *Journal of Wine Research*, 21(1), 57–75. DOI:10.1080/09571264 .2010.495854

Bannerman, C. (2011). Making Australian food history. *Australian Humanities Review*, 51. Retrieved from http://www.australianhumanitiesreview.org/archive/Issue-November-2011/bannerman.html

Barton, D., Chen, Y. & Jin, A. (2013). *Mapping China's middle class.* Retrieved from McKinsey Quarterly: http://www.mckinsey.com/industries/retail/our-insights/mapping-chinas-middle-class

Beames, G. (2003). The rock, the reef and the grape: the challenges of developing wine tourism in regional Australia. *Journal of Vacation Marketing*, 9(3), 205–212.

Brown, G.P. & Getz, D. (2006). Critical success factors for wine tourism regions: a demand analysis. *Tourism Management*, 27(1), 146–158.

Byrd, E.T., Canziani, B., Hsieh, Y., Debbage, K. & Sonmez, S. (2016). Wine tourism: motivating visitors through core and supplementary services. *Tourism Management*, 52, 19–29.

Callick, R. (2016). China's growing middle class our mega-market. *The Australian*, Apr 9, 2016. Retrieved from http://www.theaustralian.com.au/business/opinion/rowan-callick/ chinas-growing-middle-class-our-megamarket/news-story/b54c6f172420feee64f825c3 fac7de63

Chan, A.K., Denton, L. & Tsang, A.S. (2003). The art of gift giving in China. *Business Horizons*, *46*(4), 47–52.

Chang, R.C., Kivela, J. & Mak, A.H. (2010). Food preferences of Chinese tourists. *Annals of Tourism Research*, *37*(4), 989–1011.

Chen, C. (1994). 中国饮食文化的区域分化和发展趋势.The culture of Chinese diet: regional differentiation and developing trends. *Acta Geographica Sinica*, 49(3), 226–235.

Chen, Q. & Huang, R. (2016). Understanding the importance of food tourism to Chongqing, China. *Journal of Vacation Marketing*, 22(1), 42–54.

China National Tourism Administration (2014). 已正式开展组团业务的出境旅游目的 地国家(地区) List of ADS countries and regions. Retrieved from http://www.cnta.gov. cn/ztwz/cjyzt/gltl/201507/t20150708_723265.shtml

China Tourism Academy (2016). 中国出境旅游发展年度报告 *2016 Annual Report of China Outbound Tourism Development 2016*. Beijing: Tourism Education Press.

China Travel Marketing Ltd (2016). *The Chinese want to eat Chinese, right?* China Travel Marketing Ltd. Retrieved from http://www.chinatraveloutbound.com/what-are-the-food-preferences-of-chinese-travellers/

Cohen, E. & Avieli, N. (2004). Food in tourism: attraction and impediment. *Annals of Tourism Research*, *31*(4), 755–778.

Corigliano, M.A. (2002). The route to quality: Italian gastronomy networks in operation. In A.M. Hjalager & G. Richards (eds.), *Tourism and Gastronomy* (pp. 166–185). London: Routledge.

Correia R. & Brito C. (2016). Wine tourism and regional development. In: M. Peris-Ortiz, M. Del Río Rama & C. Rueda-Armengot (eds.) *Wine and Tourism* (pp. 27–40). Cham: Springer.

Destination New South Wales (2012). *China Tourism Strategy 2012–20.* Retrieved from http://www.destinationnsw.com.au/wp-content/uploads/2014/03/DNSW-China-Tourism-Strategy-2012-20.pdf

du Cros, H. & Jingya, L. (2013). Chinese youth tourists' views on local culture. *Tourism Planning & Development*, 10(2), 187–204. DOI:10.1080/21568316.2013.783732

du Rand, G.E.D., Heath, E. & Alberts, N. (2003). The role of local and regional food in destination marketing: a South African situation analysis. *Journal of Travel & Tourism Marketing, 14*(3–4), 97–112.

Duan, B., Arcodia, C. & Ma, E. (2015). Contemporary and conventional motivation of wine tourism travellers in China: an exploratory case study of Yunnan Province, China. In E. Wilson & M. Witsel (eds.) *CAUTHE 2015: Rising Tides and Sea Changes: Adaptation and Innovation in Tourism and Hospitality* (pp. 452–456). Gold Coast, QLD: School of Business and Tourism, Southern Cross University.

Er, P. (2002). 只缘口袋渐丰实—中国旅游业阔步前行 Chinese tourism industry advances in big steps, *People's Daily Overseas Edition*, 19 Oct, 2002.

Freed, J. (2016). Red envelopes and congee on hand as Whitsundays gears up for Chinese tourists. *The Australian Financial Review*, Feb 4, 2016. Retrieved from http://www.afr.com/business/tourism/red-envelopes-and-congee-on-hand-as-whitsundays-gears-up-for-chinese-tourists-20160202-gmk5fn

Frochot, I. (2003). An analysis of regional positioning and its associated food images in French tourism regional brochures. *Journal of Travel & Tourism Marketing*, 14(3–4), 77–96.

Fu, X., Lehto, X.Y. & Cai, L.A. (2012). Culture-based interpretation of vacation consumption. *Journal of China Tourism Research*, 8(3), 320–333.

Getz, D. (2000). *Explore Wine Tourism: Management, Development & Destinations.* New York: Cognizant Communication Corporation.

Gourmet Traveller (2016). *Gourmet Traveller Chinese-language edition.* Retrieved from http://www.gourmettraveller.com.au/recipes/food-news-features/2016/1/gourmet-traveller-chinese-language-edition/

Guo, Y., Kim, S.S. & Timothy, D.J. (2007). Development characteristics and implications of mainland Chinese outbound tourism, *Asia Pacific Journal of Tourism Research*, 12(4), 313–332. DOI: 10.1080/10941660701760995

Hall, C.M., Johnson, G.R. & Mitchell, R.D. (2000). Wine tourism and regional development. In C.M. Hall, E. Sharples, B. Cambourne & N. Macionis (eds.), *Wine Tourism Around the World: Development, Management and Markets* (pp. 196–225). Oxford: Butterworth Heinemann.

Harrington, R.J. & Ottenbacher, M.C. (2010). Culinary tourism: a case study of the gastronomic capital. *Journal of Culinary Science & Technology, 8*(1), 14–32. DOI:10.1080/15428052.2010.490765

Henderson, J.C. (2004). Food as a tourism resource: a view from Singapore. *Tourism Recreation Research, 29*(3), 69–74.

Horng, J.S. & Tsai, C.T.S. (2010). Government websites for promoting East Asian culinary tourism: a cross-national analysis. *Tourism Management*, 31(1), 74–85.

Hurun Report (2014). *The Chinese Luxury Traveller 2014.* Retrieved from http://up.hurun.net/Humaz/201406/20140603152517230.pdf

Jones, M.F., Singh, N. & Hsiung, Y. (2015). Determining the critical success factors of the wine tourism region of Napa from a supply perspective. *International Journal of Tourism Research*, 17(3), 261–271.

Karim, S. & Chi, C.G.Q. (2010). Culinary tourism as a destination attraction: an empirical examination of destinations' food image. *Journal of Hospitality Marketing & Management*, 19(6), 531–555.

Kawano, S., Lu, J., Tsang, R. & Liu, J. (2015). *The Chinese Tourist Boom*. The Goldman Sachs Group, Inc. Retrieved from http://www.goldmansachs.com/our-thinking/pages/macroeconomic-insights-folder/chinese-tourist-boom/report.pdf

Kong, W.H. & Chang, T.Z. (2012). The role of souvenir shopping in a diversified Macau destination portfolio. *Journal of Hospitality Marketing & Management*, 21(4), 357–373. DOI:10.1080/19368623.2011.615022

Lee, K.H. (2009). Is a glass of Merlot the symbol of globalisation? An examination of the impact of globalisation on wine consumption in Asia. *International Journal of Wine Business Research*, 21(3), 258–266.

Lee, R. & Lockshin, L. (2011). Halo effects of tourists' destination image on domestic product perceptions. *Australasian Marketing Journal (AMJ)*, 19(1), 7–13.

Lin, L. (2016). Food souvenirs as gifts: tourist perspectives and their motivational basis in Chinese culture. *Journal of Tourism and Cultural Change*. DOI: 10.1080/14766825.2016.1170841

Lin, L. & Mao, P.C. (2015). Food for memories and culture: a content analysis study of food specialities and souvenirs. *Journal of Hospitality and Tourism Management*, 22, 19–29.

Lui, V., Kuo, Y., Fung, J., Jap, W. & Hsu, H. (2011). *Taking Off: Travel and Tourism in China and Beyond*. Retrieved from The Boston Consulting Group: https://www.bcg.com/documents/file74525.pdf

Ma, C. (2016). *Chinese Travellers Lead 2015 Global Outbound Tourism*. Retrieved from http://www.chinadaily.com.cn/china/2016-01/28/content_23288004.htm

Ma, E.J., Duan, B., Shu, L.M. & Arcodia, C. (2017). Chinese visitors at Australia wineries: preferences, motivations and barriers. *Journal of Tourism, Heritage & Services Marketing*, 3(1), 3–8.

Mok, C. & DeFranco, A.L. (2000). Chinese cultural values: their implications for travel and tourism marketing. *Journal of Travel & Tourism Marketing*, 8(2), 99–114.

Oakes, T. (1999). Eating the food of the ancestors: place, tradition and tourism in a Chinese frontier river town. *Ecumene*, 6(2), 123–145.

Qiu, H.Z., Yuan, J., Ye, B. & Hung, K. (2013). Wine tourism phenomena in China: an emerging market. *International Journal of Contemporary Hospitality Management*, 25(7), 1115–1134.

Quan, S. & Wang, N. (2004). Towards a structural model of the tourist experience: an illustration from food experiences in tourism. *Tourism management*, 25(3), 297–305.

Reed, D. (2016). *Chinese Extend Lead as the World's Biggest Spenders On Foreign Travel*. Forbes. Retrieved from http://www.forbes.com/sites/danielreed/2016/01/07/chinese-worlds-biggest-spenders-on-foreign-travel/#2e8fa7ae43b3

Remy, N. & Kim, A. (2014). *Winning Today's Globe-Hopping and Shopping Chinese Luxury Consumers*. Retrieved from McKinsey Quarterly http://www.mckinsey.com/business-functions/marketing-and-sales/our-insights/winning-todays-globe-hopping-and-shopping-chinese-luxury-consumers

Sims, R. (2009). Food, place and authenticity: local food and the sustainable tourism experience. *Journal of Sustainable Tourism*, 17(3), 321–336. DOI: 10.1080/09669580802359293

South Australian Tourism Commission (2013). *Activating China: 2020 Tourism Strategy.* Retrieved from http://tourism.sa.gov.au/documents/CORP/documentMedia.ashx?A=%7B 3DEDBF6D-F230-420A-B805-A97F3E2346D9%7D&B=True

Sparks, B. (2007). Planning a wine tourism vacation? Factors that help to predict tourist behavioural intentions. *Tourism Management*, 28(5), 1180–1192.

Thorpe, M. (2009). The globalisation of the wine industry: New World, Old World and China. *China Agricultural Economic Review*, 1(3), 301–313.

Tourism Australia (2011). *China 2020 Strategic Plan.* Retrieved from http://www.tourism. australia.com/documents/Markets/China2020_Strategic_Plan.pdf

Tourism Australia (2016). *China Market Profile.* Retrieved from http://www.tourism. australia.com/documents/TASI10419_Market_Profiles_2016_China_final_copy.pdf

Tourism Australia (2017a). *There's Nothing Like Australia – Especially When It Comes to Quality Food and Wine Experiences.* Retrieved from http://www.tourism.australia. com/en/about/our-campaigns/theres-nothing-like-australia/food-and-wine.html

Tourism Australia (2017b). *Understanding the Chinese Market.* Retrieved from http:// www.tourism.australia.com/content/dam/assets/document/1/6/x/l/r/2003103.pdf

Tourism Research Australia (2014). *Chinese Satisfaction Survey.* Retrieved from http:// www.tourism.australia.com/content/dam/assets/document/1/6/w/s/y/2002066.pdf

Tourism WA (2017). *Taste 2020 – A Strategy for Food and Wine Tourism.* Retrieved from http://www.tourism.wa.gov.au/Research-Reports/Other-reports/Pages/Taste_2020_-_A_ Strategy_for_Food_and_Wine_Tourism.aspx

Vinexpo Newsroom (2017). *China is a Leading Wine Market of the Future.* Retrieved from http://www.vinexpo-newsroom.com/china-is-a-leading-wine-market-of-the- future/

WFA-Winemakers' Federation of Australia (2011). *Harnessing the Tourism Potential of Wine and Food in Australia 2020.* Retrieved from https://www.wfa.org.au/assets/ strategies-plans/pdfs/Wine_food_tourism_strategy.pdf

Willis, B. (2017). Wildlife, wine and adventure are luring high-spending Chinese tourists to Eyre Peninsula safaris. *The Advertiser*, Feb 7, 2017. Retrieved from http://www.adelaidenow.com.au/business/wildlife-wine-and-adventure-are- luring-highspending-chinese-tourists-to-eyre-peninsula-safaris/news-story/ a644d038bf9a85aef7dab3b037fd1ac2

Wine Australia (2016). Wine Australia providing insights on Australian wine. *Export Report.* Retrieved from https://www.wineaustralia.com/getmedia/07dfee9e-bc3c-4e2a- a543-d4a1bd63ab89/Export-Report-Sep-16?ext=.pdf

Woodside, A.G., Sood, S. & Miller, K.E. (2008). When consumers and brands talk: sto- rytelling theory and research in psychology and marketing. *Psychology & Marketing*, 25(2), 97–145.

World Tourism Cities Federation (WTCF) & Ipsos (2014). *Market Research Report on Chinese Outbound Tourist (City) Consumption.* Retrieved from https://www.ccilc.pt/ sites/default/files/relatorioturistachines.pdf

World Tourism Cities Federation (WTCF) & Ipsos (2016). *Market Research Report on Chinese Outbound Tourist (City) Consumption.* Retrieved from http://www.wtcf.org. cn/uploadfile/2016/1110/20161110023037669.pdf

World Tourism Organisation (2013). *China – the New Number One Tourism Source Market in the World.* Retrieved from http://media.unwto.org/press-release/2013-04-04/ china-new-number-one-tourism-source-market-world

Xiang, Y. (2013). The characteristics of independent Chinese outbound tourists. *Tourism Planning & Development*, 10(2), 134–148. DOI: 10.1080/21568316.2013.783740

Young, L. (2016). Attracting Chinese travellers to your food tourism experience. *The Culinary Tourism Alliance 2016*. Retrieved from http://www.growfoodtourism.com/attracting-chinese-travellers-to-your-food-tourism-experience/

Zhang, J. (2015). Wine-tasting Chinese tourists in Australia. *Australian Centre on China in the World*. Retrieved from https://www.thechinastory.org/2015/05/wine-tasting-chinese-tourists-in-australia/

Zhang Qiu, H., Yuan, J., Haobin Ye, B. & Hung, K. (2013). Wine tourism phenomena in China: an emerging market. *International Journal of Contemporary Hospitality Management*, 25(7), 1115–1134.

Zhou, C. & Guang, H. (2007). Gift-giving culture in China and its cultural values. *Intercultural Communication Studies*, 16(2), 81–93.

Zhou, L. & Hui, M.K. (2003). Symbolic value of foreign products in the People's Republic of China. *Journal of International Marketing*, 11(2), 36–58.

8 Food, wine and China

A tourism perspective from Western Australia

Ross Dowling

Introduction

Food and wine tourism is an Australian success story and in Western Australia food, wine and tourism have been linked for more than two decades. This chapter describes wine and tourism, and ultimately, food, wine and tourism, in Australia and Western Australia (WA) since the state hosted the First Australian Wine Tourism Conference *Wine Tourism: Perfect Partners* in 1998. Today, food and wine tourism is an important element in the visitor experience in Western Australia. In 2013 leisure travel in WA was worth $4.6 billion and an estimated 1.1 million visitors participated in food and wine activities while on holiday or on a day trip (Tourism Western Australia, 2015a).

This chapter links this growth to that of the growing number of Chinese tourists who in recent years have risen to the sixth most important market in WA by visitor numbers, and second in relation to visitor spend. It first describes tourism in Western Australia within the context of the overall international visitation by tourists to Australia. The second part examines Chinese visitors to Australia, their growing importance in relation to both visitor numbers and yield, and their key market segments. The third part examines the growing market segment of food and wine tourism in Australia and Western Australia. It examines the visitor experiences of food and wine tourists and outlines *Taste 2020*, WA's food and wine strategy. The final part of the chapter summarises the present position in regard to the importance of food and wine tourism in WA for Chinese tourists.

Tourism in Western Australia

In the year ending March 2018 there were 8.1 million international visitors to Australia (an increase of 6% over the previous year). They stayed 265 million nights in the country (an increase of 5%) and spent $41.3 billion (up by 6%) (Tourism Research Australia, 2018). The key country contributing to this growth has been China. Over the 12-month period 1.3 million visitors from China were recorded, an increase of 12% over the previous year. They stayed 50.7 million nights (up 23%) and spent $10.4 billion (up 14%) (Tourism Research Australia, 2018). Thus China is viewed as a central part of Australia's push to increase its international tourism visitation.

The state of Western Australia had a total of 10.5 million visitors in the 12 months ending March 2017 (Tourism Western Australia, 2017a). They comprised 8.1 million intrastate visitors (WA people travelling around their own state), 1.4 million interstate (visitors from other states of Australia), and 953,800 international visitors (Figure 8.1a; Tourism Western Australia, 2017a). Thus international visitation makes up 12.4% of visitors to Australia but only 9.0% of visitors to Western Australia. The majority of tourists spend time in and around the capital city Perth (61%) as well as in the more temperate southwest areas of the state (21%) (Tourism Western Australia, 2017a).

In WA the top six source markets of international visitors are the United Kingdom, Malaysia, Singapore, the United States of America, New Zealand and China (Figure 8.1b; Tourism Western Australia, 2017b). However, when examining the 'visitor spend' of the top 20 international markets to WA, China is ranked second behind the UK with a 13% increase in spend from $239 million in 2016 to $271 million in 2017 (Tourism Western Australia, 2017b).

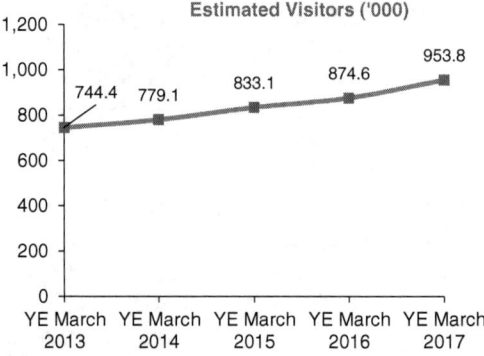

Figure 8.1a International visitors to Western Australia, 2013–2017

Source: Tourism Western Australia (2017)

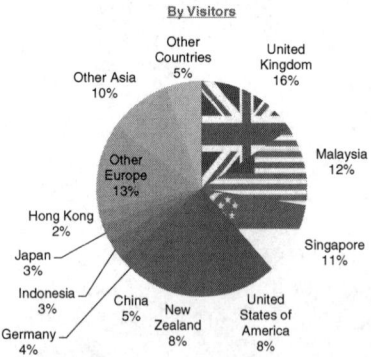

Figure 8.1b WA's international visitors by Source Market

Source: Tourism Western Australia (2017)

Chinese visitors to Australia

Australia has experienced faster international arrivals and expenditure growth from China than from any other market. Earlier this decade 542,000 Chinese visited Australia during 2011, almost 20% growth on 2010 (Tourism Australia, 2012). At that time, not only were Chinese visiting in record numbers, they were also spending at high levels. The then current overnight expenditure from China was up 15%, to more than $3.8 billion, making it the country's most valuable tourism export market. It was predicted that by 2020, Chinese visitors had the potential to add between $7 and $9 billion to the Australian economy, more than doubling their economic contribution a decade before. In order to capitalise on this opportunity, the Australian government, in collaboration with the tourism industry, started to build relationships with airlines and key industry partners.

In the first part of the decade Tourism Australia launched its *China 2020 Strategic Plan* (Tourism Australia, 2011). The plan was developed by Tourism Australia in collaboration with industry and government stakeholders and identified a number of pillars that were seen as pivotal to being competitive and winning market shares in China. The focus was set on building aviation links, understanding Chinese tourists and delivering quality tourism experiences.

Chinese visitor arrivals to Australia have increased from 100,000 in the year 2000 to one million in 2016, with an average annual growth rate of 18% since 2010 (Figure 8.2). Chinese arrivals grew two-to-three times faster in most months in 2015 than total overall arrivals to Australia. The Chinese visited Australia for holidays (53%), to visit friends and relatives (20%), and for education (13%). Under Australia's long-term tourism strategy, the value of the China market, as mentioned earlier, was originally estimated to be between $7 billion and $9 billion a year by 2020. Annual spending has already exceeded the lower end of that original range, and more recent forecasts indicate the market could be worth up to $13 billion (Tourism Australia, 2016a).

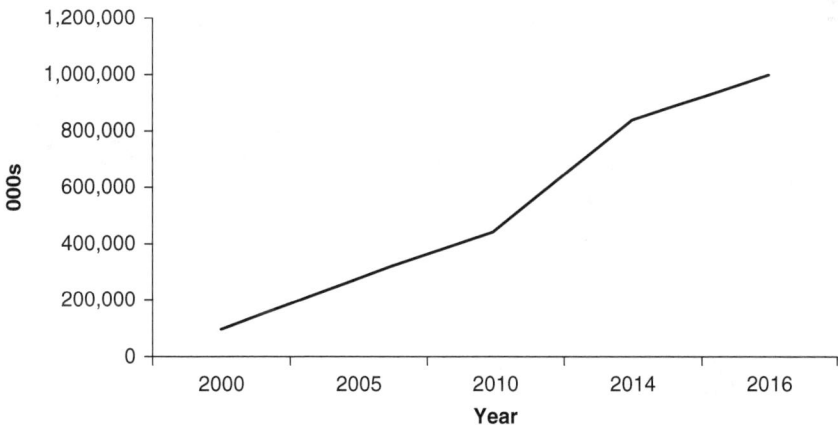

Figure 8.2 Chinese visitors to Australia, 2000–2016
Source: Tourism Australia (2016a)

Chinese free and independent travellers

Chinese visitors to Australia who are 'free and independent travellers' (FITs) are growing in number (Tourism Research Australia, 2015). FITs are defined in an Australian context as those who are not on a group tour, do not have any package inclusions and whose main purpose of holiday is for leisure. FITs have always dominated international leisure travel to Australia. However, as markets have matured and online booking capability improved, FIT travel has tended to increase in prevalence. China has dominated growth in leisure FITs to Australia since 2005 and this sector has grown by 186% in the last decade (Tourism Research Australia, 2015). In the period 2011–13, 532,000 (45%) of Chinese leisure visitors to Australia were FITs. There was low dispersal beyond Australia's main destinations (i.e. capital cities and Far North Queensland) for Asian leisure FITs generally, and Chinese leisure FITs specifically. Among Australia's top six inbound markets, China (22%) had the lowest dispersal of leisure FITs to regional areas (Figure 8.3).

Tourism Research Australia findings also show that more than one third (36%) of Chinese leisure FITs used the Internet as an information source when planning their trip to Australia, with 39% making bookings on the Internet in relation to

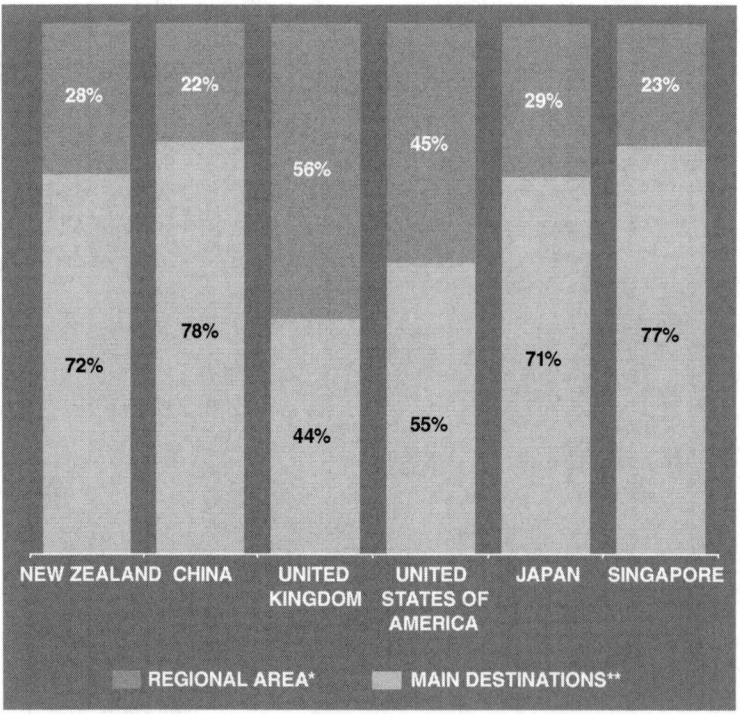

Figure 8.3 Dispersal of leisure FITs in Australia, 2011–2013

Source: Tourism Research Australia (2015b)

their trip. Of visitors who had used the Internet as an information source 29% dispersed to regional areas compared with 18% of non-Internet users. Of those visitors who had made an Internet booking, 27% dispersed to regional areas compared with 19% of non-Internet bookers. These results can be partly attributed to a much higher proportion of visitors aged less than 45 years having used the Internet for information and/or bookings than visitors aged 45 years and over (Tourism Research Australia, 2015).

Some of the key experiences Chinese visitors seek during their international holidays include a mix of natural environment, culture, history and development together with good products and services for accommodation, eating out, different cuisine options and local shopping. Importantly, though, they seek access to Chinese food, ideally once per day, particularly for breakfast (Tourism Research Australia, 2015).

In recent times Tourism Australia has conducted further research into understanding the Chinese market. The key market segments comprise holiday (53%), visiting friends and relatives (19%), education (14%) and business (9%) (Tourism Australia, 2016). A primary finding is that Australia is a major aspirational destination for the Chinese. In 2015 48% of Chinese consumers were considering travel to Australia in the next four years, with 40% intending to visit in the next two years (Tourism Australia, 2016a).

Tourism Australia devised a matrix to showcase the opportunity for Chinese tourism to Australia through identifying some key thematic appeals and experience categories. For the Chinese market, Australia performs strongest with respect

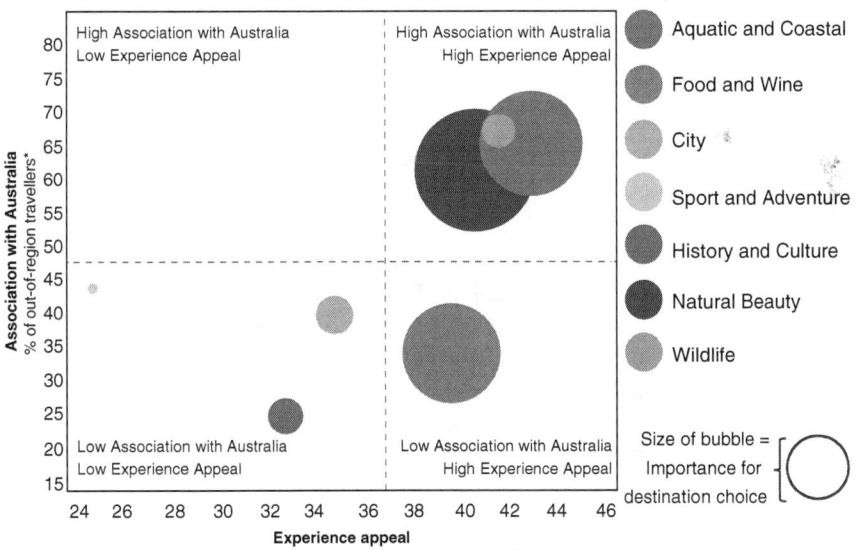

Figure 8.4 Chinese Consumers Association with Australia

Source: Tourism Australia (2016b)

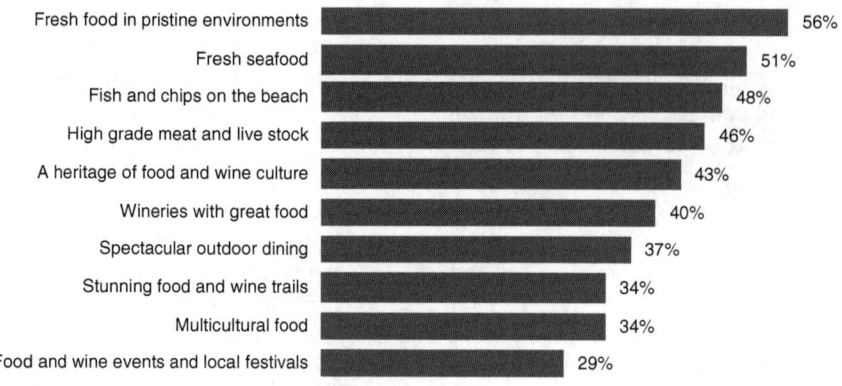

Figure 8.5 Chinese Consumers Association with Australia
Source: Tourism Australia (2016b)

to nature and wildlife. Food and wine experiences are also important amongst this market, but generate lower levels of association with Australia (Figure 8.4; Tourism Australia, 2016b).

The Chinese associate Australian food and wine with 'fresh food in pristine environments' (56%). They also associate it with 'wineries with great food' (40%), 'stunning food and wine trails' (34%) and 'food and wine events and local festivals' (29%) (Figure 8.5).

Food and wine tourism in Western Australia

Food and wine tourism is now one of the primary sectors of tourism in Western Australia (Tourism Research Australia, 2014). However, developing a wine tourism industry is not easy (Alonso et al., 2015). Research conducted by Tourism Australia across fifteen of Australia's key tourism markets showed that 'great food, wine, local cuisine and produce' is a major factor influencing holiday decision-making (38%), ranking third ahead of 'world-class beauty' and 'natural environments' (Figure 8.6; Tourism Research Australia, 2014). The research found that amongst people who have not been to Australia before, only 26% associate it with 'good food and wine' (Figure 8.7). However, those who have already visited the country ranked it second across the major markets for its food and wine experiences (60%) behind France, but ahead of Italy (third). Visitors from China, the USA, France, India, Indonesia, Malaysia, the United Kingdom and South Korea ranked Australia as their first destination for food and wine. To narrow the perception gap between those who have visited Australia and those who have not, Tourism Australia developed the idea that Australia could be the world's greatest restaurant – *Restaurant Australia* (Tourism Australia, 2014).

Thus, since 2014, Australia has set out to position itself as a food and wine destination with a culture that is different from the rest of the world and which has

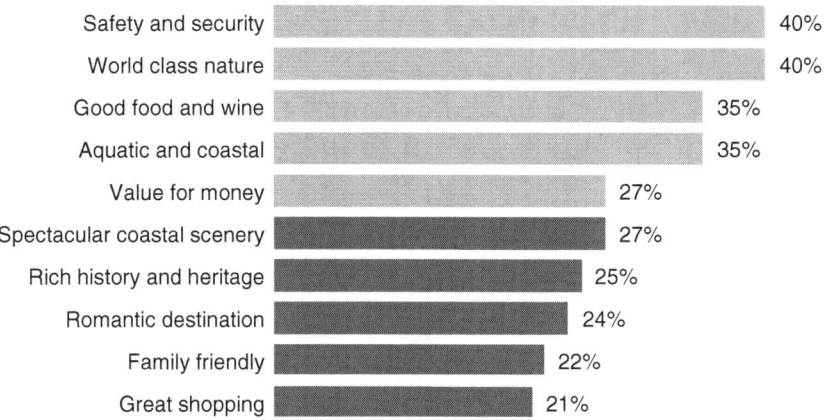

Figure 8.6 Factors that attract international tourists to Australia

Source: Tourism Australia (2016b) *Understanding the Chinese Market 2016.* Consumer Demand Project, Tourism Australia, Sydney

Figure 8.7 Visitors' perceptions of Australian food and wine

Source: Tourism Australia (2014)

three unique characteristics. They are: 'Free Thinking People' who are welcoming, open and multicultural; 'Open Places' with great weather, sunshine and areas of outstanding natural beauty; and 'Fresh Produce' which is safe, diverse and sustainable (Tourism Australia, 2014).

Western Australia has a rich heritage in agriculture and fishing and is well placed to deliver on food and wine tourism experiences through its production of a diverse range of high quality edible products (Tourism Western Australia, 2015a).

Traditionally, food and wine tourism strategies in Western Australia have focused on wine, which has a strong cultural significance for the state as vines were first planted in Western Australia in the Swan District in 1829. This was the same year that Perth was founded by Captain James Stirling as the administrative centre of the Swan River Colony. While Western Australia currently produces less than 5% of the total wine crush in Australia, the state accounts for approximately 20% of the ultra-premium segment of Australian wine sales (Tourism Western Australia, 2015a).

This emphasis on quality is reflected in all other areas of agricultural production. Western Australia is one of the most pest- and disease-free agricultural production areas in the world. Strict quarantine requirements and stringent standards in production systems ensure that this safety and quality is maintained for the benefit of local producers and consumers. Western Australia's vast geographic span provides diverse soils and climates that are suited to a variety of agricultural production. In the Ord River Irrigation Area in the north there are mangoes and other tropical crops, on the coastal sands near Perth, market gardens produce vegetables and from Perth into the southwest fruit crops are made into outstanding wines and cider (Tourism Western Australia, 2015a).

Along the state's 12,500 km coastline there are also a great number of commercial fisheries, which include rock lobster, pearling, prawns, scallops, abalone, oysters and finfish (e.g. barramundi). Western Australia is a major world supplier of lobsters and prawns and with a decline in fish stocks in other parts of the world the steadily growing aquaculture industry is successfully providing high-value products to specialist markets. The high quality of the products in each of these industries is internationally recognised, creating strong interstate and international export demand. Western Australia exports up to 80% of its agricultural production (Tourism Western Australia, 2015a). These food and wine products align with the state's tourism brand promise – *Experience Extraordinary* – and its culinary strategy incorporates all of the strengths of the state's food and wine promise within the overall tourism experience.

WA's *Signature Dish* is a competition run by the Department of Agriculture and Food Western Australia, through its *Buy West Eat Best* programme. The department works with regional food councils and 'Celebrate WA' to host regional cook-off events – culminating in a live Grand Finale cook-off as part of WA Day – to highlight the variety of top quality produce available in Western Australia. As a competition that is open to enthusiastic amateurs (not professional chefs) and supported by a social media campaign, it is an example of an event which raises awareness of the state's food culture through a significant level of community engagement. Its endorsement by prominent chefs based in Western Australia and culinary personalities generates a strong positive message about the state's food producers, their stories and regional food experiences (Tourism Western Australia, 2015a).

Food and wine tourism has been the hallmark of a number of regions of Western Australia for the past two decades. The Swan River Region of Perth and the Margaret River Region of Southwest WA have always been wine destinations, which in recent years have expanded to become wine and food destinations.

One of the first wine tourism conferences in the world was held at Leeuwin Estate in Margaret River in 1998. The First Australian Wine Tourism Conference *Wine Tourism: Perfect Partners* attracted 260 delegates from ten countries on five continents (Dowling & Carlsen, 1999). One of the observations made by the international delegates at the conference was that Australia was pioneering wine tourism and leading the world with the first global conference on the subject (Dowling, 1998) and the preparation of a National Wine Tourism Strategy (Winemakers' Federation of Australia, 1998). One year later a *Wine Tourism Strategy* was developed for Western Australia (WA Tourism Commission, 1999). The key goals of the strategy were to raise the awareness and understanding of the value-added benefit of tourism to the wine industry; establish an industry standard for wine tourism outlets and facilities; increase the skill level of employees in the wine tourism industry; and foster the links between wine, food and the Australian lifestyle. The overall goal was for the state to be recognised as a premier wine tourism destination by the year 2005.

With an emphasis on wine tourism, food tourism was overlooked for almost another decade. Then in the early 2010s food tourism became the focus of a number of conferences and events. These included the conference *Culinary Journeys – All the Tourism Ingredients* held in Manjimup in 2013, hosted by the Forum Advocating Cultural and Eco Tourism [FACET], which had hosted the First Wine Tourism Conference fifteen years earlier. Through Tourism WA, the state government sponsors and promotes a range of food and wine events in regional Western Australia, including the Margaret River Gourmet Escape, Taste Great Southern, Taste of Broome, the Truffle Kerfuffle and Cherry Harmony Festival in Manjimup, as well as many other smaller regional and themed events. Since it was launched in 2012, the Margaret River Gourmet Escape (www.gourmetescape.com.au) has quickly become the state's flagship event and the leading food and wine event in the Asia Pacific region. The event attracts famous chefs and wine critics from around the world, encouraging community vibrancy and delivering significant economic benefits and media exposure. The event has grown and is now considered to be one of the world's leading food and wine events, positioning Western Australia as a premier food and wine destination.

The visitor experience

Food and wine tourism is an important element in the visitor experience in Western Australia (Tourism Research Australia, 2014). In 2013 leisure travel in WA was worth $4.6 billion and an estimated 1.1 million visitors participated in food and wine activities while on holiday or on a day trip. Overall 55% of visitors reported that WA's food and wine offerings compared well with the offerings of other states. However, a number of key drivers in attracting visitors to WA were identified. They included that local produce and regional specialities have the broadest appeal among food and wine visitors. In addition visitors like to participate in events and those who do are likely to go to another. Therefore the opportunities for cross-promotion are high, with one of the best avenues to promote a food/wine event being another food/wine event. The most common sources of information for visitors seeking food and wine experiences in WA are friends and/or family as well as the Internet.

A survey of 750 intrastate (501) and interstate (249) visitors to WA found that 'unique or extraordinary sights' is the most influential factor for holidays in general (Figure 8.8; Tourism Research Australia, 2014). However, 'offering good food, wine, local cuisine and produce' ranks as the fourth most influential factor, with almost two-thirds of visitors rating this aspect as an important consideration when selecting a holiday destination.

Whilst 20% of all visitors to WA have no interest in food and wine experiences, a small but focused group of visitors (7%) select their travel destination base on a region's food and wine experiences. A larger number (18%) actively seek food and wine experiences when they travel and by far the largest group (54%) enjoys food and wine experiences but they do not seek them out (Figure 8.9; Tourism Research Australia, 2014).

Examination of visitors' interest in food and wine tourism experiences shows that the leading experiences are sampling *regional specialities, local produce* and *seafood* (Tourism Research Australia, 2014, p. 7). Five distinct groups of food and wine visitors were identified, each with their own expectations of food and wine experiences. They include visitors who use food and wine experiences as

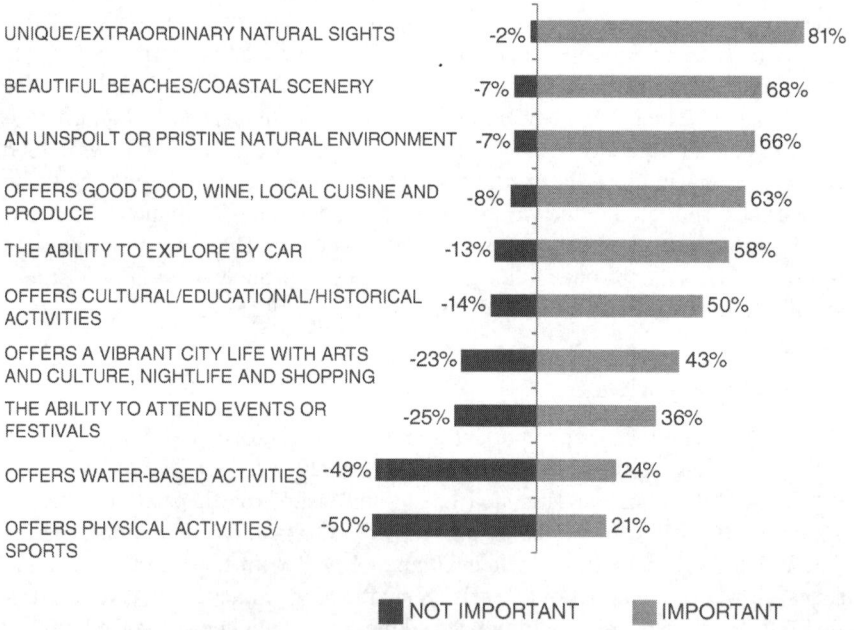

Figure 8.8 The importance of factors when choosing a holiday

Source: Tourism Research Australia (2014)

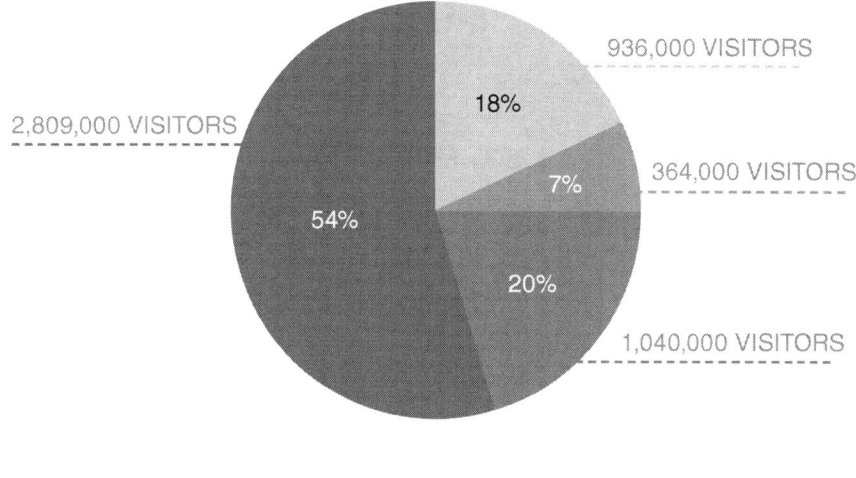

NO INTEREST IN FOOD AND WINE EXPERIENCES
ENJOY FOOD AND WINE EXPERIENCES, BUT DON'T SEEK THEM OUT
ACTIVELY SEEK FOOD AND WINE EXPERIENCES
FOOD AND WINE EXPERIENCES DETERMINE DESTINATION CHOICE

Figure 8.9 The approximate number of visitors to WA (per year) according to their interest level in food and wine experiences

Source: Tourism Research Australia (2014)

a tool for exploring or immersing themselves in a location, especially its people, culture and environment. This group, representing 39% of food and wine visitors, expresses a high interest in experiences such as authentic local produce, street food, local and farmers' markets, regional specialities and eating and drinking in spectacular surroundings. A second group of visitors, named 'traditionalists', forms 10% of the market. They are interested in 'classic' experiences associated with food and wine tourism such as fine dining restaurants, wineries, food/ wine tours and trails and events or festivals. A specialised group named 'niche experiencers', 6% of the market, are interested in more intensive food and wine experiences. They seek out hands-on cooking courses, locally caught seafood and indigenous food experiences (Tourism Research Australia, 2014).

By aligning the importance of food and wine tourism to visitors in WA's tourism regions with their satisfaction levels, the performance of each region (in terms of delivering on visitor expectations) can be seen (Figure 8.10). The bubble size represents the total number of leisure visitors (includes those who travelled for holiday and/or to visit friends and relatives) to WA in the year ending September 2013, multiplied by the proportion of the sample who visited each of the five regions.

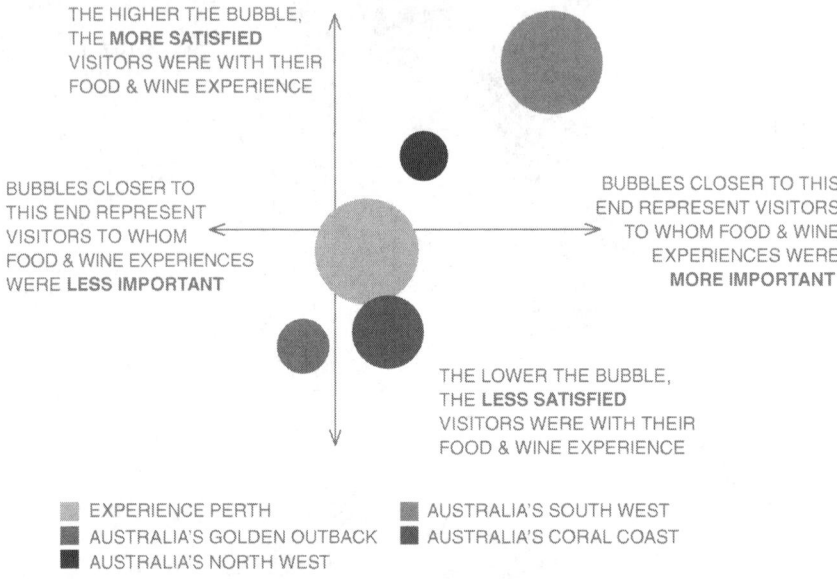

Figure 8.10 The importance of, and satisfaction with, food and wine tourism by region
in Western Australia

Source: Tourism Research Australia (2014)

Taste 2020: WA's food and wine strategy

In order to capitalise on its culinary offerings, Tourism Western Australia has developed *Taste 2020: A strategy for food and wine tourism in Western Australia for the next five years and beyond* (Tourism Western Australia, 2015a). The strategy provides a framework to align and develop new initiatives to elevate the state's tourism, wine, agricultural and fishing industries. The aim is to improve the food and wine tourism offering in the state, provide unique dining experiences and market them in the context of extraordinary and impressive locations.

The strategy aligns with Tourism Australia's 'food and wine' tourism campaign, *Restaurant Australia*, and with the state's tourism brand, *Experience Extraordinary*. It also supports the state government's *Strategy for Tourism in Western Australia 2020* and is helping to increase the value of tourism in Western Australia to $12 billion by 2020. The broad appeal of culinary tourism for all visitors ensures that it is applicable to every one of the seven key strategic pillars identified in the state tourism strategy: brand, infrastructure, business travel, Asian markets, events, regional travel and indigenous tourism. The purpose of the strategy is to enhance the positioning of Western Australia as an extraordinary destination to experience gourmet produce, fresh seafood, premium wines and boutique beverages (e.g. craft beers, cider and spirits). It also provides a cohesive framework to align and develop new initiatives to elevate the state's tourism, wine, agricultural and fishing industries.

Research undertaken for the strategy verified the importance of food and wine to the state's overall tourism offering and confirmed it as one of the fastest growing drivers of tourism worldwide. It found that almost all tourists want to try regional specialities and authentic local produce and they want to experience these in spectacular surroundings, such as a beach or by the waterfront. It also found that consumer demands of food and wine experiences can be broken down into the following five categories: natural beauty, casual dining, provenance, value for money and accessibility. 'World-class beauty and natural environments' is also in the top five most important factors when selecting a holiday destination for international visitors (along with good food, wine, local cuisine and produce).

WA's food and wine tourists

Recent consumer research shows that four out of five people who had visited Western Australia said that they were interested in culinary tourism experiences beyond the basic necessities of eating and drinking (Tourism Western Australia, 2015a). Of these people who are interested in culinary tourism experiences, approximately one in ten can be classified as a dedicated food and wine tourist. Sometimes referred to as 'gourmet travellers', these are visitors who would say that the types of culinary tourism experiences offered by a destination is an important factor in, or the main reason for, their decision to visit. The gourmet traveller is therefore an important market, accounting for almost 400,000 visitors to Western Australia each year (Tourism Western Australia, 2015a).

Gourmet travellers tend to be younger couples between 18–34 years old, who are yet to have a family, and those who are over 55 years old and whose children have already left home. For these segments, interaction and education is very important. They desire memorable experiences that increase their understanding of how to cook and prepare food, knowledge of how to appreciate good wine and experiences that help them learn more about the history and culture of the place where the food and wine is produced. The strategy also acknowledges the larger group of visitors to Western Australia who show a general interest in culinary tourism. They do not deliberately seek out culinary tourism experiences, but they enjoy them if they occur. These visitors tend to be families with young children, or singles. Instead of engaging with the heritage of the culinary culture, they experience it more incidentally at events, attractions, or as part of an overall tour. Their primary way of experiencing food and wine will be at restaurants and cafés (Tourism Western Australia, 2015a).

While gourmet travellers might engage with more niche culinary experiences, the WA research shows that no matter where the visitor comes from (intrastate, interstate, or international) and no matter what their level of interest in food and wine tourism (dedicated, interested or accidental), there are some culinary tourism experiences that have universal appeal (Tourism Western Australia, 2015a). The research also shows that international visitors are seeking Aboriginal culinary experiences in the form of either foraging for 'bush tucker' or learning about and sharing in traditional Aboriginal meals. Thus Aboriginal tours, which provide authentic local produce and also allow the visitor to experience it in-situ, are now beginning to grow in WA.

WA's Western Australian Indigenous Tourism Operators Council (WAITOC)'s 'Gourmet Experiences in WA' is an example of how new experiences can be created which match Aboriginal culture with culinary tourism. Highlighting unique food, drink and cultural experiences in Western Australia, the campaign features ten Aboriginal tourism operators. It demonstrates how tourism value can be added to an event or natural setting by enabling an exchange of culture and combining a broader experience or tour with a culinary activity.

Calendar of food and wine events

There are also a number of food and wine events, such as Taste of Perth (www.tasteofperth.com.au), the Good Food and Wine Show (goodfoodshow.com.au), and UnWined (www.wineandfood.com.au), which do not receive government sponsorship but do have the significant benefit of bringing the gourmet food and wine experience to locals and visitors. They provide a forum for showcasing the best of the state's regions and because there are generally workshops and interaction at the vendor stalls, there is an opportunity to educate the consumer on the product. This reinforces the real authority of the produce and food offerings of the state and provides an opportunity for local producers to expand their markets.

China food and wine tourism in WA

Interest in wine and wine tourism is rapidly growing amongst the Chinese (Qiu et al., 2013). Indeed China is the world's largest red grape wine consuming country (Qing et al., 2015). To capitalise on this emerging trend the WA Department of Agriculture and Food has showcased Western Australia produce in China for many years. These have included presentations at the HOFEX Food and Beverage Trade Exhibition in Hong Kong, the largest regional food and beverage trade event in Northeast Asia, held every two years (DAF, 2013). China-Hong Kong is the largest export market for Western Australian agriculture and fisheries with exports worth $1.5 billion in 2014 (DAF, 2014a). Over 90% of the food imports into Hong Kong are re-exported, one third to mainland China. Another exhibition where Western Australian producers promote their food and beverages is at the Shanghai-based 'Food and Hotel China', the country's oldest trade exhibition for the food and wine industries.

A conference aimed at strengthening business ties between Western Australia and China attracted strong international interest when it was held in 2014 (DAF, 2014b). The Western Australia-China Agribusiness Cooperation Conference enhanced relationships across the agriculture and food industries and brought together Chinese and Western Australian business and government leaders to facilitate business matching and identify key trade and investment opportunities along the supply chain. Key focus areas of the conference included premium food and wine. The importance of these products was shown by the representation from China's provinces of Zhejiang, Shanghai and Jiangsu, Shandong, Liaoning, Jiangxi, Anhui and Shaanxi.

Prospective Chinese tourists are increasingly interested in Margaret River as a destination due its reputation for good wine and food (Freed, 2016). Data from Trip Advisor's China arm revealed searches for Margaret River have risen by 525% over the last year and now make up one quarter of all searches for Perth. That compares to the 10% of Chinese visitors to Western Australia that actually visited the southwest region including Margaret River in 2016. The report noted that the Chinese outbound tourism market was not only growing fast but consumers were becoming more affluent, independent and sophisticated in their preferences. There has been a rise of interest in discovering, drinking and producing wine in China in recent years so, with Margaret River being the closest wine region to China, there is much interest in Margaret River as a travel destination for the Chinese (Freed, 2016). Further evidence of the link between Margaret River, wine and China is the 'Margaret River Wines' store which is about to open in Beijing (Figure 8.11).

China is now the biggest export destination for Australian wine, and Margaret River winemakers are hopeful for their future in a market tipped to grow to 35 million consumers within five years (Pancia, 2017). To achieve this 14 Margaret River wine producers have joined together to establish Watershed Premium Wines to sell their brands under a retail banner called Margaret River Wines (Pownall & Wilkie, 2016). The WA winemakers have set up four Chinese stores that only sell

Figure 8.11 Chinese visitors attending the Margaret River Wine Association's Grand Cabernet Auction at Eight Willows, as part of the wine region's 50th anniversary celebrations and 2017 Margaret River Gourmet Escape
Source: Tim Campbell

Margaret River wine and they have plans to open another six immediately and up to 300 within three years (Pancia, 2017).

Conclusions

Food and wine is an important tourism product for Western Australia as evidenced by the development and implementation of the *Taste 2020* food and wine tourism strategy for the state (Tourism Western Australia, 2015a) as well as its *China Strategy* (Tourism Western Australia, 2012). The principal purpose of the food and wine strategy is to strengthen WA's position as an internationally recognised food and wine destination. The aim of the China Strategy is to grow the number of Chinese visitors to WA to 100,000 by 2020 (Tourism Western Australia, 2015b). As noted previously it is currently ranked sixth by country visitor numbers (52,600 in 2017; a 12.6% increase over the previous year) but is ranked second in relation to visitor spend ($271 Million; 13% increase) (Tourism Western Australia, 2017b). Thus it is inevitable that the state's food and wine products intersect with the rising tide of Chinese visitors to Australia generally and WA in particular.

An integral part of tapping into the potential for Chinese visitations is to gain an understanding of their tourism trends. Key trends include the increasing younger age of Chinese travellers, their use of digital technologies for booking travel and the growth of travellers from second tier cities in China (Tourism Western Australia, 2016). They also seek out food and wine experiences, which are offered in many of the state's tourism regions. An example of a company that is capitalising on Chinese visitors' interest in food and wine tourism is Linkar Wines. This seven-year-old business began by exporting WA wines to China but now includes winery tours for Chinese visitors (Beyer, 2015). Thus overall this nexus between food and wine tourism and the growth of Chinese visitors is making for a perfect partnership in Western Australia's tourism supply-market match and is set to expand rapidly in future.

Acknowledgements

Much of the material and most of the figures in this chapter are drawn from the information in reports on food and wine tourism and China tourism, in both Australia and Western Australia. Thus the author wishes to acknowledge these reports by Tourism Australia, Tourism Research Australia and Tourism Western Australia. The author also wishes to thank two anonymous reviewers of an earlier draft for their constructive comments.

References

Alonso, A., Bressan, A., O'Shea, M. & Krajsic, V. (2015). Perceived benefits and challenges to wine tourism involvement: an international perspective. *International Journal of Tourism Research*, 17: 66–81.

Beyer, M. (2015). Chinese insiders provide bridge for wine exports. *WA Business News*, 20 April.

DAF (2013). *Western Australian produce on show in Hong Kong*. News Release, 5 May. Department of Agriculture and Food, Perth, Western Australia.

DAF (2014a). *Western Australian food to feature at China trade show*. News Release, 11 November. Department of Agriculture and Food, Perth, Western Australia.

DAF (2014b). *Agriculture conference draws international interest*. News Release, 9 April. Department of Agriculture and Food, Perth, Western Australia.

Dowling, R.K. (1998). The First Australian Wine Tourism Conference 'Wine Tourism: Perfect Partners'. *The Australian & New Zealand Wine Industry Journal*, 13 (3): 307–309.

Dowling, R.K. & Carlsen, J. eds. (1999). *Wine Tourism: Perfect Partners*. Proceedings of the First Australian Wine Tourism Conference, Margaret River, Western Australia, May 1998. Bureau of Tourism Research, Canberra, p. 296.

Freed, J. (2016). Why Chinese tourists are increasingly keen on Margaret River. *The Sydney Morning Herald*, 18 April (accessed 11 November 2016). http://www.smh.com.au/business/why-chinese-tourists-are-increasingly-keen-on-margaret-river-20160415-go7al

Pancia, A. (2017). Margaret River wine producers see big future in exports to China as demand grows. *ABC News Perth*, 12 May 2017. http://www.abc.net.au/news/2017-05-12/margaret-river-wine-producers-see-big-future-in-exports-to-china/8519932

Pownell, M. & Wilkie, D. (2016). Foreign investment buoys Margaret River winemakers. *Business News Western Australia*, 21 November 2016, pp. 6–7.

Qing, P., Xi, A. & Hu, W. (2015). Self-consumption, gifting and Chinese wine consumers. *Canadian Journal of Agricultural Economics*, 63, 601–620.

Qiu, H.Z., Yuan, J., Ye, B.H. & Hung, K. (2013). Wine tourism phenomena in China: an emerging market. *International Journal of Contemporary Hospitality Management*, 25:7, 1115–1134.

Tourism Australia (2011). *China 2020 Strategic Plan*. Tourism Australia, Sydney.

Tourism Australia (2012). *Knowing the Customer in China*. Research Update, Tourism Australia, Sydney.

Tourism Australia (2014). *Restaurant Australia*. Concept Magazine, Food & Wine Strategy, Tourism Australia, Sydney.

Tourism Australia (2016a). *Australia welcomes record one million visitors from China*. News Release, 13 January 2016 (accessed 11 November 2016). Tourism Australia, Sydney. http://www.tourism.australia.com/news/market-regions-greater-china-17742.aspx

Tourism Australia (2016b). *Understanding the Chinese Market 2016*. Consumer Demand Project, Tourism Australia, Sydney.

Tourism Research Australia (2013). *Understanding dispersal of Asian visitors: The International Visitor Survey data mining project*. Tourism Research Australia, Sydney.

Tourism Research Australia (2014). *Food and Wine Tourism in Western Australia: Summary*. Tourism Research Australia, Sydney.

Tourism Research Australia (2018). *International visitors in Australia: Year ending December 2017*. Tourism Research Australia, Sydney.

Tourism Western Australia (2012). *Our Direction in China 2012–2015: Summary Document*. Tourism Western Australia, Perth.

Tourism Western Australia (2015a). *Taste 2020: A strategy for food and wine tourism in Western Australia for the next five years & beyond, 2015–2020*. Tourism Western Australia, Perth.

Tourism Western Australia (2015b). *China Visitor Profile: Overnight visitor fact sheet years ending December 2012/13/14*. Tourism Western Australia, Perth.

Tourism Western Australia (2016). *China: Tourism Western Australia Marketing Forum 2016*. Tourism Western Australia, Perth.

Tourism Western Australia (2017a). *Visitation to Western Australia: Overview year ending March 2017*. Tourism Western Australia, Perth.

Tourism Western Australia (2017b). *International Visitation – fast facts year ending March 2017*. Research Team, Tourism Western Australia, Perth.

WA Tourism Commission (1999). *Western Australian Wine Tourism Strategy*. WATC, Perth.

Winemakers' Federation of Australia (1998). *National Wine Tourism Strategy Green Paper (Draft)*. WFA, Adelaide.

9 Are we China ready?

A study of Western Australian hotels and Chinese tourists' appetites

Alfred Ogle, David Lamb and Stephen Fanning

Introduction

According to Tourism Western Australia (TWA), the state's key tourism authority, the mainland Chinese inbound tourism market is presently the fastest growing and second largest market segment in Western Australia (WA) (TWA, 2016). Therefore, TWA sees Chinese travellers as an attractive market segment that needs to be nurtured. Recently, TWA posed a question to tourism operators 'Are you China ready' (TWA, 2016)? It was meant to provoke reflection by WA tourism operators on their current practices and assessment to the extent to which their products would meet the needs of ethnic Chinese inbound travellers. The Chinese love of food is well recognised (e.g. Chow & Murphy, 2008; Yang, Lai & Khoo-Lattimore, 2014; Zhang et al., 2008); moreover, extant research and anecdotal evidence suggests that while Chinese travellers wish to experience local cuisines they also desire familiar Chinese food (Liu & Hull, 2015; TWA, 2016). With the majority of Chinese travellers to WA staying in commercial accommodation, they would place high importance on the food offered by restaurants, takeaways and hotel room service (RS).

The peak body representing tourism business, industries and regions in WA is Tourism Council Western Australia (TCWA). Early in 2014, as part of its marketing strategy, TWCA joined the *China Ready and Accredited*® programme, which is a global accreditation system designed to reduce the consumer risks for Chinese travellers. According to the accreditation website, its logo designates 'product(s) and services Chinese consumers can trust' (TCWA, 2017). An accredited hotel meets the prescribed standard for quality assurance, cultural awareness, consumer protection, and respect for the Chinese free and individual travellers *(*FITs) and is highlighted in marketing paraphernalia both in China and Australia.

One of the requirements for certification of the accommodation business sector (i.e. various commercial lodging properties including hotels and resorts) was what sparked the research question vis-à-vis food: The directive to include 'Chinese instant noodles' as an item in the 'China Ready & Accredited Comfort Stay Goodie Bag' (see Appendix 9.1). Given that the *China Ready*® programme acts as a 'guide to China through "Chinese Eyes", overseen and approved by high-ranking Chinese government officials and business executives' (China Ready & Accredited®, 2016), we contend that food plays a critical role in Chinese visitor satisfaction. Consequently, we posit that this study is warranted because, based on the specificity of a food-related accreditation criterion, the availability and accessibility to food amenable to Chinese tourists' palates in a hotel is of great import.

We used a case study strategy with a positivist approach. Our independent scrutiny of RS menus based on personal experience of Chinese cuisine simultaneously through a consumer and research lens lent an ethnographic perspective to the identification of Chinese menu items. Subsequently an evaluation of the authenticity and consequently the determination of availability of items to whet our appetites and, by extension, that of the inbound Chinese visitor, could be made. This study has implications on hotel food and beverage (F&B) operations and strategic marketing for the tourism industry in general, and the hotel industry in particular.

Chinese tourism in WA

In 2012 and 2013, China was the second largest market for inbound tourism in Australia (Tourism Australia, 2014). While remaining the second largest inbound market for visitor arrivals to Australia in 2016, China was the largest market for total expenditure and visitor nights (TWA, 2017). The visitors mainly come from Shanghai, Beijing and Guangdong (Figure 9.1) which are Tier 1 cities.

While the visitation to WA is comparatively small (4%) compared to other regions (Tourism Australia, 2017), China is WA's sixth largest source market by visitor numbers and ranked second by visitor spend (Tourism WA, 2017). This expenditure trend is encouraging because in 2013 each Chinese visitor was the third highest spender amongst international travellers to WA ('State government lands Chinese air deal', 2014). Tourism WA's (2017) long-term goal for 2020 is to double the current Chinese visitor numbers to 100,000, and more than double their tourism spend to $500 million. This goal would be achievable given the

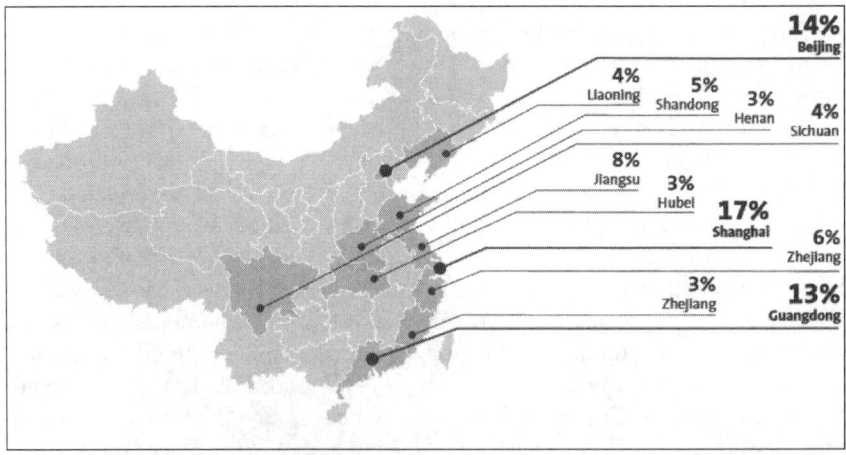

Figure 9.1 Main provinces and cities of residence of Chinese visitors to Australia
Source: Tourism Australia, 2017

growing trend among the Chinese middle class market to seek luxury accommodation and travel in WA (Acott, 2017). The vast majority of Chinese travellers to WA remain in the metropolitan area, with almost 40% staying in commercial accommodation such as hotels or resorts (Tourism WA, 2014). For such a large and growing market, hotels and resorts operating in WA need to be 'China ready'.

Dining habits of Chinese tourists in Australia

The importance of understanding the food culture and preferences of tourists for the benefit of the tourism and hospitality sectors has been noted by several researchers (e.g. Chang, Kivela & Mak, 2011; Mak, Lumbers & Eves, 2012; Reisinger & Turner, 2002). Authors such as Atkins and Bowler (2001), Mäkelä (2000) and Fieldhouse (1986) emphasised the relationships between food and culture, with culture acting as a determinant to what people eat (Atkins & Bowler, 2001; Mäkelä, 2000) and food in turn acting as a cultural symbol (Fieldhouse, 1986). In some cases, the food of an unfamiliar culture may add to the tourism experience (McKercher & Chow, 2001), whereas in other cases the cultural differences in food are not appreciated by the tourist (Cohen, 1972; Fischler, 1988). According to Tourism Research Australia (TRA) (2014), the extant research indicates that the vast majority of Chinese tourists visiting Australia fall into the latter category.

TRA (2014) found that although many Chinese tourists preferred to eat a mixture of both Chinese and Western cuisine, 80% of Chinese travellers to Australia eat Chinese food every day or most days during their time in Australia and 98% had Chinese fare at some point during their trip. Yu and Weiler (2001) supported the notion that many Chinese tourists find adjusting to Western cuisine difficult. They claimed that all of the Chinese tourists they had surveyed dined in Chinese restaurants while visiting Australia and only 50% dined in Western restaurants regularly, an activity which was rated least important amongst their respondents. Chang, Kivela and Mak (2010, p. 998) noted that whereas Chinese tourists expressed enthusiasm in trying local food when in Australia they would find it 'impossible . . . to consume local food at every meal'. Li, Lai, Harrill, Kline and Wang (2011, p. 745) found that some outbound Chinese tourists were willing to try local or new cuisines providing that the food is 'acceptable to Chinese', but expressed reservations on being able to cope without Chinese food on a daily basis. Chang et al. (2010) noted that whereas middle-class Chinese tourists were amenable to savouring local Australian cuisine in order to expand their culinary knowledge, explore the local culture, have an authentic travel experience and engage in learning/education opportunity, some may be coerced to do so because of prestige and status, reference group influence and subjective perception.

It is also indicated in the research that follows that the quality of Chinese food is related to the degree to which Chinese tourists are satisfied with their overall trip to Australia. TRA (2014) found that those tourists who rated Chinese food in Australia poorly were more likely to rate their overall experience negatively, and those rating Chinese food highly rated their overall experience positively. The implications of this behaviour may have larger ramifications: with word-of-mouth (and, by extension, word-of-mouse or e-WOM) as an influential factor in the decision-making process of prospective Chinese tourists (Simpson & Siguaw, 2008; Hsu, Kang & Lam, 2006; Law & Cheung, 2010), the availability and quality of Chinese food may ultimately impact upon the number of Chinese travellers to Australia.

In-room dining in hotels

The wide availability of Chinese cuisine in the 'Chinatowns' of all major Australian capital cities (Chinatowns across Australia, n.d.) is manifested in the myriad of stand-alone restaurants offering a variety of Chinese cuisines. Chinese visitors flock to Chinatown restaurants as they actively seek out familiar cuisine. For example Chang (2007, p. 78) found: 'with respect to food style, the most commonly preferred food style for all groups was Chinese' and 'Chinese-style food was the most common type of food, which was identified in a bad food experience' (p. 80).

Hotels, as the primary providers of commercial lodging and supporting services such as F&B (Ogle, 2009) can also capitalise on their Chinese guests' desire for Chinese cuisine. With the increasing popularity of in-room dining amongst hotel guests, optimal RS menu design is the catalyst for profitability. By testing Herzberg, Mausner and Snyderman's (1959) two-factor theory of 'hygiene' and 'motivation' factors, DeShields Jr., Kara and Kaynak (2005) confirmed that while the presence of a hygiene factor may not necessarily engender customer satisfaction, its absence often led to dissatisfaction. Ninemeier and Purdue (2008) argued that in-room dining services are merely a guest amenity that rarely generates profit for hotels. RS, therefore, could play the role of a hygiene factor, a service provided to meet guests' basic expectations. However, Suboleski (2012) asserted that RS becomes a viable revenue stream for hotels when underpinned by advanced menu engineering and clever marketing. By identifying the needs and wants of their guests, hoteliers can transform RS from a hygiene factor to a motivation factor. By catering to their Chinese guests' palates, hoteliers could effectively leverage the growing affluent inbound Chinese visitor market.

Chinese cuisine

Yan (1984) asserted that the distinguishable features that make food 'Chinese' are cooking style (stir-fry technique), *mise en place* (special ingredient cuts, seasoning, and marination), and the use of dried products.

According to Okumus, Okumus and McKercher (2007, p. 257), Chinese food is typically categorised into six sub-cuisines: Cantonese, Chiu Chow, Peking, Shanghaiese, Szechuan and Hunan. In addition, Chinese food is further differentiated as Chinese vegetarian food and Chinese festive food (Okumus et al., 2007). Wu and Cheung (2002) noted that Minnan cuisine, which originates from Southern Fujian and, together with Guangdong (Cantonese) cuisine, is the primary influence of Chinese cuisine as it is known outside of China, is not acknowledged as a regional high cuisine, which is categorised as either three, six or eight cuisines (Liang, 1985). This variability is apparent in web guides relating to Chinese sub-cuisines: Shandong, Guangdong, Sichuan (sic), Hunan, Jiangsu,

Zhejiang, Fujian and Anhui (TravelChinaGuide, n.d.); Lu (Beijing and North), Xiang, Chuan (Sichuan and West), Min (Fujian and East), Yue (Guangdong), Su (Jiangsu), Zhe (Zhejiang) and Hui (Anhui) (Zhu, 2013), and suggests that this heterogeneity might be due to the widely diverse geography, climate and produce found in a country as vast as China (Hsiung & Simonds, 2005).

Methodology

Using the hotel listing provided by Star Ratings Australia (SRA), which is an independent accommodation standards accreditation body in Australia (Star Ratings Australia, n.d.), we contacted 4- and 5-star rated hotels located in the Perth CBD by e-mail requesting a copy of their RS menu. We also approached two self-rated 5-star hotels that did not participate in the SRA scheme. Twenty hotels (out of a total of twenty-three) agreed to participate in the study and gave us

Table 9.1 Hotel sample

Hotel (n = 20)	Star rating	Affiliation	F&B operation(s)	RS after-hours operation
A	5	chain	HR; RS	22:00–06:00
B	5	chain	HR; RS	23:00–06:00
C	5	chain	HR; RS; O	23:00–05:00
D	5	IND	HR	–
E	5	chain	HR; RS	23:00–06:00
F	5	chain	HR	–
G	5	chain	HR; RS	22:30–06:00
H	5	chain	O	–
I	4.5	IND	HR; RS	23:00–06:00
J	4.5	chain	HR; RS	22:00–06:00
K	4.5	chain	HR; RS	23:00–06:00
L	4.5	chain	HR	–
M	4.5	chain	HR; RS	24H option available
N	4.5	chain	HR; RS	24H availability
O	4.5	chain	HR; RS	22:00–06:00
P	4	chain	HR; RS	22:00–06:00
Q	4	IND	HR	–
R	4	chain	HR; RS	24H pick-up service
S	4	IND	HR	–
T	4.5	chain	HR; RS	24:00–05:00

Legend:

Hotel restaurant	**HR**
Room service	**RS**
Outsourced F&B	**O**
Chain hotel	**chain**
Independent hotel	**IND**

Notes: 24-hour RS indicates a dedicated F&B operation.

their RS menus for analysis. The sample ($n = 20$) comprised of eight 5-star, eight 4½-star and four 4-star hotels.

One 4½-star hotel listed by SRA did not provide any RS facility and therefore was not included in the study. One 4-star hotel had agreed to participate but did not supply its menu, while another listed 4-star hotel outsourced its F&B and, although the supplier had initially agreed to participate, did not submit their menu.

Analysis

The sample RS menus were independently analysed by five reviewers, all academics involved in marketing, hospitality, events and services education for Chinese cuisine. Each reviewer had travelled to destinations in Asia and China including the People's Republic of China (PRC), Republic of China (Taiwan), Hong Kong SAR, Macau SAR and frequently patronised Chinese restaurants in WA. The items identified were then categorised according to a modified categorisation based on the six sub-cuisines used by Okumus, Okumus and McKercher (2007) augmented by two new categories: Pseudo Chinese and Fusion. The categories are shown in Table 9.2 below.

Each reviewer indicated his determination of whether the menus contained Chinese items/dishes and their feedback is presented in a table (Appendix 9.2).

Results

Reviewer 1

Reviewer 1 did not find any Chinese dish per se listed. He, however, noted dishes that:

a)   had Asian origins such as *nasi goreng* and Thai prawn (Hotel A)
b)   were listed under the category of "Asian Specialities" (Hotel C) and "Taste of Asia" (Hotel T)
c)   had the descriptor "Asian" in the name such as "Asian vegetables" (Hotel L), "Asian-style confit duck" (Hotel N)

Reviewer 1 observed that Hotel K offered the dish Jasmine tea poached scallops that, due to its cooking method and Chinese tea infusion, 'may appeal to a Chinese

Table 9.2 Chinese food categories

Category	Code
1 Cantonese	C
2 Chiu Chow	CC
3 Peking	P
4 Shanghaiese	Sh
5 Szechuan	Sz
6 Hunan	H
7 Pseudo Chinese	PC
8 Fusion	F

palate'. He also observed that 'green tea and jasmine tea may tempt the Chinese visitor', with reference to availability in the menu of Hotel B.

This would suggest that tea, as a beverage, is complementary to food and is an important RS menu item.

The jasmine tea scallops dish was identified as Chinese based on the usage of a Chinese tea in its preparation.

Reviewer 2

Reviewer 2 found several Asian-styled dishes and used semantic nomenclature as a basis to identify dishes of Chinese origin. He identified items:

a) using place of origin/geographic descriptors such as *Teo Chew* congee, Hainanese chicken, Shanghai pork belly (Hotel C); Shanghai rice noodles (Hotel E)
b) using cooking methods that alluded to the 'Chinese-ness' of a dish; for example, 'wok fried' (Hotels C, E & G); 'wok tossed' (Hotel G); 'stir fried' (Hotels E & N); 'steamed' (Hotels C & E). He associated Japanese culinary terms such as *sashimi* (Hotel I) and *tempura* (Hotel M) as Asian-style dishes

This reviewer listed items that were characterised as Asian in the name such as 'Asian marinated beef' (Hotel R), 'Asian vegetables' (Hotels B & L), 'Asian greens' (Hotel E), and 'Oriental vegetables' (Hotel R). In addition he highlighted Chinese/Oriental food ingredients, such as *kai lan* (Hotel E), *wasabi* (Hotel O), *kimchi* (Hotel L), *somen noodles*, shiitake mushroom (Hotel K) and satay (Hotels E & T). Lumpia spring rolls (Hotel T) were singled out as a possible Chinese dish together with 'vegetable spring rolls' (Hotel E).

Reviewer 2 observed that 'the menus offering few Chinese options often offer Chinese food as an entrée or starter only, while menus offering more options tend to include Chinese meals as mains.'

Reviewer 3

Reviewer 3 listed all dishes that explicitly or implicitly suggested Chinese or Asian flavour. He stated that there were 'almost no Chinese (and) several with no Asian, certainly no Chinese beers'. He noted that beer is classified as a food according to the Food Standards Australia & New Zealand, and should be included in the list.

Reviewer 3 indicated uncertainty about the extent to which the following dishes were Chinese:

a) Shanghai rice noodles with mixed vegetables and sesame seeds, chilli sauce (Hotel E)
b) Lumpia spring roll, pork mince, onion, carrot, soy, coriander (Hotel T)
c) Wok tossed Udon noodles with chicken, bean sprouts, spring onions and oyster sauce (Hotel G)

This uncertainty may have stemmed from the fusion (or confusion) of ingredients and preparation: Udon is a Japanese noodle served in a clear broth and also stir-fried in a soy-based sauce in a particular prefecture. However, when prepared with oyster sauce and bean sprouts a Southeast Asian flavour is evoked, and perhaps to the uninformed, this gives the implication of a Chinese characteristic. Likewise, while noodles vary greatly according to locality and ingredients, the noodle that the menu referred to is made from rice, but Shanghai noodles are a thick wheat noodle that is commonly used in the cuisines of northern China and is available in Shanghai and other cities.

Reviewer 3 observed that Hainanese chicken is served by Hotel C together with 'traditional' roast duck and 'Shanghai pork belly in Chinese wine'.

It is very unclear as to what traditional actually means given that a tradition is specific to a culture or sub-culture, and unless specified, the diner can only guess as to what style the roast duck is cooked. Pork belly is commonplace in various cuisines and apart from the reference to a Chinese city and the use of 'Chinese wine', it would be difficult to gauge the authenticity of the dish.

Reviewer 4

Reviewer 4 observed the latitude employed by Perth hotels in their RS menus in naming and describing their 'Asian' offerings. Adjectival menu item names can be misleading and perpetuate misconceptions about ethnic cuisines. Spring rolls were available (Hotels B, E & T): while the spring roll offered by Hotel B on account of no description being provided would presumably be the small deep-fried variety, Hotel E specified that the filling was vegetables but did not offer further explanation; therefore, the assumption is that it too was the small deep fried item. Hotel T offered two variants of the spring roll – the 'Lumpia spring roll' and the 'Vietnamese rice paper prawn spring roll'.

Reviewer 4 observed that Chinese Jasmine tea was available at a number of hotels (Hotels B, C, L & N).

It should be pointed out that Jasmine tea refers to tea (green, black and white) that is flavoured with jasmine.

The menu names were as follows:

Hotel B: Green tea with jasmine

Hotel C: Jasmine tea

Hotel H: Green tea

Hotel L: Jasmine green tea

Hotel N: Green tea & jasmine

Hotel Q: Green tea

Hotels H and Q referred to green tea that might be of the Japanese variant. Another observation was that only one hotel in the sample offered an Asian beer, specifically Kirin (Hotel K), which suggests an acknowledgment to the taste of Japanese guests but the absence of any Chinese beer appears to indicate otherwise.

Reviewer 4 noted that 'lumpia' is incorrectly referred to as the ubiquitous 'spring roll' served in Chinese restaurants.

Highly indigenised throughout Southeast Asia, the spring roll no longer refers exclusively to the original dish. The 'Asian' label is liberally and arbitrarily used. While possibly appropriate in Australia where the understanding of what being Asian is might be somewhat ambiguous, for an Asian traveller erroneous usage of the term may lead to confusion and even antagonism.

This reviewer had lived in Macau for an extended period and his spouse is from there. He is aware that Chinese cuisine is diverse in character and that whereas Macau residents, who are in the main of Cantonese stock, may enjoy other regional Chinese cuisine, they consider Szechuan food for example as a novelty but would not consider that as local cuisine. Hotel T highlights using Szechuan pepper as a flavouring agent in one of its side dishes. For many Chinese people, Szechuan peppers are much too piquant for their palates. Indeed some types of foods indigenous to some regions in China might be unpalatable. Zhang and Long (2014, p. 2) asserted that the 'style of dining and the language in which it (Cantonese-style Yum Char) is conducted is unfamiliar to the inhabitants of the Sichuan Province who have their own distinctive cuisine and dialect.' The reviewer also noted that the use of Chinese vegetables namely kai lan and bok choy certainly were intended to provide an Asian flavour as artichokes and Brussels sprouts would lend a European flavour to a menu.

Reviewer 5

Reviewer 5 found scarcity in Chinese dishes in the menus. Apart from some items that he could relate to his visits to China, such as dumplings and steamed rice, he found the so-called Chinese and Asian dishes to be inauthentic.
    He explained:

Normally I would look at the selected menus though a Western lens. Looking at the menus with the objectives of this research project I reflected on my experiences from two visits to China, I felt that we have let down our Chinese guests with a lack of variety, had little regard for the Chinese regional preferences, and furthermore, they implied that Asian food was 'China ready'. In short, the menus suggest that we have not catered for the Chinese tourist and this represents a short term rather than long term focus.

Table 9.3 Sample of WA hotels' RS Chinese food categorisation

Category	Freq	Remarks
1 Cantonese	5	abalone congee#; *Teo Chew** congee; steamed pork bun; Hainanese chicken+ (Hotel C); steamed fish fillet with ginger and spring onion (Hotel E)
2 Chiu Chow	0	
3 Peking	0	
4 Shanghainese	1	Shanghai pork belly (Hotel C)
5 Szechuan	0	
6 Hunan	0	
7 Pseudo Chinese	1	Jasmine tea poached scallops (Hotel K)
8 Fusion	1	Asian style confit duck Maryland (Hotel N)
9 Indeterminable Chinese	2	Chinese congee (Hotel E); traditional roast duck (Hotel C)

Notes:

\# Abalone congee could refer to Jeonbokjuk, which is a Korean porridge, or the rice congee in Cantonese style (as alluded to by Tann and Wheeler, 1980) flavoured with abalone

* *Teo Chew* is a municipality in eastern Guangdong thereby rendering it Cantonese in cuisine categorisation

\+ Hainan was until 1988 a part of Guangdong Province and, although it is now part of the Hainan Province, its cuisine would fall into the Cantonese category

Gleaning from discussions with his Chinese students, Reviewer 5 thought the heterogeneous nature of Chinese sub-regional cuisine is very evident in Chinese restaurants in Perth and that many people in the West made the incorrect assumption that a type of cuisine, for example Yum Char, which is Cantonese, is liked by all Chinese people. Segmentation of national cuisine is not well represented in the sample. This reviewer also noticed that some hotels in the sample offered European/ Australian fare, devoid of any other ethnic cuisines. While those RS menus, in his opinion, were good ones, he found it very odd that some hotels did not to any degree offer their Asian guests any choice other than what was offered to European and domestic guests.

All the reviewers felt that the availability of Chinese food in the RS menus was highly variable with some hotels offering no Chinese or Asian cuisine whatsoever (13 hotels) to some hotels that had menu categories featuring Asian flavours. Notwithstanding that some hotels in the sample are very popular with Asian travellers and this was clearly reflected in their Asian menu offerings, the choice was still very limited.

Discussion

The dearth of authentic Chinese dishes, apart from congee (Hotels C & E; Hotel C offered two types of congee – abalone and *Teo Chew*) and steamed barbeque pork bun (Hotel C), negated meaningful categorisation into the Chinese food categories.

These data suggest that the Chinese food available from Perth hotel RS menus is predominantly Cantonese in orientation. Given anecdotal data that palates across China's regions are diverse, and that Chinese people outside of Guangdong Province may not appreciate Cantonese cuisine, hotels need to be attuned to where the guests originate from. Chow and Murphy (2007, p. 61) noted that differences in travel activity preferences of Chinese outbound tourists for overseas destinations meant that the 'Chinese market should not be treated as a single homogeneous entity'. If the inbound Chinese tourists to Perth are not from Guangdong Province, Perth hotels might currently not appropriately be accommodating the palates of their Chinese guests.

The diverse interpretation of what constitutes a Chinese dish as opposed to an Asian dish possibly points to the prevalence of culinary alternative conceptions.

Culinary alternative conceptions

Spring roll

Hotel T listed an item called "Vietnamese rice paper prawn spring roll, nuoc cham" which explicitly identified the product as a particular type of spring roll distinctly different to the Chinese spring roll. Yang and Zhang's (2010) study on Canadian consumers showed that spring rolls together with rice, the hot pot, Chinese dumplings and noodles were the popular Chinese dishes. The characterisation of spring rolls as a Chinese dish is evident in the literature (e.g. Ding, Grewal & Liechty, 2005; Smith, Markandu, Rotellar, Elder & MacGregor, 2002). Smith et al. (2002, p. 1205) also refer to pseudo Chinese dishes such as 'chicken-and-mushroom soup, chicken chop suey, sweet-and-sour pork, spring roll, and fried rice'. The spring roll that the Australian consumer is probably most familiar with is the frozen product of a major food service supplier which claims '(our) "Spring Roll" has long been an Aussie favourite bought from the local takeaway shop'(Marathon, n.d.).

Lumpia is incorrectly referred to as the 'spring roll' served in Chinese restaurants in the West (Wu & Cheung, 2004), which is the Cantonese dish of '*cheun gyun*' which are shredded pork and mixed vegetables. Highly indigenised throughout Southeast Asia, the spring roll no longer exclusively refers to the original dish: lumpia is further regionally different in the Philippines (Fernandez, 2004) and called Lumpia-Shanghai in Indonesia (Wu & Cheung, 2004). 'Creolization' of Chinese food in Southeast Asia (Li, p. xiii) is evident in the literature and further confuses how Chinese food is perceived and is authenticated. Wu and Cheung (2004, p. 16) explain that Chinese food is differentiated in China 'according to regional/ethnic division – the Shanghai-style spring roll (*chunjuan* in Mandarin) is fried and served hot, while lumpia in the Minnan region or in Taiwan is non-fried and served cold.'

Spring rolls are considered as 'Thai-conventional snacks' and composed of spring roll pastry with minced pork, vermicelli, garlic, bean sprouts, carrots, Jew's ear mushroom, shiitake mushroom, soy sauce, oyster sauce and sugar (Komthong, Suriyaphan & Charoenpanich, 2012, p. 20).

When in China, Western tourists prefer sweet and sour pork, spring rolls and corn soup, all of which are widely available in Thai and Chinese restaurants abroad, but are not necessarily popular among the locals in their home countries (Cohen & Avieli, 2004).

Hainanese chicken

Hainanese chicken is known as a Chinese dish and is well known in Southeast Asia (Swaddiwudhipong, Hannarong, Peanumlom, Pittayawonganon & Sitthi, 2012). According to Tan (2001, p. 131), Hainanese Chicken Rice could have conceivably 'been invented by some Hainanese cooks in the Malay Peninsula or Singapore'. Kugiya (2010) provides a different opinion by stating that why the dish originated in the province of Hainan, it has been 'Creolised' and is claimed by Singapore as its national dish and even served on board its national carrier, Singapore Airlines. Hence, a Chinese geographical tag may not necessarily characterise a dish as being natal and indigenous to that location.

Dim Sim

One reviewer remarked that the menus appeared to primarily cater to the domestic guest as the menus were engineered to suit the palate, and by extension, the expectations of Australians who have been conditioned to Asian cuisines. Many items listed may relate to domestic guests, for example the use of the term 'dim sim' had entered the Australian lexicon which describes 'a type of deep-fried cabbage and meat filled wonton, popular in Australian Chinese restaurnts (sic) and fast food joints' which has Chinese origins (Lambert, 2010). The reviewers acknowledged the resemblance of dim sim to a *dim sum* dish called *siu mai*, which is a pork dumpling and an integral part of *yum cha*, which is peculiar to the Guangzhou province (Zhang & Long, 2014). Whilst a dim sim is immediately identifiable to Australians, it might be confusing to a tourist from southern China, or even those from Southeast Asia who are familiar with *yum cha* as it is a facsimile of a *dim sum* dish and yet is different in presentation and flavour.

Conclusion

The primary question driving this research was whether Perth hotels offered Chinese food in their RS menus. On the assumption that mainland Chinese tourists would appreciate the availability of Chinese food in their hotels based on the literature relating to Chinese tourists' dietary patterns and food-related certification criteria of China Ready® Accreditation, hotels that wish to attract this burgeoning market should provide dishes that are amenable to their target market.

In this study we found that Perth hotels do not offer a wide range of Chinese cuisine as part of the in-room product offering. It is likely that RS is viewed by hotels as an augmented product offering and not a core product offering. Our reviewers noticed that the dominant culture of one region shapes the assumptions and decision-making of WA hotels. Hoteliers need to be mindful of the peculiar needs of tourists as they evolve, including in-room dining products, which should be given more attention because of its potential to become a 'motivation' factor.

Our literature review shows that Chinese food is highly diverse and region-ally distinct. It was difficult for the reviewers to categorise the menu items that were identified as Chinese into the sub-regional food categories because those items did not clearly fall into categories, apart from one dish, *Teo Chew* Congee, which can be categorically labelled as a Cantonese dish. The inherent ambiguity in menu item naming and description demonstrates a lack of understanding by the reviewers with regard to the various regional cuisines despite their appreciation for Chinese food.

On the premises of a large healthy appetite for Chinese food and that hotels have the operational capability to serve it via RS, we recommend that hoteliers strive to be aware who their Chinese guests are in order to cater for their peculiar sub-regional palates. While the majority of Chinese tourists presently originate from Tier 1 cities as seen in Figure 9.1, the rest of China's massive market has yet to be tapped. It is, therefore, advantageous to be able to anticipate the culinary expectations of the sub-segments of the Chinese market as they emerge and grow.

The findings of this study suggest that Perth hotels are not currently 'China ready' for the mainland Chinese guests' palate particularly given the diversity of Chinese cuisines. However, it should be noted that the hospitality sector in WA is continually evolving and that hotels are continuously redesigning their product offerings to suit the changing marketplace; furthermore, great steps have been made particularly with new properties that have come on line within the past two years. We perceive that 'being Chinese ready' is actually a 'moving feast', therefore, the realignment of the in-room dining product offering should continue particularly in an increasingly competitive lodging industry and discern-ing Chinese guest. This research also suggests that in-room dining in Perth hotels may require a more thorough investigation and RS menus need to be realigned to the expectations of regional visitors and inbound tourism target markets.

Limitations

This study by reviewers who are academics in the marketing, hospitality, tourism and events fields is intended to be a catalyst for further research in the area of visi-tor palates and catering to them. This initial study involved reviewers who were not Chinese in ethnicity. However, subsequent studies involve ethnic Chinese perspectives from academic and customer viewpoints. The study focused on RS menus as an element of the broader hotel F&B offering. Whereas a number of hotels involved in this study offered food from their main F&B outlets through RS, others had dedicated RS operations albeit with a limited menu which might not contain any Chinese dishes despite possibly served in the restaurant(s). It should be noted that hotels in the sample may have a limited range of certain foods due to seasonality of inbound tourist patterns. The period of the study was not the peak inbound Chinese tourism season.

References

Acott, K. (2017). Rich Chinese eye WA holiday. *The West Australian*, Feb 6, 2017. Retrieved from https://thewest.com.au/news/wa/rich-chinese-eye-wa-holidays-ng-b88373292z

Atkins, P. & Bowler, I. (2001). *Food in Society*. London: Arnold.

Chang, R.C.Y, Kivela, J. & Mak, A.H.N. (2010). Food preferences of Chinese tourists. *Annals of Tourism Research*, *37*(4), 989–1011.

Chang, R.C.Y., Kivela, J. & Mak, A.H.N. (2011). Attributes that influence the evaluation of travel dining experience: when East meets West. *Tourism Management*, *32*(2), 307–316.

China Ready & Accredited® (2016). Retrieved from http://chinareadyandaccredited.com/about-us/

Chinatowns across Australia (n.d.). Retrieved from http://australia.gov.au/aboutaustralia/australian-story/chinatowns-across-australia

Chow, I. & Murphy, P. (2008). Travel activity preferences of Chinese outbound tourists for overseas destinations. *Journal of Hospitality & Leisure Marketing*, *16*(1–2), 61–80.

Cohen, E. (1972). Toward a sociology of international tourism. *Social Research*, *39*, 174–182.

Cohen, E. & Avieli, N. (2004). Food in tourism: attraction and impediment. *Annals of Tourism Research*, *31*(4), 755–778.

DeShields Jr., O.W., Kara, A. & Kaynak, E. (2005). Determinants of business student satisfaction and retention in higher education: applying Herzberg's two-factor theory. *The International Journal of Educational Management*, *19*(2), 128–139.

Ding, M., Grewal, R. & Liechty, J. (2005). Incentive-aligned conjoint analysis. *Journal of Marketing Research*, *42*(1), 67–82.

Fernandez, D.G. (2004). Chinese food in the Philippines. In D.Y.H. Wu & S.C.H. Cheung (eds.) *The Globalisation of Chinese Food*, 183.

Fieldhouse, P. (1986). *Food and Nutrition: Customs and Culture*. New Hampshire: Croom Helm.

Fischler, C. (1988). Food, self and identity. *Social Science Information*, *27*, 275–292.

Herzberg, F., Mausner, B. & Snyderman, B. (1959). *The Motivation to Work*. New York, NY: Wiley.

Hsiung, D-T. & Simonds, N. (2005). *The Food of China*. Millers Point, NSW: Murdoch Books.

Hsu, C.H., Kang, S.K. & Lam, T. (2006). Reference group influences among Chinese travellers. *Journal of Travel Research*, *44*(4), 474–484.

Komthong, P., Suriyaphan, O. & Charoenpanich, J. (2012). Determination of acrylamide in Thai-conventional snacks from Nong Mon market, Chonburi using GC-MS technique. *Food Additives and Contaminants: Part B*, *5*(1), 20–28.

Kugiya, H. (2010). Singapore's national dish: Hainan chicken rice, Crosscut.com. Retrieved from http://crosscut.com/2010/03/18/food/19683/Singapores-national-dish-Hainan-chicken-rice/

Lambert, J. (2010). Additions to the Australian lexicographical record III. Australian National Dictionary Centre, Australian National University. Retrieved from http://andc.anu.edu.au/australian-words/lambert-additions-corrections/third

Law, R. & Cheung, S. (2010). The perceived destination image of Hong Kong as revealed in the travel blogs of mainland Chinese tourists. *International Journal of Hospitality &Tourism Administration*, *11*(4), 303–327.

Li, X.R., Lai, C., Harrill, R., Kline, S. & Wang, L. (2011). When East meets West: an exploratory study on Chinese outbound tourists' travel expectations. *Tourism Management*, *32*(4), 741–749.

Li, Y-Y. (2001). Foreword. In D.Y.H. Wu & C-B. Tan (eds.) *Changing Chinese Foodways in Asia*. Hong Kong: The Chinese University Press.

Liang, S.-C. (1985). *Ya-she Tan-chi* [My Experience of Food]. Taipei: Jiuge.

Liu, C. & Hull, J.S. (2015). ADS tour operators' perspective of the Chinese tourism market and sustainable strategies for developing the Auckland city destination. *International Journal of Tourism Cities*, *1*(3), 254–268.

McKercher, B. & Chow, S.M.B. (2001). Cultural distance and participation in cultural tourism. *Pacific Tourism Review*, *5*(1), 23–32.

Mak, A.H.N., Lumbers, M. & Eves, A. (2012). Globalisation and food consumption in tourism. *Annals of Tourism Research*, *39*(1), 171–196.

Mäkelä, J. (2000). Cultural definitions of the meal. In H.L. Meiselman (ed.) *Dimensions of the Meal: The Science, Culture, Business, and Art of eating* (pp. 7–18). Gaithersburg, ML: Aspen Publication.

Ninemeier, J. & Purdue, J. (2008). *Discovering Hospitality and Tourism: The World's Greatest Industry*. Upper Saddle River, NJ: Pearson Prentice-Hall.

Ogle, A. (2009). Making sense of the hotel guestroom. *Journal of Retail & Leisure Property*, *8*(3), 159–172.

Okumus, B., Okumus, F. & McKercher, B. (2007). Incorporating local and international cuisines in the marketing of tourism destinations: the cases of Hong Kong and Turkey. *Tourism Management*, *28*(1), 253–261.

Reisinger, Y. & Turner, L.W. (2002). Cultural differences between Asian tourist markets and Australian hosts: Part 1. *Journal of Travel Research*, *40*(3), 295–315.

Simpson, P.M. & Siguaw, J.A. (2008). Destination word of mouth: the role of traveller type, residents and identity salience. *Journal of Travel Research*, *47*(2), 167–182.

Smith, S.J., Markandu, N.D., Rotellar, C., Elder, D.M. & MacGregor, G.A. (1982). A new or old Chinese restaurant syndrome? *British Medical Journal* (Clinical research ed.), *285*(6349), 1205.

State Government lands Chinese air deal (2014). Retrieved from http://www.media statements.wa.gov.au/Pages/StatementDetails.aspx?listName=StatementsBarnett&Sta tId=8511

Suboleski, S.D. (2012). *Room Service Principles and Practices: An Exploratory Study* (unpublished doctoral dissertation). Las Vegas: University of Nevada.

Swaddiwudhipong, W., Hannarong, S., Peanumlom, P., Pittayawonganon, C. & Sitthi, W. (2012). Two consecutive outbreaks of food-borne cholera associated with consumption of chicken rice in northwestern Thailand. *Southeast Asian Journal of Tropical Medicine & Public Health*, *43*(4), 927–932.

Tann, S.P. & Wheeler, E.F. (1980). Food intakes and growth of young Chinese children in London. *Journal of Public Health*, *2*(1), 20–24.

TravelChinaGuide (n.d.). Chinese Food. Retrieved from http://www.travelchinaguide. com/intro/cuisine_drink/cuisine/

Tourism Australia (2012, March). *Knowing the Customer in China: Research Update*. Australia: TA.

Tourism Australia (2013). *China Market Profile*. Australia: TA. Retrieved from http:// www.tourism.australia.com/documents/Markets/MarketProfile_China_May14.pdf

Tourism Australia (2014). *China Market Profile*. Australia: TA. Retrieved from http:// www.tourism.australia.com/documents/Markets/MP-2013_China-Web.pdf   Retrieved from http://www.tourism.australia.com/documents/Markets/China2020-Building_the_Foundations.pdf

Tourism Australia (2017). *China Market Profile*. Retrieved from http://www.tourism.australia.com/content/dam/assets/document/1/6/x/g/p/2002921.pdf

Tourism Research Australia (2014). *Chinese Satisfaction Survey*. Australia: TRA. Retrieved from http://www.tra.gov.au/documents/Chinese_Satisfaction_Survey_FULL_REPORT_FINAL_24JAN2014.pdf

Tourism Western Australia (2012). *Our Direction in China*. Western Australia: TWA. Retrieved from http://www.tourism.wa.gov.au/Industry/Documents/TWA_China_Strategy_Summary.pdf

Tourism Western Australia (2013). *China Visitor Profile*. Western Australia: TWA. Retrieved from http://www.tourism.wa.gov.au/Research-Reports/Facts-Profiles/Documents/International%20Visitor%20Profiles/China%20fact%20sheet%20YE%20December%202013.pdf

Wu, D.Y.H. & Cheung, S.C. (2002). The globalisation of Chinese food and cuisine. *The Globalisation of Chinese Food*. Honolulu, HI: University of Hawaii Press.

Yan, M. (1984). *Chinese Cooking: Step-by-Step Techniques*. New York: Random House.

Yang, C.L., Khoo-Lattimore, C. & Lai, M.Y. (2014). Eat to live or live to eat? Mapping food and eating perception of Malaysian Chinese. *Journal of Hospitality Marketing & Management*, 23(6), 579–600.

Yang, S. & Zhang, Y. (2010). The research of the differences between Chinese and Western diet cultures. *Cross-Cultural Communication*, 6(2), 75–83.

Yu, X. & Weiler, B. (2001). Mainland Chinese pleasure travellers to Australia: a leisure behaviour analysis. *Tourism, Culture & Communication*, 3(2), 81–91.

Zhang, X., Dagevos, H., He, Y., Van der Lans, I. & Zhai, F. (2008). Consumption and corpulence in China: a consumer segmentation study based on the food perspective. *Food Policy*, 33(1), 37–47.

Zhang, Y. & Long, M. (2014). The role of Yum Cha (Cantonese morning tea) in the integration process among interprovincial migration in China. *Leisure Studies*, (ahead-of-print), 1–8, DOI: 10.1080/02614367.2014.962587.

Zhu, W. (2013). Eight regional cuisines of China. Retrieved from http://www.theworldofchinese.com/2013/05/eight-regional-cuisines-of-china/

Appendix 9.1

China Ready & Accredited® Standard

The detailed criteria for certification of each business sector and level of accreditation are set out below.

FOOD AND BEVERAGE

Bronze Accreditation

- Company employs at least one certified CRA individual member (who is likely to engage face to face with Chinese consumers).
- Food and beverage menus translated into Chinese in a meaningful way.
- Company to provide access to the China Ready & Accredited® 24/7 Helpline.

- At least one web image has been supplied in high resolution (jpg or png format) in 667 x 510 size to feature on the China Ready & Accredited® website.

Silver Accreditation

- Website/webpage of the company is to be translated into Chinese in a meaningful way.

ACCOMMODATION

Silver Accreditation

- Company employs at least one CRA individual member (who is likely to engage face to face with Chinese consumers).
- Company to employ at least one person who is fluent in the Chinese language or company to provide access to a 24/7 Chinese interpreter phone service.
- Company to provide access to the China Ready & Accredited® 24/7 Helpline.
- Website/webpage of the company is to be translated into Chinese in a meaningful way.
- All essential information, products, services and facilities translated into Chinese in a meaningful way.
- Company to provide to Chinese customers on arrival:

  - toothbrush
  - toothpaste
  - Chinese instant noodles
  - a copy of translated products/services/facilities and essential safety information

- Signage, noticeboards and/or banners with the China Ready & Accredited® logo are located in prominent and visible locations.
- At least one web image has been supplied in high resolution (jpg or png format) in 667 x 510 size to feature on the China Ready & Accredited® website.

TRANSPORT & GETTING AROUND

Bronze Accreditation

- Company employs at least one CRA individual member (who is likely to engage face to face with Chinese consumers).
- All essential information, products, services and facilities to be translated into Chinese in a meaningful way.
- Company to provide access to the China Ready & Accredited® 24/7 Helpline.
- Waiver/Liability forms translated by a NATTI accredited translator.
- At least one web image has been supplied in high resolution (jpg or png format) in 667 x 510 size to feature on the China Ready & Accredited® website.

Silver Accreditation

- Website/webpage of the company is to be translated into Chinese in a meaningful way.

Appendix 9.2

Hotel (n = 20)	Hotel Star Rating	Affiliation	F&B Operation	Additional Charges	Reviewer 1 (UK; Male; Senior Lecturer)	Reviewer 2 (UK; Male; Honours Candidate) Menu	1 Chinese	1 Generic Asian	Reviewer 3 (AU; Male; Lecturer/Chef)	Reviewer 4 (MAL; Male; Lecturer)	Reviewer 5 (AU; Male; Lecturer)
A	5	Chain	HR; RS		No discernible Chinese dishes are offered						
B	5	Chain	HR; RS	$5.00 service fee	No discernible Chinese dishes are offered, but two Asian dishes are listed in the room 'in the 'in dining menu' (Nasi Goreng and Thai prawn and Rice). Green tea and Jasmine tea may tempt the Chinese visitor.	All Day	– –	4 – Taste of Asia plate – Sweet miso fish w/ – Asian veg – Nasi goreng – Thai prawn	Tastes of Asia Beef satay, spicy wings, vegetable samosa ans spring rolls served on Asian slaw Thai prawn and rice noodle salad with nam jim dressing	Spring rolls? (as part of the Tasts of Asia starter). Green Tea with Jasmine is available as a beverage (although not explicitly a Chinese Jasmine Tea)	This menu treats Asians as a homogeneous group and as a result can offend Asian tourists especially in relation to religious conventions.
C	5	Chain	HR; RS; O		Several Asian inspired breakfast items are offered, whereas 'A La Carte' is very restrictive, although there are a number of Asian dishes described as 'Asian specialities' in the all-day dining menu.	Breakfast 5	– Fish ball & vermicelli soup – Wok fried bee hoon etc. – Steamed BBQ pork bun – Abalone congee – Teo Chew congee	1 – Nasi lemak	BREAKFAST: Asian Breakfast Nasi Lemak coconut flavoured rice with chicken curry & condiments; Fish ball flavoured with vegetable vermicelli soup; Wok fried bee hoon, prawn, fish cake, sprouts, soy sauce, and spring onion; Steamed barbeque pork bun; Abalone congee with condiments; Teo Chew congee with condiments Specialities Fish ball with vegetable vermicelli soup; Hainanese chicken with fragrant jasmine rice; Barbeque pork served with fragrant jasmine rice; Barbeque pork served with egg noodles or steamed rice and bok choy; Traditional roast duck with egg noodle or steamed rice and bok choy; ?Wok fried bee hoon, prawn, fish cake, sprouts, soy sauce, and spring onion; Shanghai pork belly in Chinese wine withsteamed rice LATE NIGHT MENU ?Nasi goreng - spicy fried rice, char grilled chicken satay, ground peanut sauce, fried egg	Steamed barbeque pork bun & Abalone congee with condiments (Asian Breakfast). Jasmine tea. Hainanese chicken with fragrant jasmine rice [NB: this is not a Chinese dish per se]. Barbeque pork served with egg noodle or steamed rice and bok choy. Traditional roast duck with egg noodle of steamed rice and bok choy?, Shanghai pork belly in Chinese wine with steamed rice? (All Day Dining)	Steamed barbeque bun appears to be authentic. The Chinese style food offering in the Breakfast menu is limited but again it should be because the restaurant is French. This also applies to the All Day Dining menu. The Late Night Menu is devoid of Chinese dishes.
						All Day 6	– Fish ball & vermicelli soup – Hainanese chicken – BBQ pork & egg noodles – Roast duck & egg noodles – Wok fried bee hoon etc. – Shanghai pork belly	2 – Nasi goreng Penang keow teow			
						Late	– –	1 – Nasi goreng			

D	5	IND	HR		No Chinese dishes are offered, other than steamed broccoli as a side dish.
E	5	Chain	HR; RS	Breakfast 1 Chinese congee All Day 6 – Shanghai rice noodles – Wok fried prawns – Pork knuckle **SIDES** – Prawn crackers – Kai lan – Steamed jasmine rice 11 – Chicken & corn soup – Veg spring rolls – Mixed satays – Nasi goreng – Singapore rice noodles – Red curry chicken – Stir fried chicken – Thai green chicken curry – Beef rending – Steamed fish fillet w/ginger **SIDES** – Asian greens BREAKFAST: Mains Chinese Congee With selection of condiments With Chicken $x $x + 3 JOE'S ORIENTAL DINER RICE AND NOODLE DISHES ?Shanghai rice noodles with mixed vegetables and sesame seeds, chilli sauce WOK DISHES Stir fried chicken with dried chilli, cashew nuts and steamed rice Steamed fish fillet with ginger and spring onion, topped with soya sesame sauce Chinese Congee (In-Room Dining; Breakfast available 24 hours). Chinese jasmine tea available. Crispy vegetarian spring rolls? Shanghai rice noodles with mixed vegetables and sasame seeds, chilli sauce? Braised pork knuckle with star anise, cinnamon, fennel seeds, cloves and Peppercorns in soy sauce? Stir fried chicken with dried chilli, cashew nuts and steamed rice? Steamed jasmine rice (All Day Menu side dishes)	Amongst the various menus offered Joe's oriental dinner offers a small selection of rice and noodle dishes, which are not available as room service. Chinese congee is authentic but it would be limited and would not be sufficient variety for a multiple night stay. There are some rice and noodle dishes that appear to be Western versions of Asian food.

(continued)

(continued)

Hotel (n = 20)	Hotel Star Rating	Affiliation	F&B Operation	Additional Charges	Reviewer 1 (UK; Male; Senior Lecturer)	Reviewer 2 (UK; Male; Honours Candidate) Menu	1 Chinese	1 Generic Asian	Reviewer 3 (AU; Male; Lecturer/Chef)	Reviewer 4 (MAL; Male; Lecturer)	Reviewer 5 (AU; Male; Lecturer)
F	5	Chain	HR		No evidence whatsoever of dishes that might appeal to a Chinese visitor.	All Day	– –	1 – Beef rendang		Steamed rice (All Day Menu side dishes)	
G	5	Chain	HR; RS		Nothing on either the all-day menu or night menu is fashioned towards meeting the needs of a Chinese diet.	All Day 2 – Wok tossed Udon noodles SIDES – Wok fried vegetables	2 – Green Thai chicken curry – Laksa etc.		All Day Menu (Mains) ?Wok tossed Udon noodles with chicken, bean sprouts, spring onions and oyster sauce	Wok fried vegetables (All Day Menu sides).	Wok tossed Udon noodles.
H	5	Chain	O		Nothing on either the breakfast menu or lunch menu is fashioned towards meeting the needs of a Chinese diet.					NB: Asian flavours evoked eg. Crispy nori with tuna tartare and toasted seame seeds, Chilli prawn with roast capsicum, salsa verde. Green tea is available.	
I	4.5	IND	HR; RS	$5.95 service delivery fee	There is no evidence of any of the room service menus catering for the needs of potential Chinese culinary tastes.	All Day 2 – Pork belly – Lemon pepper squid	2 – BBQ Beef sashimi salad – Soft shell crab.			Pork Belly: Twice cooked slow braised ans roasted glazed belly of porrk with stir fried bok choy, cashew and ginger carrot puree (Entrée). Lemon Pepper Squid: Crispy Lemon pepper rice flour dusted calamari served with Asian salad and nam jin dressing? (Snacks).	

		Type	Service	Price	Comments	Meal			Dish	Notes
J	4.5	IND	HR; RS		There is no evidence of any of the room service menus catering for the needs of potential Chinese culinary tastes.	All Day	— —	1 —	Beef salad	NB: Asian flavours evoked eg. Beef Salad: beef skewers, rice noodles, mixed greens, asian herbs, vietnamese dressing (Light Meals)
K	4.5	Chain	HR; RS		Jasmine tea poached scallops may appeal to a Chinese palate (Entrée), otherwise there is very limited choice in the other menus and no option within the Sleep Walker menu.	All Day	— —	2 —	Beef sirloin w/ Thai aromats; Poached chicken, quail eggs, shitake mushroom, somen noodles	NB: Chinese flavours evoked: Jasmine tea poached scallops: minted pea puree, prosciutio, shiso and baby basil, Twice cooked pork belly: with glutinous black rice, orange and chilli, tatsoi salad (From The Restaurant). Seasonal steamed vegetables (From the Restaurant side dishes) NB: Kirin (Japanese) beer available
							—		Jasmine tea scallops	
						Late	1 —	Pork belly		
							1 —	Jasmine tea scallops		
L	4.5	Chain	HR	$5.50	There facility does not offer room service and there is very little of any menu items catering for the needs of potential Chinese culinary tastes. That is, other than Asian vegetables sautéed in teriyaki sauce.	All Day	1 ENTRÉE — S&P Squid	6 ENTRÉE — Pumpkin soup — Thai seafood salad **MAIN** — Pan fried Barra w/ Asian veg **SIDES** — Asian veg — Kimchi — King oyster etc.	Vegetable Fried Rice; Saffron Rice; King Oyster & Shiitake Mushroom Sauteed in Soy Ginger (Lunch & Dinner Sides). Jasmine Green Tea is available.	Asian vegetables sauteed in teriyaki sauce whilst Asian in essence is a using a Japanese theme.

(continued)

Hotel (n=20)	Star Rating	Affiliation	F&B Operation	Additional Charges	Reviewer 1 (UK; Male; Senior Lecturer)	Reviewer 2 (UK; Male; Honours Candidate)			Reviewer 3 (AU; Male; Lecturer/Chef)	Reviewer 4 (MAL; Male; Lecturer)	Reviewer 5 (AU; Male; Lecturer)
						Menu	1 Chinese	1 Generic Asian			
M	4.5	Chain	HR; RS	$4.00 tray charge – all a la carte orders (5:30 to 9:30 pm)	In room dining is offered at this facility, but there is no evidence on the menu items listed that would meet the needs of Chinese culinary tastes.	All Day	2 ENTRÉE – S&P Squid BAR – S&P Squid	1 BAR – Tempura prawns		NB: Chinese flavours involved e.g. Salt & Pepper Squid: coated in sea salt and Szechuan pepper served with garden salad, lemon and aioli (A la Carte).	
N	4.5	Chain	HR; RS?	$4.00 Service charge (food)	Within the 'IN Room Dining Menu', Asian vegetables are offered as a stir fry dish and within the "classics Menu', Asian style confit duck is available.	All Day	1 – Asian confit duck	1 SIDES – Stir fry veg		Asian Style Confit Duck: Confit duck maryland with stir fry wonton noodle, bok choy, kai lan, fresh coriander and soy sesame dressing (Classics). Green Tea & Jasmine is available.	Asian style but uses terminology Maryland which would be confusing to a Chinese visitor because it is an American dish
O	4.5	Chain	HR; RS	$4.50 tray charge	The "IN ROOM DINING MENU" on offer from 10pm until 6am. there are no discernible Asian inspired dishes	All Day	1 – –	1 – Cuttlefish w/ wasabi		NB: Asian flavour evoked e.g. Cuttlefish: Spiced crispy cuttlefish with wasabi mayonnaise (In-Room Dining).	
P	4	Chain	HR; RS	$6.00 tray charge (orders < $10.00; $5.00 Surcharge on PHs	Room service at this facility is offered from 11am until 10 pm, but there is no options for Chinese palates	All Day	1 – S&P Calamari	1 – Thai beef salad			

Q	4	IND	HR	There is no evidence of any of the menu items catering for the needs of potential Chinese culinary tastes			Green tea is available	There are no Chinese dishes and this is acceptable because a the Swedish theme of the hotel.
R	4	Chain	HR; RS	The '24/7 Food and Drink' menu offers limited choice of menu items and no choice of Asian inspired menu items.	All Day	1 **ENTRÉE** – S&P Squid	3 **ENTRÉE** – Tempura prawns – Mini chicken satay **SIDES** – Asian marinated beef	NB: Asian flavours evoked e.g. Asian Marinated Beef (Main Course)
S	4	IND	HR	There is no evidence of any of the menu items catering for the needs of potential Chinese culinary tastes	All Day	1 **ENTRÉE** – Pork belly	1 – Skin on barra w/ oriental veg	Twice cooked pork belly: accompanied with local scallop and cauliflower puree? Premium Market Skin On Barramundi: with steamed king prawns & Oriental vegetables and tomato reduction

(continued)

(continued)

Hotel (n = 20)	Star Rating	Affiliation	F&B Operation	Additional Charges	Reviewer 1 (UK; Male; Senior Lecturer)	Reviewer 2 (UK; Male; Honours Candidate)				Reviewer 3 (AU; Male; Lecturer/Chef)	Reviewer 4 (MAL; Male; Lecturer)	Reviewer 5 (AU; Male; Lecturer)
						Menu	1 Chinese	1 Generic Asian				
T	4.5	Chain	HR; RS		This room service based menu offers a number of Asian inspired dishes within the "Taste of Asia" section of the menu and all are either with meat or fish. There is one discernible Indonesian dish 'Nasi goreng', a 'Vietnamese rice paper prawn spring roll, nuoc cham' and an 'Indian style buttered chicken with rice and poppadums'. The 'Overnight Menu' offers no Asian inspired dishes.	All Day Late	2 **ENTRÉE** – Grilled squid salad **MAIN** – Lumpia spring rolls	3 – Chicken satay – Viet spring roll – Nasi goreng 2 – Chicken satay – Nasi goreng		Taste of Asia ?Lumpia spring roll, pork mince, onion, carrot, soy, coriander	Lumpia spring roll, port mince, onion, carrot, soy, coriander? Vietnames rice paper prawn spring roll, nuoc cham (Room Service, Tast of Asia). Use of Szechuan pepper in sides item.	Taste of Asia menu does not represent Chinese cuisine and is stereotyping Asian cuisine.

General Remarks

The reviewer's Chinese friends informed him that the culinary preferences differ according to geographic location.

"Another pattern which I have noticed is that the menus offering few Chinese options often offer Chinese food as an entree or starter only, while menus offering more options tend to include Chinese meals as mains".

"Almost no Chinese. Several with no Asian. Certainly no Chinese beers".

The hotels appear to be catering to domestic guests.

Lumpia is incorrectly referred to as the "spring roll" served in Chinese restaurants. Highly indigenised throughout SEA, the spring roll no longer exclusively refers to the original dish. The "Asian" label is frequently used.

The reviewer observed a lack of variety of food items and little regard for the Chinese palate, regional preferences. He surmised from his analysis that Perth hotels have not catered for the Chinese tourist and this represents a short term rather than long term focus.

Legend

Chain	Chain hotel
IND	Independent hotel

Htl Rest	HR
Rm Svc	RS
Outsourced	O

10 Preference for Australian premium wines in China

The effect of tourism

Graham Ferguson, Isaac Cheah and Sean Lee[1]

Introduction

Getting novice consumers of premium wine, in new markets, to consider brands outside of the most recognised, dominant wine regions is a challenge for premium wine marketers. This is particularly the case for premium Australian wine brands competing in the China market where French wines dominate (Cohen et al., 2017; Lee et al., 2015; Wu, 2017). The China market represents a tantalising opportunity for Australian premium wine exporters (Cohen et al., 2017; Cohen, Corsi & Lockshin, 2015; Maguire and Lim, 2015) because of market potential but the challenge of being accepted by novice wine consumers may dissuade some brands from entering when they can enter much more advanced wine markets in Europe or the USA (Galbraith and Gao, 2016).

Premium wine, like most premium products, offers consumers the opportunity to showcase their social status and to enjoy their success (Cohen, Corsi & Lockshin, 2015; Hall & Mitchell, 2007). But the quality of premium wine is difficult to evaluate and even experienced consumers rely on the recommendations of experts to choose the right products and the right brands (Danner et al., 2016). Consumers new to wine feel even more uncertainty because they do not know where to find reliable assessments and do not have a body of experience to rely on. Bourdieu (1984) would say they are lacking the domain-specific cultural capital (DSCC) required to gain advantage within the premium wine domain. DSCC in this context would be comprised of resources such as knowledge, experience and a network of trustworthy advisors (Maguire & Lim, 2015). In this situation, the novice consumer chooses the most recognised, dominant brands because they represent the lowest risk of damaging status or having a poor experience (Erdem & Swait, 1998; Maguire & Lim, 2015). When these new consumers are part of new markets the effect is even more pronounced because the market is comprised largely of consumers and suppliers with limited DSCC who all begin by relying on dominant brands (Maguire & Lim, 2015). As a consumer's DSCC grows, they rely less on dominant brands because they become more aware of situations where second tier brands can be substituted, that second tier brands offer similar experiential outcomes at a lower cost, and that more specialised or unique brands can be used to signal personal identity (Stiehler et al., 2016).

This leads to a hierarchy of brand cues used by consumers of premium products in new markets. Everyone aspires to the most prestigious brand (e.g. Lafite) but few can afford it for every circumstance (Huang & Lockshin, 2017; Maguire & Lim, 2015). Therefore, they rely on other extrinsic brand cues. At the highest level, this will usually be wine brands that are high priced from a dominant country of origin (COO) (e.g. expensive French wines), then, as DSCC grows, wines may be chosen from smaller wine regions (e.g. Burgundy), and lastly individual wine brands (Cohen, Corsi & Lockshin, 2015; Maguire & Lim, 2015). This is consistent with findings that when brand names are unfamiliar, consumers rely on COO cues (e.g. Chaney, 2000) and even more so for wines where the *terroir* or geographic origin is integral to the wine (Lockshin et al. 2000; Moulard, Babin & Griffin, 2015). Therefore, COO is especially important in determining dominant premium wine brands in new markets. In many product categories, there is one country that is considered better than all the others: Germany for motor vehicle engineering, Japan for electronic goods and Italy for leather goods. Similarly, premium wine consumers in new markets such as China rely on the dominant COO brands (such as France and Italy) for good wine (Muhammad et al., 2014).

During the establishment of a premium wine market, expensive brands from the dominant country will have an advantage because they pose the lowest risk to the consumer (Carsana et al., 2017). In this situation, the COO provides a halo effect (Batra et al., 2000; Reardon, Vianelli & Miller, 2017). Premium brands from other countries or brands positioned as less premium can only be successful in the premium market as consumers develop their DSCC, reducing the risk of having a bad experience, and as the DSCC of the market overall increases, reducing the risk to status (Carsana et al., 2017). Therefore, premium brands from the non-dominant wine country must work hard to increase the consumer's DSCC and the DSCC of the entire market. Marketers of these brands must increase recognition of their brand while clearly positioning it at the premium end of the market in terms of quality and status (Muhammad et al., 2014).

Tourism offers wine marketers the opportunity to rapidly increase DSCC by exposing consumers from a new market to wine drinking in an established wine culture; and to build brand specific knowledge by getting tourists to experience premium local brands (Sørensen & Jensen, 2015). While tourists account for only a small proportion of most nations, they also tend to be more affluent, more willing to engage in new experiences, more worldly, and therefore leaders of consumer trends (Horner & Swarbrooke, 2016). Winery tours and experiences are a major part of tourism activities in countries with strong wine cultures and gastronomic experiences are a growing trend in many countries (Carlsen & Boksberger, 2015; Kim & Bonn, 2016). The key for premium wine brands is to ensure that consumers have a memorable, premium experience at the cellar door, can link the brand strongly to that experience (to aid recall), and can access the wine brand when they return home (Carlsen & Boksberger, 2015; Kim & Bonn, 2016).

China is an important opportunity for Australian premium wine exporters (Cohen et al., 2017; Cohen, Corsi & Lockshin, 2015; Maguire & Lim, 2015).

China's wine imports in the first months of 2017 amounted to 407.37 million litres with a net worth of approximately US$1.74 billon (Wang, 2017). In 2016, Australian wine ranked second behind France in volume (approximately 80 million litres) and logged a 40% increase from the previous year's figures (Wu, 2017). It is the top-ranked destination by value for Australian bottled wine exports to Asia and the third largest market globally, after the United Kingdom and the United States (Galbreath & Gao, 2016). However, Australian wine exporters have little direct influence over Chinese consumers, as wines are sold through distributors and retailers who refuse to dedicate the same amount of space to Australian wines as they allocate to France or Italy (Cohen et al., 2017; Lee et al., 2015). Direct mass advertising is expensive in China's fragmented media landscape (Gou et al., 2014) and is likely to be inappropriate to communicate premium positioning. But there is substantive opportunity for Australia's premium wine brands to target Chinese tourists, to build brand awareness, preference and to become advocates. Of course, there are many types of tourists. The current chapter focuses on long-term tourists who visit a nation for an extended period of time and therefore have a much greater chance to become immersed in the culture and the local brands. For the purposes of this study, long-term tourists were defined as individuals who stayed in Australia for longer than three months (based on the Australian Department of Immigration's maximum stay of three months for a visitor visa). These long-term tourists encompassed both individuals with long-term holiday visas (Visa subclass 600) or student visas.

The next section of the chapter discusses the market for premium wine in China, the concepts of COO, DSCC, brand cues and long-term tourism. A qualitative study then compares the perceptions of premium wine consumers in China to long-term Chinese tourists in Australia. The results of this study are used to test some propositions posited about the importance of COO and the potential of tourism to provide opportunities for Australian premium brands. Implications for Australian premium wine brands are discussed.

Premium wine consumption in urban China

Wine has a favourable image in China, however most Chinese consumers have little wine knowledge or appreciation (Jin, 2004; Maguire & Lim, 2015). The Chinese white spirit culture dominates alcohol consumption in China but younger consumers from the large urban areas are motivated to drink wine for reasons relating to identity and social status, while older generations focus primarily on the potential of wine as a social catalyst (Liu & Murphy, 2007; Maguire & Lim, 2015). Therefore, young consumers who purchase premium red wine to show their individual social status form an important part of the China market (Li et al., 2011; Maguire & Lim, 2015).

In China, where consumers are less accustomed to the complexities of different wines, the perception of an origin and the presentation of origin information in a retail environment can have a great influence on wine sales (Chaney, 2002; Maguire & Lim, 2015). A recent research study by Li et al. (2011) demonstrated that Chinese consumers have a high awareness of alternative brands in established product markets, including foreign brands.

Chinese wine consumers are typically market followers, not leaders (Hu et al., 2008). Chinese wine consumers need to feel safe in the knowledge that the wine they are drinking gives out a desirable image, as opposed to 'losing face' by choosing the wrong wine or admitting their knowledge of wine is limited (Chaney, 2002; Maguire & Lim, 2015). 'Face saving' or '*mianzi*' is one of the important factors that will be considered during decision-making. Being conscious of *mianzi*, directly translated as face, is a key Chinese characteristic; people are conscious of what other people think about them (Graham & Lam, 2003; Sun, D'Alessandro & Johnson, 2014; Zhang, 1996). In this case, being able to afford an expensive wine, especially one that is imported, would show off to others that the individual has a high social status and has succeeded financially (Maguire & Lim, 2015). Liu and Murphy (2007) also found that drinking red wine suggests good social image, elegance and grace, all of which indicate good *mianzi*. Thus, it is important for wine brands to use their packaging and marketing to convey the status of their wine (Williamson et al., 2017).

Hall and Lockshin (2000) found that the intrinsic and extrinsic attributes used to choose a wine brand are related to the wine consumption situation. Therefore, different consumption situations amplify or mute the importance of different wine attributes (Lockshin et al., 2017). Liu and Murphy (2007) conducted in-depth interviews with 15 consumers in Guangzhou and revealed that Chinese consumers tend to purchase wine and consume it during important business situations, or during important holiday celebrations (e.g. Chinese New Year). Similarly, Balestrini and Gamble (2006) reported that buying of wine in Shanghai was generally tied to an 'occasion'.

Domain-specific cultural capital

Cultural capital could be broadly defined as 'the knowledge' of the upper class, an asset that the upper class can use to achieve outcomes that others cannot achieve (Bourdieu, 1984). The idea of DSCC extends cultural capital to mean that people can have the resources in a specific domain even though the capability may not extend to other domains (Kipnis et al., 2014; Tan, 2017). For many products, consumers require a certain amount of cultural capital in order for the right signals to be sent and to enable interpretation of the signals (Kipnis et al., 2014; Nella & Christou, 2014). Domain-specific cultural capital refers to the resources that enable an individual to interpret a cultural code within a specific domain. There is an argument that a person cannot engage in a culture until they have enough cultural capital i.e. consumption literacy (Maciel & Wallendorf, 2016; Wallendorf, 2001).

The role of brands

Brands represent a strong extrinsic cue for consumers (Carsana et al., 2017; Kelley, Hyde & Brewer, 2015). For products that are new, vary a lot or affect social status, brands can communicate a product's positioning in the market and therefore provide a short-cut and more certainty for the consumer (Carsana et al., 2017; Erdem & Swait, 1998). Brands on premium products are especially important as signals of social status (Audrin et al., 2017; Beneke & Zimmerman, 2014; Holt, 1995).

The current study focuses on emerging markets for premium wine. Emerging markets have been established but are early in the growth phase (Winterhalter et al., 2017). In the case of China, the market seems likely to grow and to become substantial. Emerging markets are characterised by novice consumers that have limited experience of the product existing in a whole market of relatively novice consumers (Kipnis, Broderick & Demangeot, 2014). This enhances the consumer's reliance on extrinsic indicators to guide their product selection. Consumers rely on extrinsic factors generally for products that are high in credence properties as these products are difficult to assess prior to, or after, consumption. Wine is one of those products where even experienced consumers rely on the ratings of wine experts, the recommendations of others, prior experience and 'luck' to find good wines (Bruwer, Chrysochou & Lesschaeve, 2017). Michman and Mazze (2006) argue that novice consumers of luxury goods mitigate the substantive perceived risk by seeking recommendations from friends and by purchasing known brands. The consumer's choice depends on the perceived consequences of making the wrong choice, and their confidence in themselves and the retailer to help them make the right choice (Michman & Mazze, 2006).

The China premium wine market is characterised by preference for overseas brands with a small but substantial volume of local brands (Maguire & Lim, 2015). Local brands typically cater to the low end of the market because they have limited brand story and little status as premium wines (Zhang et al., 2013). The premium end of the market is dominated by international brands that are established in the international market and therefore represent lower status risk for consumers (Maguire & Lim, 2015).

Established premium wine brands are attracted to emerging markets because they represent growth opportunities that they cannot gain in mature markets and if the emerging market has a premium consumer class then there is the potential to charge premium prices (Williamson et al., 2017; Zeng & Szolnoki, 2016). Dominant global brands launch in the emerging market far easier than non-dominant brands because even consumers with limited DSCC can recognise or easily find out about the dominant brands.

Country of origin becomes the dominant brand cue

Many studies have indicated that country of origin (COO) bears a significant influence on consumer perception and decision-making (Phau & Prendergast, 1998; Phau & Suntornnond, 2006; Josiassen, 2010; Cheah & Phau, 2015). More specifically, the COO effect refers to how a country's image (e.g. workmanship, innovation and technological advancement) is projected onto the products of the producing country and how it has a substantial impact on the consumers' product judgements and purchase intentions (Aichner, 2014; Maheswaran, 1994;

Papadopoulos & Heslop, 2003). Studies have shown that the COO influences consumers' product evaluations by signalling product quality when they are unable to detect the true quality of a country's product (Huber & McCann, 1982). In particular, the COO can be an indicator of quality when it is difficult to assess by other objective means (Aichner, 2014; Ahmed & d'Astous, 2001; Cheah & Phau, 2015).

As a general consensus, products produced in less developed countries tend to have a less positive image than products from more developed countries (Akdeniz & Kara, 2014; Wang & Chen, 2004). For example, Siu and Chan (1997) found that Chinese consumers in Hong Kong perceived American products to be prestigious, Japanese products to be innovative and Chinese products to be cheap. In addition, past research also indicated that Chinese consumers weigh a product's COO heavily, and perceive a product made outside of China as a strong positive stimulus or attribute to consider while making selection and purchasing decisions (Zhang, 1996). Akdeniz and Kara (2014) found that consumers in developing countries tend to have even more positive perceptions of foreign products if the consumers are favourable to the culture of the country being considered, are susceptible to normative influence or are considering products with high social signalling value.

Specific to wine, Lockshin et al. (2000) indicated that quality-conscious consumers process various perceived signals of quality, mainly of an extrinsic nature, such as price, producer, brand, vintage, region, awards, ratings and recommendations. Studies have suggested that wine is a product associated to a *terroir* or even a country, as such the wine's *terroir* or country of origin is generally viewed as an extrinsic cue (i.e. that the grapes' geographic origin, as a physical place, prompts the consumer to purchase a wine) (Quintal, Thomas & Phau, 2015; Moulard, Babin & Friffin, 2015). Furthermore, wine, it can be argued, may have to conform to 'regional standards', that is it may have to show a typicality of origin to be considered a quality wine (Byrd et al., 2016; Jackson, 1994; Pratt & Sparks, 2014).

Wine consumers were also found to utilise intrinsic cues to aid in the choice process. Williamson et al. (2016) found that quality perception of wine was based on intrinsic cues such as physical attributes including colour, aroma, and taste, alcohol content and wine style which relate to the product and the processing method. However, the average wine consumer is more likely to rely on extrinsic cues such as price or origin when making quality assessments (Lockshin & Rhodus, 1998). More specifically, the importance of extrinsic cues for wine quality judgement occurs especially when consumers have little knowledge and/or feel that some risk remains in the purchase decision for a product (Maciel & Wallendorf, 2016; Michman & Mazze, 2006).

Literature affirms that consumers rely on country-of-origin information in situations when the brand name is unfamiliar (Chaney, 2000; Coskun & Burnaz, 2016). A product's origin is considered one of the more important types of information necessary to assist in consumers' decision-making. When consumers are not familiar with a country's product, they will tend to use a country's image as a sort of 'halo effect' in the product evaluation (Elliot et al., 2010; Lee & Lockshin, 2012). On the other hand, when consumers are familiar with the product, country images serve as summary constructs to reinforce perceptions (Maheswaran, 1994;

Reardon, Vianelli & Miller, 2017). This is especially the case when consumers are faced with several hundred wines and competing brands and are therefore likely to use short cuts to form their choice set. The use of COO information also helps to reduce dissonance in the purchase process (Keown & Casey, 1995; Lee et al., 2017; Reardon, Vianelli & Miller, 2017).

Proposition 1: Consumers with low DSCC, in new markets, will rely more on COO brand cues to direct their product choices.

Tourism builds DSCC

Studies in tourism have highlighted that past experience of a travel destination affects tourists' knowledge of the destination and its products (e.g. Dodd et al., 2005; Horner & Swarbrooke, 2016; Kersetter & Cho, 2004; Sørensen & Jensen, 2015). Through visitation to various travel destinations, tourists gain greater insight into the intricacies of the culture and its cultural products (Horner & Swarbrooke, 2016; Kersetter & Cho, 2004; Sørensen & Jensen, 2015). For instance, wine tourists, who visit wine regions acquaint themselves with the local winemaking heritage and crafting methods which, in turn, increase their cultural capital within the domain of wine consumption (Ye, Zhang & Yuan, 2014). This familiarity, prior knowledge and cultural capital then impacts on their information search behaviour (Alba & Marmorstein, 1987; Chevalier, Maury & Fouquereau, 2014; Kersetter & Cho, 2004).

In the context of wines, a wine tourist's experiences at a destination imbues them with a certain degree of expertise which allows them to be more discerning when purchasing and consuming wine (Bianchi, Drennan & Proud, 2014). As suggested by Reardon, Vianelli and Miller (2017), higher familiarity with a country's product and knowledge is often associated with reduced reliance on COO cues. Increased short- and long-term tourism (including studying abroad) has led many people to be more receptive to the taste and use of red wine, because of their exposure to wine drinking cultures and exposure to red wine on social occasions (Maguire & Lin, 2015). Being integrated into the local wine drinking culture exposes the tourist to wine consumption situations and brands that are an integral part of culture (as per Douglas & Isherwood, 2002).

Proposition 2: Consumers from novice markets who experience the product over an extended period in mature markets reduce their reliance on COO brand cues.

Method

The current study explored the most recent premium wine purchase situations of Chinese consumers who have not toured outside of China, to the most recent purchase situations of Chinese consumers temporarily residing in Australia (long-term tourists). Snowball sampling was used to identify these hard to find consumers which enabled the researchers to gain access to participants that qualified as young consumers of premium wine from a seed group of three premium wine consumers (as per Noy, 2008). The researchers knew the seed participants, two females and one male, aged 27–35, who recommended the subsequent participants. The resulting sample was comprised of five purchasers of premium wine who live in China and six living temporarily in Australia. For the purposes of this study, the respondents must have purchased 10 or more bottles of premium wine in the previous 12 months. Premium wine was described as bottles costing more that AU$60 in Australia or 300 Yuan in China. The respondents were all aged 35 years or younger, seven of the participants were female and four were male, all of the respondents had bachelor degrees or higher levels of education, and all came from major cities in China.

In-depth interviews were undertaken with each participant to identify at least two recent premium wine purchase situations and to uncover the process they went through to choose premium wine in each situation. In-depth interviews allowed the researchers to explore the complex and abstract processes used by Chinese premium wine consumers (Schmidt, 2010), by fostering open conversation and encouraging interviewees to express their ideas, inner thoughts and personal views (Herz, 2015). Interviews were recorded in Mandarin in order to allow the respondents to feel as comfortable as possible and to facilitate open discussion in the means easiest to the participants and then translated into English. Interviews took up to 30 minutes and were conducted face-to-face with respondents in Australia, and by telephone with respondents in China.

The resulting transcripts were analysed with a grounded theory approach to reveal the situations in which young Chinese consumers purchase premium wine and to identify important themes within those purchase situations. The purchase situations described by each participant were synthesised to derive representative 'cases' that depict a typology of purchase situations that confront young Chinese consumers and to explore the way the decisions are made in each of the typologies. A second round of analysis sought to derive themes that address the research propositions. To ensure validity and reliability a second coder re-coded three of the interviews to the same themes with almost complete agreement.

Results

Participants residing in China each described the circumstances around their most recent purchases of premium wine. The participants purchased premium wine for special occasions: specifically, to give as a gift, to be perceived as a good host, to celebrate important occasions with family and friends, and to celebrate extravagantly when out partying. Participants residing in China tended to talk about the wines they chose as being 'French' with very little discussion of the wine itself or specific brands:

When I think of wine, my first thought is French wine.

(Male, 26 years, China)

This is consistent with arguments that premium wine consumers in China are relatively novice consumers of wine and are not (yet) as concerned about buying wine for taste and experience but are concerned about appearing to choose the right wine. Participants focused on mitigating risk by choosing international wines from the right country (usually a French brand) and ensuring that the price is expensive enough to signal the right meaning:

> *If it is a gift to friends I will choose a more expensive wine.*

(Male, 26 years, China)

Despite this need to purchase expensive wines, participants also emphasised getting good value for money:

> *I would buy a French wine if I had the right price, but I only know the brand Lafite, which is very famous and expensive, and I would not buy it. But if there were other French wines with a good price and good wines, I would think about that wine whether it is worthy to buy it or not.*

(Female, 24 years, China)

Participants residing in China were also wary of fake wines:

> *Several times I was going to buy wine from the Internet to buy some foreign brand wine but lots of fake products are selling online, you cannot see the product and taste it. You just have the picture and the reputation from the wine. It may lead to buying the fake products which wastes lots of money.*

(Female, 24 years, China)

Participants temporarily residing in Australia also described situations in which they purchased premium wine. The respondents talked about purchasing premium wine for consumption with others, as gifts, and to spoil themselves.

The most obvious attribute of these interviews was that participants liked and consumed wine more:

> *We eat traditional Western food with wine. Western food always has wine.*

(Male, 28 years, Australia)

This was even the case for participants who hadn't drunk wine in China:

> *When I was in China, there were not many times when I had wine. Even at business dinners or official parties, we normally drink spirits. But having been here a while, there have been a lot of opportunities to drink wine. Maybe because of different wine cultures, or because my friends like to drink wine when we have a party, so I have accepted the taste of wine.*

(Male, 35 years, Australia)

Consumers residing in Australia discussed their wine choices in much more depth and could articulate why they chose a particular brand and product for a particular situation. They were also much more likely to prefer Australian premium wines and also were more knowledgeable about pairing wine with food:

> *I think the price and quality of the wine are important when I buy wine, and I choose different wine for different types of food. For instance, Merlot is perfect for seafood for its strong taste; I'd choose light fruit wine for Chinese food. I think the wine produced in South Australia and Western Australia are of high quality because the grapes from these two sunshine states are good.*
>
> (Female, 34 years, Australia)

They all discussed the importance of choosing a brand that tastes good, that matches the consumption situation and is a brand that is recognised by their friends. Friends were a major influence on brand choice. For example, they visited wineries and built stories about their experience:

> *One of my friends recommended me to buy this wine. So, I drove to Margaret River to buy this brand before returning home. We spent only three-hour driving from Perth by car; I think it is not too far.*
>
> (Male, 28 years, Australia)

However not all participants in Australia showed this attachment to a brand or talked as specifically. They tended to talk about overall perceptions of quality rather than specific product attributes (e.g. taste). They tended to talk about what other people expect (e.g. gift recipients, my friends, etc) and they tended to rely on the perceptions of what is a good brand (e.g. people expect this type of brand) and getting them for a good price:

> *I usually buy Penfolds from Australia for around $120 a bottle. This is a well-known brand in China and it is from Australia, the price is relatively cheaper here, so it seems like a bargain to me.*
>
> (Female, 27 years, Australia)

Typology of wine consumption situations

Type 1: Premium wine as a gift

Gift-giving was the most common reason for purchasing premium wine for all participants. Despite the prevalence of gift-giving for business purposes in China, both sets of participants described only premium wine gift-giving to family and friends for weddings, birthdays and celebrations:

> *If a good friend has a birthday or relatives get married, I will buy red wine as a gift for them.*
>
> (Female, 24 years, China)

> *I don't buy wine often and only do so as birthday gifts for friends, and it's a good gift because we can share it.*

> (Male, 35 years, Australia)

While the Australian participants talked about giving Australian wines, the China participants talked about French wines as the dominant description of the brand they chose:

> *I do not have a particular interest in a wine brand, as long as the price can be appropriate, but most times I will choose French brand because they do good promotions.*

> (Male, 26 years, China)

Participants residing in Australia also preferred Australian wines as gifts in Australia but French wines for gifts in China:

> *I have a good impression of Australian wine after I tasted it. But if I go back to China, I may choose wine which is made in France. I feel French wine is better and more popular in China and it is a more selective choice.*

> (Male, 28 years, Australia)

Although not always:

> *If I send wine to people living in China, I will choose the most famous brands, such as South Australia's Penfolds and Henschke.*

> (Female, 34 years, Australia)

> *Because most of my family members live in China, they do not know much about Australian wines, but I would like to introduce them to the brands, origin, and taste of the wines when we share them together.*

> (Male, 35 years, Australia)

They also acknowledge that French wine appeals a little differently to Australian wine:

> *France's packaging design is delicate, while Australian red wine does not pay too much attention to packaging but pays more attention to the quality of wine.*

> (Male, 35 years, Australia)

The price of the wine was also a strong indicator of its propriety as a gift. All of the interviewees bought more expensive wine for gifts than for other uses. As mentioned by an interviewee:

> *The price must be slightly higher and must look better. Probably the price is between 600–1000 Yuan.*

> (Female, 24 years, China)

As such, Chinese *mianzi* of face dictates that well-known foreign wines are chosen to ensure that the gifts are perceived in a favourable light. The risk of choosing the wrong wine led one participant to note that she avoids wine as a gift:

> *Normally, I will not buy wine. If it is as a gift, I do not like to send wine. I feel it is too weird as I do not have the knowledge about premium wines. What if I buy the wrong one and give it to other people? That is not good.*
>
> (Female, 25 years, China)

All of the China respondents preferred imported wines bought at reputable retailers as gifts, perhaps because of scepticism about fake products:

> *If you buy fake wine as gift to send your friends, it is bad.*
>
> (Female, 24 years, China)

The COO of the wines chosen as gifts depends on the expectations of the recipient. Two of the China respondents gave French wine as gifts to match the expectations of the gift recipients. But not all participants purchased French wines, with one participant in China gifting Penfolds Australian wine because a key member of their friend group has extensive experience of wine and considers that this is the best Australian brand and they are much better value for money:

> *When I buy wine as a gift I choose the Penfolds brand. Because my friend said this brand of wine is cost-effective and the quality of this brand is better than others at the same price.*
>
> (Female, 24 years, China)

Type 2: Good hosts provide good wine

Being a good host, especially to women, was a common reason given for purchasing premium wine:

> *If I choose to buy wine that may only be given as a gift to someone else, or as a host I invite guests to eat food at home, women and most people like to drink red wine.*
>
> (Female, 24 years, China)

> *Most people who don't drink are women. They like wine instead of spirits when they must drink.*
>
> (Male, 24 years, China)

Type 3: Celebrations with family and friends

Participants in China and Australia bought premium wine for important festivals celebrated with family or friends. These festivals included Chinese New Year and Spring Festival which are predominantly family or close-friend occasions in Chinese culture. They are popular celebrations that often involve serving the best fare for the day and offerings are tailored to the tastes of the guests. The choice of what is served also serves a social function to impress their guests (face):

> *We'd buy expensive wine for the Chinese New Year or other traditional festivals because some people don't drink white spirit.*
>
> (Male, 26 years, China)

Only one participant argued that Chinese brands met their needs for special occasions. That participant argued that while Chinese brands did not have the luxury reputation of French brands, the best wines from Zhangyu, a large Chinese wine brand, are reliable and better value for money. Her rationale was as follows:

> *I have had this brand of wine before, thought it was pretty good, and the brand is very famous in China.*
>
> (Female, 24 years, China)

In Australia, the focus is on sharing important celebrations with friends and loved ones:

> *On Valentine's Day, I'd buy expensive wine because I'd have dinner with my girlfriend and the wine shows I have a good taste.*
>
> (Male, 28 years, Australia)

Premium wines were also purchased to share with others by participants living in Australia:

> *A few months ago, I bought myself a bottle of wine, just to drink it with my friend at a normal dinner, no special reason.*
>
> (Female, 34 years, Australia)

Health was also a key reason for purchasing premium wines to share with family. Recently, in China, there has been concern about the negative health effects of the dominant alcoholic drink, white spirit (Chinese *bai jiu*). Chinese consumers are increasingly aware that wine provides a healthier alternative to traditional Chinese spirits. Therefore, some participants referred to the health benefits as a reason for purchasing premium wine. This was particularly so for those young consumers residing in Australia purchasing wine to take back to their aging relatives:

> *I bought a red wine when I left for China and shared it with my parents, because I think it's more healthful than white spirit. My parents are pretty old and I don't want them to drink white spirit too often.*
>
> (Male, 28 years, Australia)

Type 4: Partying with friends

Several respondents purchased premium wine when out celebrating with friends. This is often linked to spoiling oneself and showing status. For example, in China:

> *It was completely impulse consumption; the waiter told me that the brand is good so I bought it, and I thought 'it is my birthday, I can buy a little expensive wine'. I did know it was expensive, it is a luxury brand. The other reason I bought that wine was because I didn't want to lose face.*
>
> (Female, 25 years, China)

And in Australia:

> *It was a Penfolds, Australian made. Around 150 Australian dollars. The party was quite nicely planned with a huge comfortable room and premium food. So, it is polite to bring premium wine as a gift. And all the attendees are relatively rich, so it is better to bring wine that they usually drink or at least know.*
>
> (Female, 26 years, Australia)

What happens when long-term tourists make visits back to China?

Participants residing in Australia also disseminate Australian wine consumption in China when they return:

> *I decided to return to China to celebrate Spring Festival with my family, that is why I bought wines as gifts for relatives and my parents.*
>
> (Male, 28, years Australia)

> *When I return home, I will still buy Australian well-known brands, such as Penfolds.*
>
> (Female, 34, Australia)

The effect of DSCC on COO, region and brand choices

There was a clear difference between the DSCC of the China-based participants compared to the Australia-based participants. This was evident in terms of general knowledge of wines, consumption situations, as well as Australian wine regions and brands. The source of this domain-specific cultural capital in Australia is exposure to a wine drinking culture, memorable experiences at wineries, memorable experiences with family and friends, and repeat experience of the same brand:

> *I often buy wines that are produced in Western Australia's Margaret River, and sometimes there are South Australian wines. My favourite brand is Gralyn. Recently I have felt that Kay Brothers is also good.*
>
> (Female, 34, Australia)

Table 10.1 Premium wine purchase situations and dominant reason for brand choice

ID	Respondent	Preferred premium wine brand	Gifts	To be a good host	Celebrate with family and friends	Partying
China						
1	Male 26 Employed	Wines from France (especially Bordeaux)	Face	Face (business dinner)	Face	
2	Female 24 Employed	Zhangyu brand ('the top China brand')	Value for money	Value for money		
3	Female 25 Employed	Wines from France	Face (wine is too risky as a gift)			Face
4	Female 24 Employed	Penfolds brand	Confidence in brand Value for money		Confidence in brand	Spoil myself
5	Female 29 Employed	Wines from France (especially Burgundy)	Purchase the recipient's preferred brand		Taste and providence	
Australia						
6	Female 26 Student	Moss Wood brand	Face		I identify with the brand Share my knowledge	Face
7	Female 27 Student	Penfolds brand	I identify with the brand	Face Value for money Spoil myself	Face Value for money Spoil myself	Spoil myself
8	Male 28 Student	Voyager Estate brand in Australia ('Australia's premier brand')	Face	Face I identify with the brand	Face	
9	Male 35 Employed	Penfolds brand in China ('I don't buy premium wine in Australia')	Face (for friends in China)		Health (when return home to China) Face (in China) Value for money	
10	Female 34 Employed	Kay Brothers brand in Australia Penfolds brand in China	Purchase the recipient's preferred brand I identify with the brand		I identify with the brand Prefer the taste Share my preference	Spoil myself
11	Male 35 Employed	Gralyn brand	I identify with the brand Prefer the taste		I identify with the brand Share my preference	Spoil myself

*Some respondents discussed purchase of non-premium brands – these are not included in the table.

Past experience in a mature wine market appeared to be the main contributor to knowledge of and preference for Australian wines. Chinese consumers residing temporarily in Australia appeared to have a significant preference for Australian wines. This may be attributed to their consumption of wine in Australia during their stay.

> *In China, DSCC is built in similar ways but there is a lot less wine drinking culture and a lot fewer experts to provide substantial information about a brand. My friend teaches us a lot of knowledge about wine, including how to choose the wine from taste, origin and so on.*
>
> (Female, 24 years, China)

Country of origin

The results also revealed that French wines are preferred by Chinese consumers. They preferred French wines in general although many were unable to clearly state the name of the brand and cited regions instead:

> *At the mention of wine my first thought is French wine but I do not care too much about the origin of wine, because the taste seems the same to me. If the price is right, I will try it no matter the brand or country, but I still tend to prefer France-made.*
>
> (Male, 26 years, China)

However, French wine was reported to have a better reputation in China and therefore prevailed as the first choice for most:

> *I also have a good impression for Australian wine after I tasted it. But if I go back to China, I may choose a wine which is made in France. I feel French wine is better and more popular in China, and it is also viewed as a more selective choice.*
>
> (Male, 28 years, Australia)

Regions and brands

Regions and brands were not mentioned by participants residing in China but were mentioned by nearly all of the participants in Australia. This reflects the increased focus of the participants in Australia on specific wines and brands:

> *Moss Wood winery is based in the Margaret River region, WA. This region is one of the best regions in the world.*
>
> (Female, 26 years, Australia)

Discussion

The social status of a premium wine appears to be crucial to the Chinese consumer and this is most commonly ensured through choosing the leading COO and high price. Signalling this status through choice of brands is inseparable from culture (Douglas & Isherwood, 2002). The participants residing in China acknowledged that they do not know wine well but preferred French wines and wines that were higher priced. Specific brands, regions and wine attributes were hardly mentioned. The results indicate that many Chinese premium wine consumers are using COO branding as the mechanism to choose their wine. Using COO as an indicator of quality is unsurprising given that consumers in developing countries tend to have positive perceptions of products from more developed countries, especially for high social value products (Batra et al., 2000). As well, those consumers are seeking to mitigate high levels of risk associated with high prices, product variability, low consumer confidence due to inexperience and dealing with strong peer opinions (Michman & Mazze, 2006). However, over-reliance on COO appears to be because the consumers lack domain-specific cultural capital (DSCC) in the wine market. Therefore, they do not yet understand the differences and qualities of regions within countries or specific brands. They also give little indication of the type of wine that they prefer other than red wine. These findings seem to apply irrespective of the reason for purchasing the wine. Gifts in particular seem to reflect the general lack of DSCC amongst the consumers except to say that the consumers recognise that premium wine comes from France (COO).

It is apparent that young Chinese wine consumers do not have wine knowledge in terms of brands, regions, types, taste and consumption situations. So external factors become an important consideration when a customer selects wines. French wine was the earliest brand to enter into the Chinese market. Furthermore, consumers are often exposed to luxury French wines in movies and TV series, such as Lafite, and that influences consumer perceptions that French wine is of a higher grade compared to other COOs. Penfolds, the most famous Australian brand, is also very famous in China. The name of the wine in Mandarin means running-to-riches. The Chinese easily accept that the brand name has a positive meaning. Yet, its popularity is still eclipsed by French wines that have a strong market hold in China.

For those young Chinese consumers residing in Australia we see a different level of DSCC and a different set of behaviours. These consumers can express a far greater level of knowledge of wines, like drinking wine themselves, and can express preference for brands, region, type of wine and consumption situation.

> *I want to drink Chardonnay tonight, so I will decide Burgundy or Margaret River Chardonnay. If I chose Margaret River region, then I will choose the brand maybe Leeuwin Estate Art Series Chardonnay or Moss Wood Chardonnay.*
>
> (Female, 26 years, Australia)

This would indicate that exposure to the Australian market and to the marketing and the experience of the wine industry in Australia rapidly extends the ability of the consumer to choose premium wines by region and by brand (although still usually well-known brands).

Implications

From the results it is apparent that wineries from countries other than France need to connect with the Chinese to build their knowledge and preference for their wines. It is crucial that this is achieved prior to their return to China. According to Lee et al. (2015, p. 64), 'Chinese tourists in Australia are more likely to enjoy than be disappointed with their tourism experience, hence, there is a greater propensity of them forming positive perceptions of Australia's products.' One key way of addressing this is to ensure that the wine experience is a prominent part of their Australian experience. Previous literature shows that the tourist experience acts like a halo to colour the perceptions of the country's products (Elliot et al., 2010; Lee & Lockshin, 2012). While this might mean a risk that product perceptions may suffer due to less positive tourism experiences, countries with a well-established tourism industry such as Australia are well positioned to take advantage of this approach.

There are a number of ways in which this can be achieved. For short-term tourists visiting Australia, stronger collaboration with Chinese tour companies both onshore and offshore would help ensure that tourists are channelled into the wine regions of Australia. Promotional campaigns within China should aim at creating greater awareness of the various Australian wine regions highlighting both the intrinsic and extrinsic value of visiting these regions. Consumers who are more aware of these regions will more likely participate in tours of these regions. As a consequence, when Chinese tourists return to China and want to buy a bottle of wine, Australian wine brands will be forefront in their minds. Furthermore, these returning tourists can become effective word-of-mouth advocates for the brands. For longer-term tourists such as individuals studying or working in Australia on a temporary basis, a greater effort to engage these tourists in wine consumption is required to build a base in the Chinese market. These temporary residents will in future serve as brand advocates and opinion leaders when they return to their home country. Given the prolonged time spent in Australia, it is possible to educate these tourists in greater depth with regards to the intricacies of wine consumption and the uniqueness of Australian wines through memorable experiences in the wine regions. As these tourists are already onshore, it also makes them a more accessible target market. Wine regions can embark on publicity and public relations activities such as wine tastings and education programmes amongst the local Chinese community to engage these tourists. Further, offering tour packages to wine regions to Chinese companies in Australia may go a long way in consolidating emotional connections with these regions.

Comparing short- to long-term tourism

However, a study by Lee et al. (2015) found that the 'tourism effect' decayed over time; that is tourists' perceived image and likelihood of purchasing Australian wine eroded over time, after they had returned to China. To be exact, the study found 'about 12 months upon their return, there were no differences between visitors and non-visitors in their perceptions of and willingness to buy Australian wine' (Lee et al., 2015, p. 65). This, however, suggests that Australian brands have a year to capitalise and take advantage of the favourable experiences and positive predispositions that Chinese tourists will have formed towards Australian wines during their time spent in Australia. This will also mean that Australian wine exporters will need to continue to engage and market to these Chinese consumers after they have returned to the mainland in order to maintain these positive associations or at least slow the decline of the initially formed positive perceptions. Furthermore, and in this case, the use of social media and other forms of direct marketing and communications to reinvigorate the memories of visitors would be worthwhile pursing.

The ultimate goal is to educate Chinese consumers to increase DSCC. Knowledge of French wines is more salient in China but they do not know the value of Australian wines. In Australia, education can come through the creation of memorable experiences that increase involvement, and subsequently, facilitate learning. However, offshore, this needs to be addressed at a more macro level by Australian wine brands, regions and the Australian government. The great advantage for Australian products is that Australia offers so many good tourism experiences (Galbreath & Gao, 2016). Wineries just have to make sure tourists have a chance to encounter their wine. It is imperative to begin building the DSCC of key influencers in the China market. A market push approach through advertising, sales promotions and publicity may produce positive results in terms of creating greater brand awareness and promote purchases. However, significantly improving perception of the wine amongst the general consumers, particularly those purchasing for gifts and special occasions, will require a concerted market pull approach. Given that the recognition of the quality and prestige of Australian brands is currently low, a possible strategy would be pulling these consumers to the wine region and working on creating a memorable experience while they are there. This will serve as a two-pronged approach with the regions' appeal serving as an impetus for travel to the region at which these consumers may be educated about the quality and prestige of Australian wines.

Chinese consumers consider more extrinsic attributes such as the region, brand and price. The lack of wine-specific knowledge makes it difficult for a majority of Chinese consumers to build up a standard of wine quality in terms of taste, flavour and aroma. While education is a long-term strategy to increase purchase and perceptions of Australian wines, more short-term strategies can aim at satisfying this need for extrinsic qualities. This could involve the development of more gift-ready packaging targeted at Chinese consumers. For instance, high quality and luxurious wooden gift boxes and packaging will offer these Chinese tourists an attractive souvenir which could be used for personal consumption or gifting. These gift boxes may not only serve as an appealing packaging for the product but also a reminder of their memorable experience in Australia for many years to come. Within the boxes, Chinese language information booklets on the wine production process and the region may also work towards further educating these consumers on Australian wines.

While it is possible for Australian wine brands to mass advertise, the highly fragmented Chinese medial landscape, where each city has its own local and provincial newspapers, radio and TV stations, on top of nationwide media such as CCTV and *China Daily* would place significant costs on such wine promotional methods. There is, however, potential to promote Australian wines and wine regions through publicity in China. Research has shown that Chinese consumers are highly susceptible to trends and TV dramas (Yang, 2012). Thus, movie-induced tourism may be a means of pulling more Chinese tourists into specific wine regions in Australia. In 2014, the Tourism Authority of Thailand initiated working agreements with the Korean Tourism Organisation to promote the filming of Korean movies and TV dramas in Thailand (Yonhap News Agency, 2016). Similar co-branding collaborations may draw more Chinese consumers into Australia allowing for more engagement and education of these potential markets. Furthermore, other collaborations between Australian wine regions and Australian organisations in China may also help bring Australian wines into China. For instance, consulate events catering local Australian wines will help boost the prestige perceptions of these wines. Sponsorship of major Chinese events will also consolidate the luxury perceptions of products from Australian wine regions.

Yet, while it is often cited as the main goal of local Australian stakeholders, it is important to understand that New World wineries are subject to the proverbial glass ceiling when it comes to developing the positioning of prestige. The luxury that is associated with Old World wineries in France, Spain and Italy is one that has been honed over decades of winemaking heritage. Without heritage, there is no history and without history, there is no provenance. To attempt to craft heritage at a New World winery will be a difficult task. As such, Australian wineries and wine regions should focus on delivering the short-term needs of Chinese consumers while working on a long-term plan to educate these consumers on the relative strengths of Australian wines. Instead of attempting to reproduce the aura of the Old World wine regions, Australian wine regions should embrace their New World spiritedness and vibrancy as a unique selling proposition for its wines. Through the continued promotion of wine regions as tourist destinations, greater awareness, familiarity and education amongst short- and long-term tourists may be achieved. This in turn, will gradually accumulate in the development of memorable experiences, place attachment and loyalty.

Overall, a concerted effort is required to create greater brand awareness, educate consumers and adapt products to satisfy the Chinese market. The creation of memorable experiences with Australian wines and wine regions will create market mavens and influencers amongst the Chinese market. This can be facilitated through the creation of Chinese-targeted products such as gift packages and wine tours. The responsibility for developing the wine regions go beyond the local winery and requires a strong collaboration between individual sellers, wine region authorities and even the Australian government. Through both short- and long-term strategies, it is simply a matter of time before Chinese consumers are familiar with, educated about and partial towards Australian wines.

The current study is constrained by its limited sample size, by focusing on young consumers and by including only long-term tourists. For these reasons, it may not encompass the full range of premium wine purchase situations by Chinese consumers nor the full range of drivers of purchase. However, for this specific sample it does provide some useful insights that can be expanded upon. Further, the suggestions derived from the study can be tested more specifically in future studies.

Conclusion

The current chapter shows that consumers of premium wine from an emerging market primarily rely on COO branding to inform their wine choice; that as they get more experienced and knowledgeable non-dominant COO branding, region branding and winery branding become more influential. This process is used to highlight strategic options for non-dominant COO brands, region brands and winery brands to enter the emerging market. Evidence is also provided to show that long-term tourists in a strong wine producing country develop stronger knowledge of wines from a country, the wine regions and local brands. A strategy of inducting tourists into the local wine culture is explored as a way to achieve the intended outcomes. Lastly, the study explores the implications for national and regional wine branding along with implications for individual wine brands including targeting tourists as part of an Australian wine brand's overall marketing strategy. It is effective, it is cost-efficient and it offers an alternative platform for wine brands to nurture long-term consumers.

Note

1 All three authors are from Curtin University, Australia and would like to thank attendees at the 2016 'Food, Wine and China: A Tourism Perspective' Symposium, organised by Curtin University's Tourism Research Cluster, for stimulating this chapter. Author contacts: g.ferguson@curtin.edu.au; i.cheah@curtin.edu.au; sean.lee@curtin.edu.au.

References

Ahmed, S.A. & D'Astous, A. (2001). Canadian consumers' perceptions of products made in newly industrialising East Asian countries. *International Journal of Commerce and Management*, 11 (1): 54–81.

Aichner, T. (2014). Country-of-origin marketing: a list of typical strategies with examples. *Journal of Brand Management*, 21(1), 81–93.

Akdeniz Ar, A. & Kara, A. (2014). Emerging market consumers' country of production image, trust and quality perceptions of global brands made in China. *Journal of Product & Brand Management*, 23(7), 491–503.

Audrin, C., Brosch, T., Chanal, J. & Sander, D. (2017). When symbolism overtakes quality: materialist consumers' disregard product quality when faced with luxury brands. *Journal of Economic Psychology*, 61, 115–123.

Balestrini, P. & Gamble, P. (2006). 'Country-of-origin effects on Chinese wine consumers'. *British Food Journal*, 5(108), 396–412.

Batra, R., Ramaswamy, V., Alden, D.L., Steenkamp, J.-B.E. & Ramachander, S. (2000). Effects of brand local and non-local origin on consumer attitudes in developing countries. *Journal of Consumer Psychology*, 9(2), 83–95.

Belk, R.W. (1988). Possessions and the extended self. *Journal of Consumer Research*, 15(2), 139–168.

Bourdieu, P. (1984). *Distinction: A Social Critique of the Judgement of Taste*. Cambridge, MA: Harvard University Press.

Bruwer, J., Chrysochou, P. & Lesschaeve, I. (2017). Consumer involvement and knowledge influence on wine choice cue utilisation. *British Food Journal*, 119(4), 830–844.

Byrd, E.T., Canziani, B., Hsieh, Y.C.J., Debbage, K. & Sonmez, S. (2016). Wine tourism: motivating visitors through core and supplementary services. *Tourism Management*, 52, 19–29.

Carlsen, J. & Boksberger, P. (2015). Enhancing consumer value in wine tourism. *Journal of Hospitality & Tourism Research*, 39(1), 132–144.

Carsana, L. & Jolibert, A. (2017). The effects of expertise and brand schematicity on the perceived importance of choice criteria: a Bordeaux wine investigation. *Journal of Product & Brand Management*, 26(1), 80–90.

Chaney, I.M. (2002). Promoting wine by country. *International Journal of Wine Marketing*, 14(1), 34–40.

Cheah, I. & Phau, I. (2015). Effects of 'owned by' versus 'made in' for willingness to buy Australian brands. *Marketing Intelligence and Planning*, 33(3), 444–468.

Chevalier, A., Maury, A.C. & Fouquereau, N. (2014). The influence of the search complexity and the familiarity with the website on the subjective appraisal of aesthetics, mental effort and usability. *Behaviour & Information Technology*, 33(2), 117–132.

Cohen, J., Corsi, A.M. & Lockshin, L. (2015). China: a'\ 'show system' approach for better marketing of Australian wine in China. *Wine & Viticulture Journal*, 30(4), 62–63.

Cohen, J., Corsi, A.M., Lockshin, L., Lee, R. & Bruwer, J. (2017). China: this isn't the time to pat ourselves on the back. *Wine & Viticulture Journal*, 32(2), 28–29.

Coskun, M. & Burnaz, S. (2016). Exploring the literal effect of COO for a new brand: a conjoint analysis approach. *Journal of International Consumer Marketing*, 28(2), 106–120.

Danner, L., Ristic, R., Johnson, T.E., Meiselman, H.L., Hoek, A.C., Jeffery, D.W. & Bastian, S.E. (2016). Context and wine quality effects on consumers' mood, emotions, liking and willingness to pay for Australian Shiraz wines. *Food Research International*, 89, 254–265.

Douglas, M. & Isherwood, B. (2002). *The World of Goods: Towards an Anthropology of Consumption* (6). New York: Routledge.

Elliot, S., Papadopoulos, N. & Kim, S.S. (2010). An integrative model of place image: exploring relationships between destination, product and country images. *Journal of Travel Research*, 50(5), 520–534.

Galbreath, J. & Gao, G. (2016). WA winee: building an economic future with China. *Bankwest Curtin Economic Centre Research Report*, 1(16) 1–128.

Graham, J.L. and Lam, N.M. (2003), The Chinese negotiation. *Harvard Business Review*, 81(10), 82–91

Hall, C. M. & Mitchell, R. (2007). *Wine Marketing*. London: Routledge.

Hall, J. & Lockshin, L. (2000). Using means-end chains for analysing occasions not buyers. *Australasian Marketing Journal*, 8, 45–54.

Holt, D.B. (1995). How consumers consume: a typology of consumption practices. *Journal of Consumer Research*, 22(1), 1–16.

Huang, A. & Lockshin, L. (2017). Wine pricing: understanding consumer response to price changes for high-priced wine brands. *Wine & Viticulture Journal*, 32(3), 61–62.

Huber, J. & McCann, J. (1982). The impact of inferential beliefs on product evaluations. *Journal of Marketing Research*, 9, 324–33.

Hu, X., Li, L., Xie, C. & Zhou, J. (2008). The effects of country-of-origin on Chinese consumers' wine purchasing behaviour. *Journal of Technology Management in China*, 3(3), 292–306.

Jackson, R.S. (1994). *Wine Science*. London: Academic Press Ltd.

Jin, W. (2004). The forecast of wine markets in China. *SINO-Overseas Grapevine and Wine*, 4, 69–74.

Josiassen, A. (2010). Young Australian consumers and the country-of-origin effect: investigation of the moderating roles of product involvement and perceived product-origin congruency. *Australasian Marketing Journal*, 18(1), 23–27.

Kelley, K., Hyde, J. & Bruwer, J. (2015). US wine consumer preferences for bottle characteristics, back label extrinsic cues and wine composition: a conjoint analysis. *Asia Pacific Journal of Marketing and Logistics*, 27(4), 516–534.

Keown, C. & Casey, M. (1995). Purchasing behaviour in the Northern Ireland wine market. *British Food Journal*, 97(1), 17–20.

Kim, H. & Bonn, M.A. (2016). Authenticity: do tourist perceptions of winery experiences affect behavioral intentions? *International Journal of Contemporary Hospitality Management*, 28(4), 839–859.

Kipnis, E., Broderick, A.J. & Demangeot, C. (2014). Consumer multiculturation: consequences of multicultural identification for brand knowledge. *Consumption Markets & Culture*, 17(3), 231–253.

Lascu, D.N. & Babb, H.W. (1995). Market preference in Poland: importance of product-country of origin, in E. Kaynak and Y. Eren (eds.) *Technology and Information Management for Global Development and Competitiveness*, Istanbul, 216–22.

Lee, J. K., Ahn, T., Lee, W.N. & Pedersen, P.M. (2017). Managing sports brands in a global consumer market: country-of-origin fit in cross-border strategic brand alliances. *South African Journal for Research in Sport, Physical Education and Recreation*, 39(1), 81–96.

Lee, R. & Lockshin, L. (2011). Halo effects of tourists' destination image on domestic product perceptions. *Australasian Marketing Journal*, 19(1), 7–13.

Lee, R., Corsi, A.M., Lockshin, L. & Cohen, J. (2015). They came, they like and they buy: turning tourists into long-term customers. *Wine and Viticulture Journal*, 30(6), 64–65.

Li, J.G., Jia, J.R., Taylor, D., Bruwer, J. & Li, E. (2011). The wine drinking behaviour of young adults: an exploratory study in China. *British Food Journal*, 113(10), 1305–1317.

Liu, F. & Murphy, J. (2007). A qualitative study of Chinese wine consumption and purchasing: implications for Australian wines. *International Journal of Wine Business Research*, 19(2), 98–113.

Lockshin, L., Corsi, A.M., Cohen, J., Lee, R. & Williamson, P. (2017). West versus East: measuring the development of Chinese wine preferences. *Food Quality and Preference*, 56, 256–265.

Lockshin, L.S. & Rhodus, W.T. (1993). The effect of price and oak flavour on perceived wine quality. *International Journal of Wine Marketing*, 5(2/3), 13–25.

Maciel, A.F. & Wallendorf, M. (2016). Taste engineering: an extended consumer model of cultural competence constitution. *Journal of Consumer Research*, 43(5), 726–746.

Maguire, J.S. & Lim, M. (2015). Lafite in China: media representations of 'wine culture' in new markets. *Journal of Macromarketing*, 35(2), 229–242.

Maheswaran, D. (1994). Country of origin as a stereotype: effect of consumer expertise and attribute strength on product evaluations. *Journal of Consumer Research*, 21, 354–65.

Michman, R.D. & Mazze, E.M. (2006). *The Affluent Consumer: Marketing and Selling the Luxury Lifestyle*. Westport, CT: Greenwood Publishing Group.

Moulard, J., Babin, B.J. & Griffin, M. (2015). How aspects of a wine's place affect consumers' authenticity perceptions and purchase intentions: the role of country of origin and technical *terroir*. *International Journal of Wine Business Research*, 27(1), 61–78.

Muhammad, A., Leister, A.M., McPhail, L. & Chen, W. (2014). The evolution of foreign wine demand in China. *Australian Journal of Agricultural and Resource Economics*, 58(3), 392–408.

Nella, A. & Christou, E. (2014). Segmenting wine tourists on the basis of involvement with wine. *Journal of Travel & Tourism Marketing*, 31(7), 783–798.

Papadopoulos, N. & Heslop, L.A. (2003). Country equity and product-country images: state-of-the-art in research and implications. In S.C. Jain (ed.) *Handbook of Research in International Marketing*. Northampton, MA: Edward Elgar.

Phau, I. & Prendergast, G. (1998). Tracing the evolution of country of origin research: in search of new frontiers. *Journal of International Marketing and Exporting*, 4(2), 71–83.

Phau, I. & Suntornnond, V. (2006). Dimensions of consumer knowledge and its impacts on country of origin effects among Australian consumers: a case of fast-consuming products. *Journal of Consumer Marketing*, 23(1), 34–42.

Pratt, M.A. & Sparks, B. (2014). Predicting wine tourism intention: destination image and self-congruity. *Journal of Travel & Tourism Marketing*, 31(4), 443–460.

Reardon, J., Vianelli, D. & Miller, C. (2017). The effect of COO on retail buyers' propensity to trial new products. *International Marketing Review*, 34(2), 311–329.

Siu, W.-S. & Chan, C.H.-M. (1997). Country of origin effects on product evaluation: the case of Chinese consumers in Hong Kong. *Journal of International Marketing and Marketing Research*, 22(2), 115–22.

Stiehler, B.E., Caruana, A. & Vella, J. (2016). Using an aesthetics and ontology framework to investigate consumers' attitudes toward luxury wine brands as a product category: evidence from two countries. *International Journal of Wine Business Research*, 28(2), 154–169.

Sørensen, F. & Jensen, J.F. (2015). Value creation and knowledge development in tourism experience encounters. *Tourism Management*, 46, 336–346.

Sun, G., D'Alessandro, S. & Johnson, L. (2014). Traditional culture, political ideologies, materialism and luxury consumption in China. *International Journal of Consumer Studies*, 38(6), 578–585.

Tan, C.Y. (2017). Conceptual diversity, moderators and theoretical issues in quantitative studies of cultural capital theory. *Educational Review*, 69, 600–619.

Wall, M., Hofstra, G. & Liefeld, J. (1991). Impact of country-of-origin cues on consumer judgements in multi-cue situations. *Journal of the Academy of Marketing Science*, 19(2), 105–13.

Wallendorf, M. (2001). Literally literacy. *Journal of Consumer Research*, 27(4), 505–511.

Wang, C. & Chen, Z. (2004). Consumer ethnocentrism and willingness to buy domestic products in a developing country setting: testing moderating effects. *Journal of Consumer Marketing*, 21(6), 391–400.

Wang, N. (2017). China's wine imports continue upward trend. Retrieved from https://www.thedrinksbusiness.com/2017/08/chinas-wine-imports-continue-upward-trend/

Williamson, P.O., Mueller-Loose, S., Lockshin, L. & Francis, I.L. (2017). More hawthorn and less dried longan: the role of information and taste on red wine consumer preferences in China. *Australian Journal of Grape and Wine Research*, Early view.

Winterhalter, S., Zeschky, M.B., Neumann, L. & Gassmann, O. (2017). Business Models for Frugal Innovation in Emerging Markets: The Case of the Medical Device and Laboratory Equipment Industry. *Technovation*, 66–67, 3–13.

Wu, S. (2017). 2016 China wine import figures round up: Australia grows by 40%. Retrieved from https://www.decanterchina.com/en/news/2016-china-wine-import-figures-round-up-australia-grows-by-40

Ye, B.H., Zhang, H.Q. & Yuan, J.J. (2014). Intentions to participate in wine tourism in an emerging market: theorisation and implications. *Journal of Hospitality & Tourism Research*, 1096348014525637.

Zhang Qiu, H., Yuan, J., Haobin Ye, B. & Hung, K. (2013). Wine tourism phenomena in China: an emerging market. *International Journal of Contemporary Hospitality Management*, 25(7), 1115–1134.

Zhang,Y. (1996). Chinese consumers' evaluation of foreign products: the influence of culture, product types and product presentation format. *European Journal of Marketing*, 30(2), 1–17.

11 China as an export market

The case of Western Australian wine

*Jeremy Galbreath, Grace Gao, Louis Geneste,
Kristina Georgiou, Niki Hynes and Paull Weber*

Introduction and background

China represents a significant export market opportunity for wine producers (Camillo, 2012). In fact, research demonstrates that China is now one of the top five wine consuming countries in the world (Galbreath et al., 2015), while predictions suggest that China could represent the largest market for wine consumption in the world within the next three decades (Camillo, 2012; Thorpe, 2012). However, exporting wine to China can be very challenging. Challenges include tariffs (although the Chinese government is currently reducing tariffs on wine for some exporting countries), cultural differences, language barriers, differences in consumer palates, access to networks, superstitions and peculiarities (e.g. the symbolism of different colours), and a population that is relatively new to drinking wine (Bretherton & Carswell, 2001; Camillo, 2012; Jenster & Cheng, 2008; Rabobank International, 2010).

To address these challenges, some research has focused on success factors for the export of wine to China. In their study, Bretherton and Carswell (2001) present a general view of ways that Western wine producers can increase their export success to the Chinese market. The approach they take is based on the '4Ps': price, product, promotion and place. The authors offer suggestions regarding the 4Ps; however, their suggestions rely on the existing literature and are framed within a generic framework aimed at any Western producer of wine. Further, they did not survey or interview any individuals in their study to gain insight nor is there specificity to a specific wine-producing region. Generic prescriptions and lack of data collection leave room for deeper research.

In an effort to expand insight into China as an export market for wine, this chapter focuses on wine producers in the state of Western Australia (WA). Located in nine distinct geographic indicator (GI) regions, WA wine production represents approximately five per cent of Australia's overall production volume, which perhaps reflects the nature of the many small and boutique producers in the state (Galbreath et al., 2015). Yet, the wine industry in WA delivers 12 per cent of Australia's wine by value and 25 per cent of value in the speciality and super-premium wine categories (Galbreath et al., 2015). Alternatively, WA wine exports lag behind the country average almost 5 to 1, which is estimated to cost their economy nearly AUD$100 million annually (Galbreath et al., 2015). Exacerbating this lag in overall exports is the fact that only around one quarter of WA wine producers export to China, which represents one of the biggest opportunities for wine sales in the world (Camillo, 2012; Galbreath et al., 2015; Thorpe, 2012).

Building on the work of Bretherton and Carswell (2001), we explore wine exports to China. This is done by focusing on a specific wine-producing state and by interviewing a variety of participants, including industry stakeholders, producers who export wine, and producers who do not engage in exporting to China. By capturing different viewpoints, perspectives are built that result in deeper insights into the issues of exporting wine to China.

Research questions

From the above discussion, the following research questions are formulated:

- To what extent is China a viable export market for wine?
- What barriers or risks are faced when exporting wine to China?
- Are there any specific success factors for exporting wine to China?
- How can wine producers increase their ability to export wine to China?

Methods

Sample

Following the guidelines of Lincoln and Guba (1985) for purposeful sampling, participants were chosen who would be most able to provide insight on the research aims. I relied on my contacts in the WA wine industry to solicit participants. A snowball technique was used and 32 participants were secured, with relatively equal distribution among industry stakeholders, wine exporters and those who do not export (six industry stakeholders and 26 producers). Table 11.1 demonstrates the mix of participants.

As for producers, Figure 11.1 illustrates the number of producers by GI region. The Great Southern has the largest representation with six producers, followed by Margaret River and Pemberton with four each, Blackwood Valley, Geographe, Manjimup, and Swan Valley with three each, Peel with two and Perth Hills with one. Figure 11.2 represents graphically the location of producers.

With respect to annual case production, Figure 11.3 shows a range, from less than 999 cases to over 250,000 cases, annually. The most common production range is 20,000 cases annually.

Regarding export volume, Figure 11.4 illustrates that 15 out 26 producers reported their overall export volume.

As for exports to China, Figure 11.5 shows that 13 of 26 producers export to China.

Table 11.1 Participants in WA wine export study

Type of participant	Affiliation	Coding designation	Exports to China?
Stakeholder	Industry association	S1	NA
Stakeholder	Industry association	S2	NA
Stakeholder	State government	S3	NA
Stakeholder	Industry association	S4	NA
Stakeholder	Supplier	S5	NA
Stakeholder	Distributor (China)	S6	NA
Producer	Margaret River	P1E	Yes
Producer	Margaret River	P2NE	No
Producer	Swan Valley	P3NE	No
Producer	Great Southern	P4E	Yes
Producer	Great Southern	P5E	Yes
Producer	Margaret River	P6E	Yes
Producer	Perth Hills	P7E	Yes
Producer	Swan Valley	P8NE	No
Producer	Great Southern	P9NE	No
Producer	Pemberton	P10E	Yes
Producer	Great Southern	P11E	Yes
Producer	Manjimuo	P12E	Yes
Producer	Pemberton	P13E	Yes
Producer	Geographe	P14E	Yes
Producer	Pemberton	P15NE	No
Producer	Blackwood Valley	P16NE	No
Producer	Pemberton	P17NE	No
Producer	Geographe	P18NE	No
Producer	Great Southern	P19NE	No
Producer	Peel	P20E	Yes
Producer	Geographe	P21NE	No
Producer	Blackwood Valley	P22E	Yes
Producer	Peel	P23NE	No
Producer	Blackwood Valley	P24E	Yes
Producer	Manjimup	P25NE	No
Producer	Swan Valley	P26E	Yes

'E' = exporter to China

'NE' = non-exporter to China

Figure 11.6 shows the price points of producers (AU$) for both the domestic and Chinese markets. The prices range from the lower end (AU$7–AU$10) to the premium end (higher than AU$30).

Finally, two scatterplots are provided. Figure 11.7 illustrates the relationship between annual production volume and export volume to China, while Figure 11.8 demonstrates the relationship between export volume to China and export value to China.

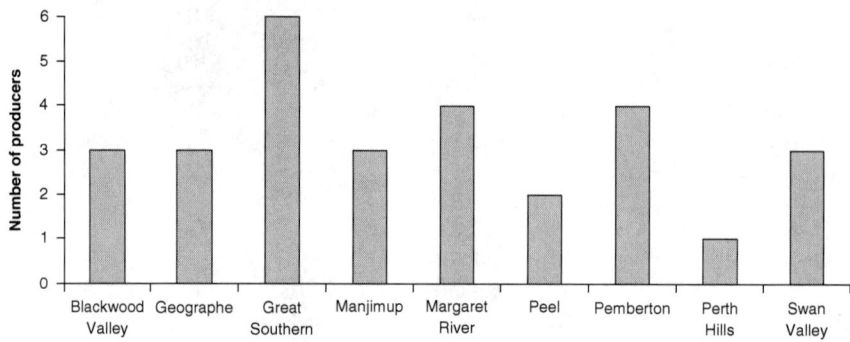

Figure 11.1 Producers by GI region (by number)

Note: Some producers reported being based in one location but operating in more than one region

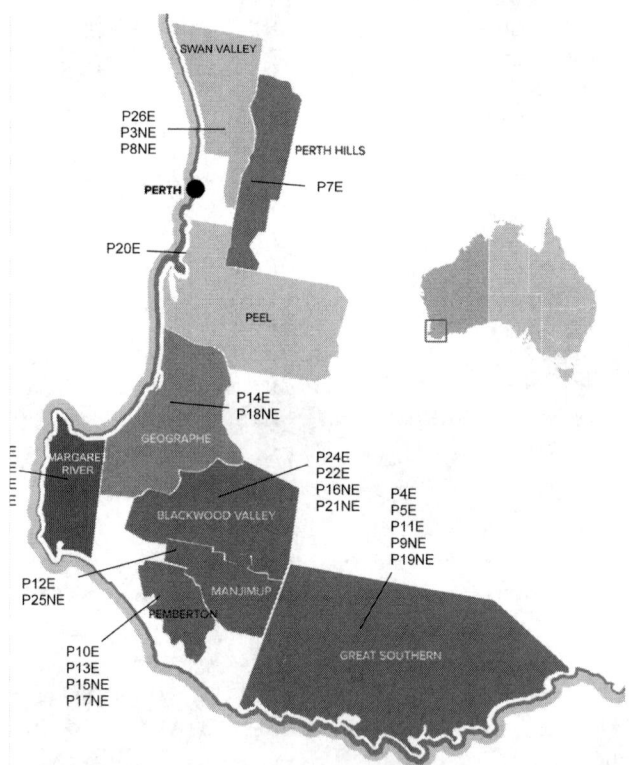

Figure 11.2 Producers by GI region (by location)

Note: Some producers reported being based in one location but operating in more than one region (only base location is reported)

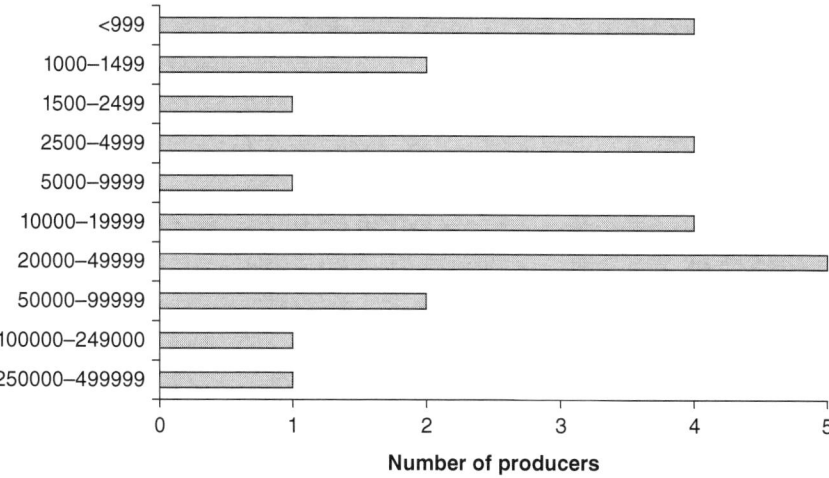

Figure 11.3 Number of producers by annual case production

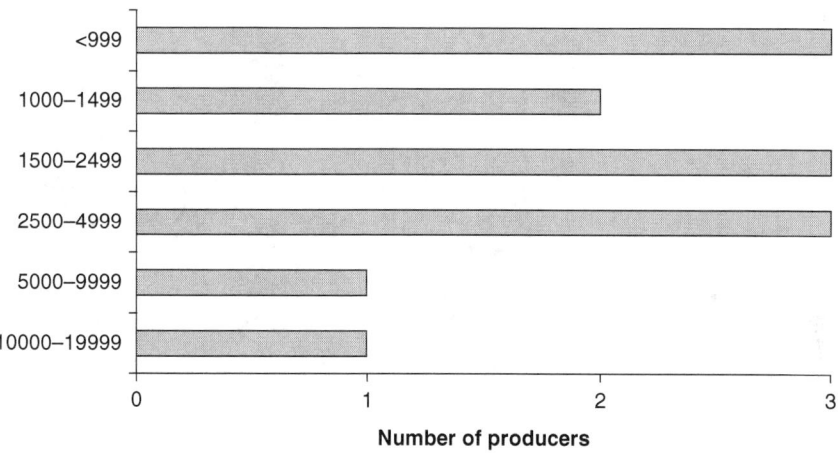

Figure 11.4 Overall export case volume (by number of producers)

Data collection

To explore the objectives of the study, semi-structured, open-ended interviews were used. Semi-structured interviews involve gathering rich and multi-layered information, allowing a few prepared questions to form the skeleton of the interview, with additional questions emerging during the interview process (Hoggart et al., 2002). By pre-determining some questions, the comparability of responses is increased and the interviewer's effects and biases reduced (Kitchin & Tate, 2000).

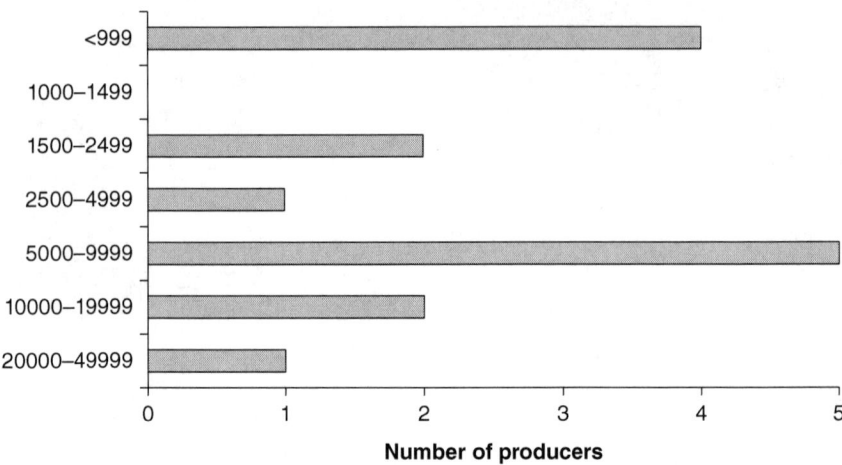

Figure 11.5 Export case volume to China (by number of producers)

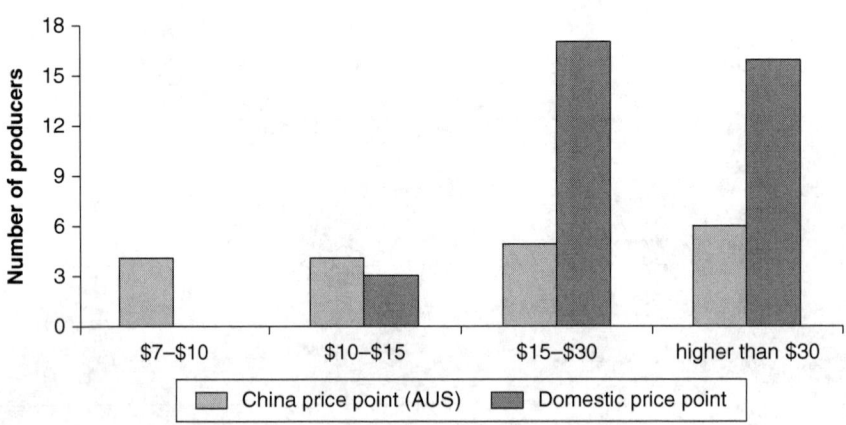

Figure 11.6 Price point domestic and Chinese markets (AU$)

Around 10 of the interviews were conducted face-to-face and at a place that suited the participant's work commitments. The remainder of the interviews were completed over the phone. All interviews were recorded digitally with permission. Interviews were conducted from February to June 2015 with government-related employees, industry policy makers and strategists, managing directors, CEOs, winemakers, viticulturists, marketers, suppliers and distributors. All participants demonstrated a high level of professionalism in their understanding of the Chinese market, and its specific impacts on the WA wine export opportunity. The average length of an interview was 37 minutes, with some lasting more than an hour.

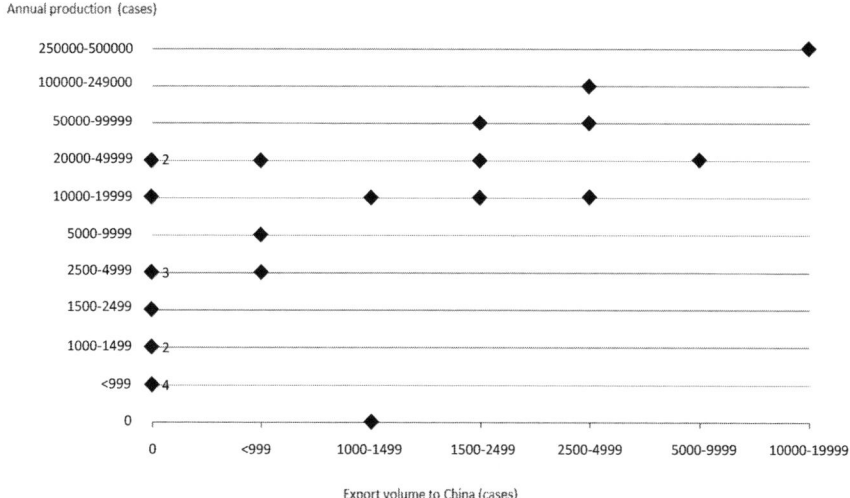

Figure 11.7 Annual production x export volume to China
Note: Number of duplicates are indicated next to the diamonds

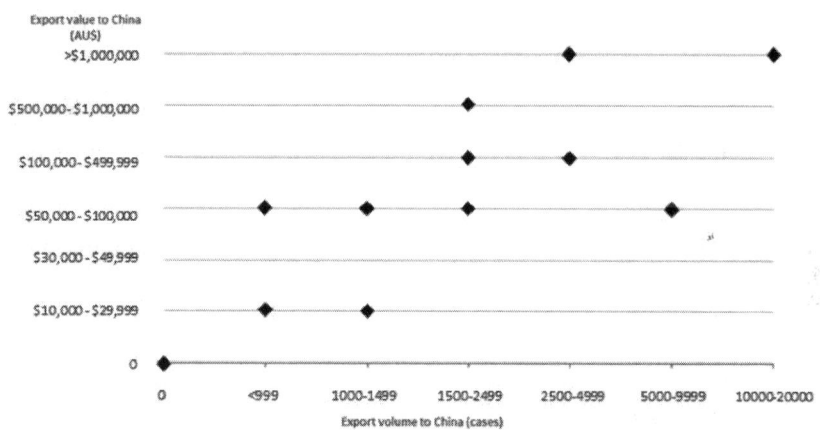

Figure 11.8 Export value to China x export volume to China (AU$)

A careful orthographic transcription was made of the interviews to accurately reproduce the semantic content of what each participant said.

A team of researchers, experienced in qualitative methods, undertook an analysis of the transcripts. Thematic analysis was conducted using an interpretive approach (Braun & Clarke, 2006). Themes are developed through the careful iterative and reflexive examination and re-examination of the raw interview data (Braun & Clarke, 2006).

The interpretive approach to thematic analysis attempts to determine the substance of the themes and their broader meanings and implications (Patton, 1990).

Following the inductive coding process (Bryman & Burgess, 1994), a close reading of the transcripts was undertaken to become familiar with the content and to gain an understanding of the details in the text. After the initial reading, each participant's transcript was entered into Excel to assist with content analysis. Initial codes were generated across the corpus of interview data, based on the actual words or terms used by the participants, using a system of in vivo coding, or coding taken directly from the participants' discourse. In this step, first-order categories were derived and we reflected on the coded files by re-reading the interview transcripts, coding for more in vivo words.

In the second step of the process, first-order codes were examined for relationships between and within the passages, which facilitated assembling them into first-order categories (Sharma & Vredenburg, 1998). As displayed in Figure 11.9, several first-order categories emerged. In the third step, analysis was undertaken to look for links and relationships among first-order categories so that they could be collapsed into distinct clusters (Platt, 1981), or second-order themes. Here, a recursive approach rather than a linear one was employed, namely iteration between first-order categories and emerging patterns in the data until conceptual second-order themes emerged (Eisenhardt, 1989). Fourth, second-order themes were organised into final themes that reflected the overarching dimensions that emerged from the data. Lastly, where there was discrepancy in the coding of any transcript, these were resolved until 100 per cent agreement was reached among the researchers.

Findings

Based on the interviews, three key themes emerged from the data (Figure 11.9). The first reflects the realities of the barriers and risks of exporting to the Chinese market. The other emergent themes reflected more of the participants' viewpoints and perspectives on WA export success to China, including new business models and product-related features. Please note that in the findings section, to protect the privacy of the participants, a coding designation in parentheses is used at the end of each quote (see Table 11.1 for more information).

Theme 1: Realities of barriers and risks

From a stakeholder perspective, the size of the wine industry in WA is seen to inhibit greater export efforts, particularly against other regions in Australia, such as South Australia, who leads the nation in exports, and specifically, exports to China.

As some stakeholders commented:

> So we are a large region in area, but we're not a massive producer either . . . Moving into China, they're supporting large volume producers, basically (S4).

Wine producers offered a similar perception with respect to size.

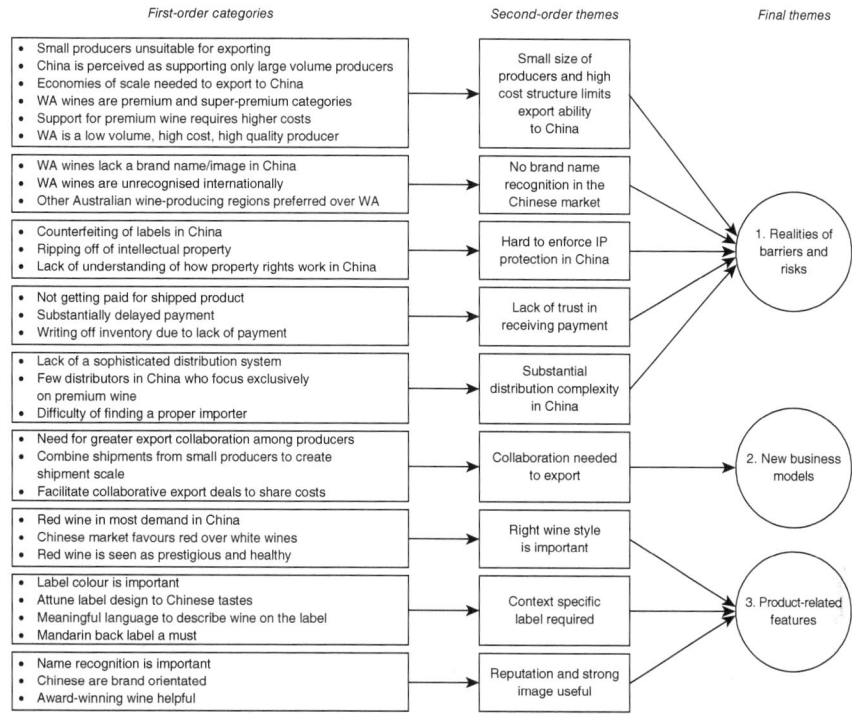

Figure 11.9 Themes emerging from the data

Non-exporters to China commented:

> So to make it work [in China] is quite difficult because you have to have the volumes and the economies of scale (P8NE).
>
> Seems the bigger people and the mid-size wineries can do it but the smaller ones just haven't got the volume (P17NE) (this participant was an exporter to China but has since pulled out).
>
> We're small producers so often we don't have great volumes of any one wine that we can present so that's often a factor that blows in (P25NE) (this participant was an exporter to China but has since pulled out).

Current exporters to China agreed:

> Well I reckon if you're doing less than 5,000 cases, forget trying to export [to China] (P24E).

Closely related is the type of product produced (e.g. bulk wine versus premium wine) and its associated cost structure. For example, stakeholders perceive that China is more interested in cheaper wines in volume, which appears to put WA at a disadvantage, both on price and cost.

Stakeholders commented that:

> [As for barriers to exporting to China] price point is one of them (S1).
>
> Western Australia, across the board, is not a low cost producer . . . It can't do it (S2).
>
> So it's really related to what the Chinese market is after and how much they are willing to pay for it. Moving into China, they're supporting large volume producers, basically . . . [where WA] . . . are premium wine producers (S4).

Perceptions of stakeholders are backed up by producers who export to China and those who previously exported, but longer do so.

Non-exporters to China stated that:

> When you crunched the numbers and then started adding in your real expenses, we were actually losing money on it (P15NE) (this participant was an exporter to China but has since pulled out).
>
> Our business model works on small volumes large margins, rather than large volumes small margins . . . then I looked at the pricing structure and what other wineries that were compatible to us were selling their wine into China for, and there's just no margins in it (P15NE) (this participant was an exporter to China but has since pulled out).
>
> Most export deals rely upon high volume, low margin, which is not really where [WA is] at (P25NE) (this participant was an exporter to China but has since pulled out).

Similarly, exporters to China suggested that:

> The major reason [for our lack of exports to China] is because our cost of goods is so high . . . [while] . . . There is little appreciation of value in high price wines [in China] (P1E).
>
> The only problem we have is our costs; it's so hard to produce here we have to be careful of where we go and what we do because you can't compete with the Chilean wines and South African wines [on costs] (P5E).

Another barrier that appears to be affecting WA wine exports to China is the infatuation with prestige. The Chinese tend to gravitate towards 'brand image'. This is reflected in both stakeholder and producer perceptions as there is a concern that WA wines, in general, do not have a known 'brand' in China.

Stakeholders suggested that:

> Even a region with the gravitas of Margaret River internationally doesn't fit that [reputational] bill [in China] right now (S1).
>
> [When you look at things internationally] you begin to realise just how unrecognised we really are (S1).

Non-exporters to China tended to agree:

> [China is] not a market we know, it's not a market we control, we're a no-name brand, there's no reason for them to buy our wine because they don't know who we are, so it's just quite difficult (P8NE).

Exporters to China had similar views:

> The Chinese market has always been talking about the eastern states and the Barossa and the reds and all that and rarely [are they talking about] the WA wines (P4E).
>
> It's just the Chinese prefer South Australian wine because they're better known, they've been around a lot longer (P10E).

Three other factors that can be seen as 'barriers' to entry to China. First, counterfeiting is seen as a particular risk.

Stakeholders had this to say:

> With the problems with counterfeiting, with the gravitas that there is in guaranteed place of origin and authenticity, I think you've got to be pretty consistent (S1).

Non-exporters to China held a similar view:

> Probably one of the risks unfortunately . . . is the duplicating of labels, ripping off labels (P2NE).

Exporters to China commented:

> Even though you protect your intellectual property, people absolutely rip off your intellectual property all the time (P1E).
>
> The problems are counterfeiting . . . we investigated the legal implications because we had no idea about China's legal system. How do property rights apply to wine? What about label integrity? All that sort of stuff (P6E).

Second, receiving payment for purchased product is also seen as a risk.

Stakeholders commented:

> The risks of actually getting paid and how you set up all those risk aversion things so that you're going to get paid if things go wrong (S3).
>
> Certainly first and foremost and most fundamental is just payment (S6).

Non-exporters to China expressed a similar perspective:

> Outstanding invoices and when things don't get paid and they have to be written off, it makes the entire exercise a very negative experience (P15NE) (this participant was an exporter to China but has since pulled out).

We got a few [cases] in but, because we're small, they strung out payment and it took 6, 7, 8, 9 months to get paid. We haven't got the clout over there (P17NE) (this participant was an exporter to China but has since pulled out).

Exporters to China demonstrated the same concern:

Certainly . . . payment [is an issue], we don't do anything until the money is in the bank. We don't even pack the container or anything (P7E).

The other thing too is you've got to be careful when you're dealing with overseas. I got caught a few times with not being paid so our policy is pay for it before it leaves (P10E).

Third, distribution in China is seen as potentially the biggest barrier among stakeholders and producers alike.

Stakeholders believed that:

There's a lot of distribution complexity here [and] one of the major challenges and you ask this question further down, but one of the major challenges here really is finding a good partner that's reliable, that really genuinely is capable of distributing (S6).

Non-exporters to China commented:

They've had people [in China] buying – going from buying 2,000 cases to 50,000 cases to 100,000 cases and then, the next year, saying we don't need any wine this year. Because they haven't been able to – because they're so new at it – they haven't been able to manage their stocks (P1E).

Getting access to those [premium] segments and the other side of that is there are very few Chinese distributors that focus in that area and that's probably one of the biggest problems (P2NE).

Exporters to China also expressed distribution concerns:

This is the hardest thing for people to be able to cope with. That you're selling to a guy who's a trader. He sells to everyone he knows. That's how they sell. That's what they do; they trade in China. He may have two restaurants, but 500 guys buying his wine. That's the reality (P1E).

In particular, distribution, good distribution is so hard to get. In all of China there wouldn't be more than four companies that can do it on a national scale (P6E).

I've had various different people in China selling our wine, so it's always that struggle, you've got to find someone to import it and these people have tried and got out of the business or whatever, I don't know, so it's always about trying to find an importer, it's never about trying to find the buyer, it's always about trying to find the importer (P11E).

Theme 2: New business models

WA production is mainly based on low volume, higher price point premium wines (Wines of Western Australia, 2014). As noted in Theme 1, a low volume, higher price point strategy is one that imposes a higher cost structure, which is believed to inhibit exporting ability among some producers. In light of such a strategy, some participants believe that WA should consider new or alternative business models where producers build 'exporting consortiums' or otherwise seek to combine or share the costs of exporting. For example, stakeholders commented:

> So if you've got ten producers of a similar scale across, say, three [WA] regions that all are interested in Singapore, why wouldn't you facilitate for them the ability to collaborate to do something (S1)?
> [as for producers collaborating] I'd really like to do that. I've had somebody offer me the concept in the last couple of months. Then it's just a question of engaging with the producers to see if they really want to do it (S4).

Views on collaboration were not limited to stakeholders. Some of the producers also intimated that some level of collaboration may be necessary to crack the export market to China. For example, such views are reflected in the following comments:

> We have a situation where my view is that a lot of the smaller wineries that are really struggling should collaborate, they should get together (P2NE).
> Unless we can get like a cooperative going where we can find somebody [in China] that's prepared to buy individual lots that will then make up a 40 foot container, it's not very feasible I don't think (P18NE).

Theme 3: Product-related features

The third emergent theme clearly focused on product features as a mechanism for success in exporting to China. That is, for those WA producers who do choose to export wine to China, participants views suggested that there needs to be attention given to a variety of produce features. First, there is consideration of wine style.

Stakeholders suggested:

> So what we find is definitely the [Chinese] market favours red wine over white right now probably to the magnitude of perhaps ten to one (S6).

Similarly, non-exporters to China commented:

> Definitely reds and it's more of a prestige thing (P15NE) (this participant was an exporter to China but has since pulled out).
> They want sweet red wine mainly. Red wine is perceived as being of extreme health benefit to them (P16NE).

Currently it's all about red . . . yes, it's mainly red styles (P25NE) (this participant was an exporter to China but has since pulled out).

Exporters to China held the same perspective:

Most Asian countries . . . love their red wine, they think it's more superior and important and everything like that (P5E).

You want to keep it simple, just to one or two wines, red wines initially and sweeten one up and try it make it to the Chinese taste (P7E).

They're generally reds. They love that for a number of reasons, health reasons. They think red wine is the way to go (P26E).

Second, to be successful, participants suggested that there is a need to ensure that packaging, and particularly the wine label, meets Chinese requirements. Labels are important because many Chinese consumers are unfamiliar with Australian wine (Liu & Murphy, 2007), and labels are their first 'point of contact' with respect to making a purchasing decision (Rabobank International, 2010).

Non-exporters to China suggested:

I think it's a very important area, the labels . . . You know, the red colours and the gold colours and the happy New Year colours (P3NE).

You'd have to possibly rebrand it or re-label it or have some sort of a Chinese lucky number on the label and away they go (P18NE).

Exporters to China went further:

Words that are meaningful to them that are descriptors of wine, information and back labels, which are all in Chinese (P1E).

For the top end wines the label packing is not so much of a thing, but on the lower end wines it is . . . So the lower end wines need to have a nice colourful label, a good catchy label and things like that (P4E).

Yes and the packaging has to [represent] good luck and wealth and that type of thing, that's what they're in to and once you understand all those traditions and those types of things you're a long way down the track especially if you can deliver (P12E).

Third, recognition or acclaim is seen as a requirement to export to China because the Chinese tend to value prestige. This is evident in the following quotes.

Stakeholders held the view that:

How are you going to get traction [in China]? Name and reputation, brand platform, product differentiation, that's how (S2).

Non-exporters to China agreed:

> It takes ratings, no doubt about it, it takes building a brand in China because people are brand-orientated (P2NE).

Exporters to China corroborated others' perspectives:

> We actually have to build our reputation first. Because reputation is really important . . . perceptions are really important in China (P1E).
>
> But the other thing is brand recognition, just gaining that recognition on an international scale can open up real opportunities and I guess there are a couple of ways to do that and one is to participate in international competitions and try to gain recognition that way which can then drive your export opportunities (P13E).
>
> If you've won awards and got good write-ups in magazines etc. (P14E).

Discussion

Given the growth and opportunity for wine in China, wine producers are taking notice and making efforts to enter the market there (Bretherton & Carswell, 2001; Camillo, 2012; Yu et al., 2009; Jenster & Cheng, 2008; Liu & Murphy, 2007). However, the challenges of exporting wine to China are many (Bretherton & Carswell, 2001). Relying on interviews with 32 industry stakeholders and wine producers in Western Australia, this chapter presented a qualitative study examining wine exporting to China. The findings both corroborate, and extend, previous research.

First, previous research suggests that product features are important. Studies have identified that wine style, and particularly red wine, is favoured by the Chinese (Camillo, 2012; Liu & Murphy, 2007). Red wine is favoured because of a positive image: drinking red wine denotes symbolism such as health, happiness, good taste, an image of wine knowledge, and good luck. The participants in the current study also acknowledge that to be successful as an exporter of wine to China, selling red wine is a key. As one participant (S6) noted, red is preferred over white wine in China on a magnitude of ten to one. Packaging has also been previously identified as an important product feature. In their study, Yu et al. (2009) find that a properly designed back label, one that conveys 'taste descriptions', is the most important information that can be provided. The current study demonstrates similar findings. For example, 'words that are meaningful to them that are descriptors of wine, information and back labels, which are all in Chinese (P1E).'

Second, whether seen as a marker of success or barrier to entry, brand image/name has been identified as an important characteristic of exporting and selling wine in China. In their study, Yu et al. (2009) find that brand name rates highly as a wine attribute among both ordinary consumers and university students. Camillo (2012) finds that Chinese consumers of wine pay attention to brand and that brand image influences their purchasing decision. Our study corroborates these findings. For example, as one participant (P8NE) noted, in China they are a 'no-name brand' and that consumers do not know who they are. Similarly, another participant (P10E) suggested that South Australian brands are the brands that are known in China and that wine from WA is simply not known as a brand. Such views tended to reflect a concern or barrier to entry rather than an opportunity to exploit for these particular participants.

Third, previous research has reflected on distribution in China and finding importers who are capable and trustworthy. For example, Bretherton and Carswell (2001: 27) call distribution in China 'difficult', while Galbreath et al. (2015) suggest there are 20,000 wine importers in China, making navigation of the distribution system complicated. Participants in the present study concur. One (S6) suggests that a major challenge of exporting to China is finding a distribution partner that is 'reliable' and 'genuinely capable of distributing'. Another noted that wine distribution in China 'is the hardest thing for people to be able to cope with (P1E)'. In-country distribution, then, remains a critical aspect that needs to be mastered to be a successful exporter of wine to China.

Fourth, some new evidence emerged. For example, enforcing intellectual property (IP) (e.g. a brand name or label design) can be difficult in China (Kambill et al., 2006; Schotter & Teagarden, 2014). IP protection is difficult because IP laws can be unclear, uncertain, or defined differently than in the West (Kambill et al., 2006; Schotter & Teagarden, 2014). Several of the participants acknowledge issues with IP protection in China. One participant (P6E) acknowledged that counterfeiting is a problem and that there is a lack of understanding of how property rights apply to wine, such as label integrity. Yet, in China, some view copying not as an infringement of IP rights, but as a way to honour past works of masters (Kambill et al., 2006). In fact, one participant said, 'The Chinese copy, they copy labels. I always joke that if they copied my label I'd be laughing if they spread it all over the country, it would be great marketing (P12E).' However, most participants view IP as problematic and lack of IP protection as a risk of selling wine in China.

Fifth, business transactions in the West are based on a fundamental principle of 'immediate' or direct trust. In essence, in the West, individuals are given the benefit of the doubt and are considered trustworthy until something is done to break trust (De Cremer, 2015). In China, trust is gained or awarded only after one has proven oneself worthy of it (De Cremer, 2015). In other words, 'guilty until proven innocent'. Largely, differences in trust between the West and China are believed to be a result of history and culture. In the case of this study, new evidence emerged from several participants in that they stated that they had difficulty getting paid for wine shipped. Some never got paid. Others didn't receive payment for several months. While not directly noted by any of the participants, trust could be a factor here. That is, unless a wine producer has built solid relationships in China and has demonstrated reliability in conducting business transactions over a period of time, there may be reluctance on the part of Chinese importers or buyers to grant immediate favour and trust (De Cremer, 2015), creating risk in the supply chain.

Lastly, there is recognition that given the nature of the wine industry in WA – mainly small producers, mainly premium wine production – export opportunities to China are perceived to be out of reach to many due to limitations of size, scale, resources and higher cost structures. Factors such as lack of size, scale, or resources are common perceptions among small- and medium-sized enterprises in terms of export challenges (Paul et al., 2017). However, some new evidence that emerged in our study demonstrated that participants did recognise a need for new business models to ramp up, or in some cases facilitate, the ability to export wine to China. New business models were largely viewed in terms of collaboration and collaborative arrangements. For example, some views suggested that rather than individual wine producers in a given region (or regions) trying to 'go it alone' in terms of exporting, an export consortium could be established where each producer shares the costs of filling containers and shipping their wine to China. Others suggested that individual small producers tend to struggle in terms of developing healthy export markets and that instead they should collaborate, sharing cost and expense as an alternative approach to exporting. The ability to leverage networks and collaborative arrangements is increasingly seen as important to the export success of small businesses (Paul et al., 2017).

Conclusion

The Chinese market is said to represent unprecedented opportunities for wine producers. Yet, navigating the Chinese wine export market is difficult and challenging. While some have investigated the Chinese wine consumer to better understand the landscape (Camillo, 2012; Liu & Murphy, 2007; Yu et al., 2009), relatively little research has studied the perspective of the wine producer. This chapter presents a study that did explore the perspective of the wine producer in the context of the Chinese export market.

The findings are particularly important to wine producers who are small and who produce premium wine. More specifically, after interviewing 32 participants in Western Australia ranging from industry stakeholders to wine producers, the results identified three key themes. First, exporting to China is difficult and one is faced with several barriers and risks. These include overcoming the cost and time of cracking the Chinese market, gaining brand image/name recognition, protecting IP, ensuring payment for services delivered, and navigating a very complex distribution system. Second, because of the cost to export and the limited resources of most small producers, the development of business models that include cost-sharing and collaborative export strategies may help open up more affordable opportunities to export to China. Third, to export successfully, wine producers are likely best to focus on red wine, develop context-specific wine labels, and work towards creating award-winning wines or wines that demonstrate strong reputations.

With this contribution the authors hope to provide researchers in the field with additional insights into the wine export market to China. We invite future contributions with the aim of moving beyond research that focuses on consumers of wine, to additional studies that focus on wine producers, including those in different countries and perhaps those that are large in size. In this way, the gap between academic knowledge and field practices can be reduced.

References

Braun, V. & Clarke, V. 2006. Using thematic analysis in psychology. *Qualitative Research in Psychology*, 3: 77–101.

Bretherton, P. & Carswell, P. 2001. Market entry strategies for Western produced wine into the Chinese market. *International Journal of Wine Marketing*, 13: 23–35.

Brown, E. 2015. Barossa winery makes deal to send two million bottles of wine to China. Available at http://www.abc.net.au/news/2015-05-28/barossa-winery-makes-two-million-bottle-deal-to-china/6503434 (accessed 11 August 2016).

Bryman, A. & Burgess, R.G. 1994. *Analysing Qualitative Data*. Routledge: London.

Camillo, A.A. 2012. A strategic investigation of the determinants of wine consumption in China. *International Journal of Wine Business Research*, 24: 68–92.

De Cremer, D. 2015. Understanding trust, in China and the West. Available at https://hbr.org/2015/02/understanding-trust-in-china-and-the-west (accessed 29 April 2017).

Eisenhardt, K.M. 1989. Building theories from case study research. *Academy of Management Review*, 14: 517–554.

Galbreath, J., Gao, G., Geneste, L., Georgiou, K., Hynes, N. & Weber, P. 2015. *WA Wine Exports: Building an Economic Future with China*. Perth, Western Australia: Bankwest Curtin Economics Centre.

Hoggart, K., Lees, L. & Davies, A. 2002. *Researching Human Geography*. London: Arnold.

Jenster, P. & Cheng, Y. 2008. Dragon wine: developments in the Chinese wine industry. *International Journal of Wine Business Research*, 20: 244–259.

Kambill, A., Long, V.W.-t. & Kwan, C. 2006. The seven disciplines for venturing in China. *MIT Sloan Management Review*, 47: 85–89.

Kitchin, R. & Tate, N.J. 2000. *Conducting Research in Human Geography: Theory, Methodology and Practice*. London: Chapman.

Lincoln, Y. & Guba, E. 1985. *Naturalistic Inquiry*. Beverly Hills, CA: Sage.

Liu, F. & Murphy, J. 2007. A qualitative study of Chinese wine consumption and purchasing implications for Australian wines. *International Journal of Wine Business Research*, 19: 98–113.

Patton, M.Q. 1990. *Qualitative Evaluation and Research Methods*. Newbury Park, CA: Sage.

Paul, J., Parthasarathy, S. & Gupta, P. 2017. Exporting challenges of SMEs: a review and future research agenda. *Journal of World Business*, 52: 327–342.

Platt, J. 1981. Evidence and proof in documentary research: 2: some shared problems of documentary research. *The Sociological Review*, 29: 31–52.

Rabobank International. 2010. *Project Tannin: The Chinese Grape Wine Market*. Adelaide: Grape and Wine Research Development Corporation and Rabobank International.

Schotter, A. & Teagarden, M. 2014. Protecting intellectual property in China. *MIT Sloan Management Review*, 55: 41–48.

Sharma, S. & Vredenburg, H. 1998. Proactive corporate environmental strategy and the development of competitively valuable organisational capabilities. *Strategic Management Journal*, 19: 729–753.

Thorpe, M. 2009. The globalisation of the wine industry: New world, Old World and China. *China Agricultural Economic Review*, 1: 301–313.

Wines of Western Australia. 2014. *Western Australia Wine Industry Strategic Plan 2014–2024*. Claremont, Western Australia: Wines of Western Australia.

Wu, S. 2015. Younger and more sophisticated drinkers mark a new era for wine in China – study shows. Available at https://www.decanterchina.com/en/index.html?article=1143 (accessed 11 August 2016).

Yu, Y., Sun, H., Goodman, S., Chen, S. & Ma, H. 2009. Chinese choices: a survey of wine consumers in Beijing. *International Journal of Wine Business Research*, 21: 155–168.

12 Great Wall or red carpet?

Challenges and opportunities for Australian wines in China

Piyush Sharma

Introduction

A recent study funded by the Bankwest Curtin Economics Centre (BCEC) at Curtin University in Australia provides some important insights into the challenges and opportunities facing Australian wine exporters in the Chinese market, based on interviews with 26 wine producers across multiple Western Australian (WA) regions and six industry stakeholders (Galbreath et al., 2016). According to this report, the export performance of the Australian wine industry had been under a tremendous strain until 2015 due to overproduction, high value of the Australian dollar and the increasing competition from New World wine producers. As a result, most wine producers had become unprofitable with declines in their revenues from exports.

Notwithstanding the above dismal scenario, the recent weakening of the Australian dollar combined with the demand for wine in the Chinese market and a rise in the popularity of Australian wines as well as the recent signing of the China Australia Free Trade Agreement (ChAFTA) bodes well for the future of Australian wines in China, its most important export market. However, there are still many major challenges in 'cracking' the Chinese wine market due to its complex and fragmented marketing and distribution infrastructure coupled with diverse and continuously changing consumer tastes and preferences.

One of the key opportunities in China is the growing demand for premium wines, a segment that has traditionally been dominated by the Old World countries of origin such as France and Italy. However, more recently New World wines from Australia and the USA have begun to find a foothold in this segment, while making strong forays into the mass end of the wine market in China at the same time. However, it is difficult to penetrate the Chinese wine market because of its complicated and underdeveloped distribution infrastructure. According to some reports there are more than 20,000 wine importers and distributors in China who use a wide range of retail as well as food and beverage outlets that sell wine to the end customers (Corsi et al., 2016).

The unique habits and preferences of the Chinese consumers also pose a strong challenge for wine exporters from Australia and other Western countries. Specifically, most Chinese cannot read or write English and they have unique perceptions and preferences for different colours (Corsi et al., 2016). In addition, most Chinese consumers are not very familiar with the history of specific regions that the imported wine is produced in and hence they tend to rely on the brand name or price as the signal of quality (Corsi et al., 2016). Similarly, Chinese consumers have not been used to having grape wine with their food and instead, they have preferred to have either domestic rice wines or other alcoholic beverages such as beer and Moutai (distilled Chinese liquor). As a result, most Chinese dishes have not been paired with the imported varieties of wines, especially the Australian wines (Corsi et al., 2016). Therefore, Australian wine exporters to China need to adapt their products, packaging, pricing and promotions as well as use the most appropriate communication strategies to suit the unique tastes and preferences of their target Chinese customers.

Another key challenge for Australian wine exporters is the differences in the extent to which the various wine regions of Australia are recognised among Chinese wine consumers. For example, regions such as Margaret River in Western Australia have very little global recognition compared to others like Barossa, Yarra and Hunter Valley (Corsi et al., 2016). Hence, there is a clear need for the Australian wine industry to engage with their overseas trade partners in a more meaningful and sustained manner, in order to create, promote and sustain a unique identity for the major wine brands from each of its wine regions in the minds of their target consumers, especially in a large market like China with a huge potential for growth in demand.

In this context, the relatively smaller regional wine producers from Australia have found it particularly useful to join hands with each other on their own or under the umbrella of a national association (e.g. Wine Australia) in order to attain a critical mass and a collective voice in the highly competitive global marketplace. This has also helped them increase their competitiveness in the export markets through collaborative and cooperative efforts, by sharing costs (e.g. shipping, distribution, warehousing and marketing) as well as information (e.g. market intelligence, rules and regulations, government restrictions etc.) (Corsi et al., 2016).

This chapter addresses all these challenges and opportunities faced by the Australian wine industry in China, by exploring the changing consumer attitudes and lifestyles, increasing consumer awareness and growing preference for branded products coupled with improving penetration and repeat purchase rates for wines in China. This chapter concludes by exploring the impact of the recently signed China Australia Free Trade Agreement (ChAFTA) on the demand for Australian wines in China.

The Chinese wine market and Australia

Global trade in wine has doubled in the last 15 years to about 28.3 billion Euros per annum and about 40% of all wine is now exported, up from 25% in early 2000s (OIV, 2016). Consumers in the new wine markets like China are typically fuelling this growth in the global demand for wine, with the Chinese domestic wine industry worth US$7.3 billion per annum in 2014. China already has the second largest vineyard area with 1.97 million acres although it is only the eighth largest wine producer (1.12 billion litres) in the world. However, quite interestingly, China is the fifth largest wine consumer (1.58 billion litres) and the No.1 red wine consumer (1.40 billion litres) in the world (Corsi et al., 2016). In view of the gap between its domestic wine production and overall consumption figures, it is not surprising to note that China is also the No.1 importer of wine in the world, to meet its growing demand.

The Chinese domestic wine market is dominated by local players such as Great Wall, Changyu and Dynasty. Among the countries exporting wines to China, France is the biggest exporter followed by Australia. Australian wine exports to China increased by 8% to 40 million litres worth about A$224 million in 2014 (Wine Australia, 2016). Australia's market share of the bottled wine imports into China also remained strong at about 18% by value and 12.6% by volume in 2014. Australian wines also have a relatively high average value of bottled imports (US$6.83/litre), which is the highest among the top five importing countries (Corsi et al., 2016).

Challenges for Australian wine in China

Low awareness levels

After a period of rapid growth in popularity and initial offtake in consumption, Australian wines seem to have hit a plateau with Chinese consumers. According to the China Wine Barometer (Corsi et al., 2016), the awareness of all foreign brands in China dropped during the 2013–14 period, with the Australian wines dropping by almost 10%, from 76% to 66%, while France remained at the highest level with 97% awareness (China Wine Barometer, 2015).

During the same period (2013–14), the awareness of the origins of most imported wines also dropped, however, the awareness of Barossa Valley in Australia fell the most (-23%) compared to its counterparts from other countries (e.g. Bordeaux in France, Sicily in Italy or Ningxia in China) but similar (-21%) to those from Napa Valley (China Wine Barometer, 2015). These findings further highlight the importance of keeping a close eye on the ever-changing consumer awareness and preferences in a vastly complex market like China.

More recent figures about the awareness of various Australian regions of origin (in China) show that most existing regions (e.g. Barossa Valley, McLaren Vale, Margaret River, Yarra Valley and Hunter Valley) have fairly low awareness levels (below 50%) and even these lower levels of awareness actually declined significantly during the period March–October 2014 (China Wine Barometer, 2015). These are clearly alarming signs for Australian wine exporters. However, on a more positive note, new regions such as Mornington Peninsula, Coonawarra, Clare Valley and Langhorne Creek made an impressive entry into the minds of Chinese consumers, presenting new opportunities to the wine producers in these regions.

Once again, it is extremely important for the Australian wine producers and trade associations to closely monitor these changes in the awareness levels because these reflect the outcomes of ongoing promotions and marketing efforts of their global competitors in the highly competitive Chinese market.

Low penetration and repeat purchase levels

Although awareness levels are reasonably good indicators of market presence of various brands, these figures may not always accurately represent their market performance. Hence, we next look at the market penetration and repeat purchase levels of all the countries of origin in the Chinese wine market. 'Penetration' is defined in this context as the number of buyers of any given product attribute over the total number of shoppers and 'repeat purchase rate' is defined as the buyers of any given product attribute conditional on being a previous buyer of the same attribute.

According to China Wine Barometer (2015), France and China lead all the other countries by a huge margin, with penetration rates of 48% and 30% coupled with repeat purchase rates of 36% and 55% respectively. These two are followed by Italy and Australia with the same but much lower penetration rate of 4%; however, Italy leads with a repeat purchase rate of 28% compared to 16% for Australian wines. Interestingly, Chilean, Portuguese and Californian wines have similar repeat purchase rates as Australia, despite their lower penetration rates at about 2%, which possibly represents a few major cities in China.

In fact, wines from two regions – Ningxia in China and Bordeaux in France – have the highest and very similar market penetration of 26% and 24% respectively, however the repeat purchase rate of 50% for Ningxia is double that of 25% for Bordeaux. Clearly, the Chinese consumers seem to favour their domestic wine for regular consumption, whereas the more prestigious foreign brands could be used for special occasions such as festivals, weddings and for gifting purposes during Chinese New Year and other holidays. Most foreign brands from the other regions seem to have quite low market penetrations (< 10%), yet those from Sicily, Barossa Valley and Napa Valley have quite high repeat purchase rates (18–23%). From these findings, it appears that even though fewer Chinese consumers may buy the wines from these regions, many of them are quite loyal and are willing to purchase these wines again.

Finally, Cabernet Sauvignon seems to be a clear favourite among Chinese consumers with a market penetration of 33% and repeat purchase rate of 38% (China Wine Barometer, 2015). All the other types of wines have significantly lower penetration levels (≤ 5%); however, Grenache, Merlot, Riesling and Malbec have reasonably high repeat purchase rates (18–24%).

Complex and fragmented distribution network

One of the major challenges for any foreign wine producers in entering the Chinese market is to understand and navigate the hugely complex and fragmented distribution network, which ranges from hypermarkets and speciality wine stores in large cities such as Beijing and Shanghai to supermarkets, department stores and grocery stores in other cities (China Wine Barometer, 2015). In addition, there are many wholesalers as well as online retailers and discount stores with a national or regional distribution reach, which further complicate the distribution network. Therefore, it is nearly impossible for the wine exporters from Australia to penetrate all the diverse channels in the Chinese wine market. Instead, it may be more prudent for them to identify and focus on a few key market segments, such as hypermarkets or specialised wine shops or even duty free outlets for the premium brands and possibly online retailers for the regular brands. As their brands find greater acceptance and repeat purchase, Australian wine exporters may be able to move up or across the distribution channel in order to extend their reach to market segments that they were unable to target in the beginning.

Chinese wine consumers – a closer look

A recent study (Qing et al., 2015) shows that Chinese consumers buy wine primarily for two main reasons; for self-consumption or gifting to others. Next, both these motivations are discussed in more detail.

Self-consumption vs. gifting segments

According to China Wine Barometer (2015), mostly younger and well-educated middle to upper income consumers, consisting mainly of executives and professionals, tend to buy wine for self-consumption at home or outside. They tend to spend about RMB67 per month on an average, which is typically the price of one bottle of reasonable quality wine. Hence, it seems that these consumers probably consume wine only once every month for which they buy a fresh bottle of wine every time, either to be consumed at home or with a meal in a restaurant. They seem to prefer buying wine in supermarkets or speciality wine stores rather than online or from overseas/duty-free shops. Interestingly, their main purposes for consuming wine are health, enjoyment and cosmetic reasons. In contrast, those buying wines for gifting purposes tend to be older and higher income consumers who spend almost double the amount on average (RMB127 per month), which means that they either buy a bottle of expensive wine (e.g. French or Italian wine) or more than one bottle of good quality but inexpensive wine.

A close look at these two segments and the positioning of various countries-of-origin of wines in the minds of Chinese consumers (as discussed earlier) shows that Australian wines are likely to be preferred by the self-consumption segment rather than the gifting segment. This is because Australian wines tend to be associated with fun, enjoyment and health, especially among younger Chinese consumers, whereas wines from the traditional producers such as France and Italy tend to be associated with class, heritage and history. However, it is not as if Australian wines do not have any competition in the self-consumption segment, because they compete with wines from new producing regions, such as California, New Zealand

and South Africa, which are all very popular with younger Chinese consumers. To maximise their penetration in this segment, Australian wines would need to increase their width as well as depth of distribution as well as point-of-sale promotions, to generate trials and repeat buying (China Wine Barometer, 2015).

Super premium segment

With the recent spurt in income levels in China, the top 100 richest Chinese hold US$376 billion (2014), which has more than doubled (121%) since 2009 (Forbes, 2014). China also boasts 2.4 million millionaires (2014), up +189% since 2009 (Forbes, 2014). In fact, Ultra-High Net-worth investors put 15–20% of wealth into collectibles, e.g. jewellery, art and fine wine (Barclays, 2012). These rich Chinese business families and social elite have very high expectations from fine wine (Masset et al., 2015), which includes its ability to improve in the bottle, show ageing potential, that it should emanate from a well-known wine growing region, have a long-standing history and reputation, should be awarded high scores by experts and hold high standing in an official classification (Corsi et al., 2016). Unfortunately, Australian wines continue to struggle in this super premium segment because they find it nearly impossible to compete with the relatively better established wines from traditional producing countries such as France, Italy and Spain.

Chinese outbound tourists segment

A recent report by the China Tourism Research Institute shows a rapid rise in Chinese outbound tourism, with the number of Chinese tourists travelling overseas crossing the 120 million mark in 2015, an increase of 25% over 2013 (CTRI, 2016). In fact, Chinese outbound tourists now outnumber even those from the US (68 million), most of whom travel only to Canada and Mexico (55%). South Korea, Taiwan, Japan, Hong Kong and Thailand remain the top five destinations for Chinese outbound tourists, followed by France, Italy, Switzerland, Macau and Germany. Chinese outbound tourists spend about US$300 billion during their overseas trips and their main objectives for overseas tourism include sightseeing, leisure and shopping (WTCF, 2015). However, most Chinese tourists continue to prefer travelling in large package tour groups and they show very low levels of awareness, enthusiasm or interest towards wine tourism (Hussain, Simeon & Sayeed, 2015).

Some researchers argue that the current lack of awareness and interest for wine tourism among Chinese tourists may well present an opportunity for Australian wine producers because the number of Chinese tourists coming to Australia has also grown quite rapidly in recent years, with arrivals in 2015 growing by 22% over 2014 (Becken & Scott, 2016). Moreover, even though Australia only attracts 1.5% of the total number of Chinese outbound tourists (ranked 15th among all destinations), China is now Australia's largest source of inbound visitors, after excluding visitors from New Zealand. Chinese tourists spent about A$6.2 billion in Australia in 2015 (average of A$6,489 per person), which represents 21% of all money spent by overseas travellers (Corsi et al., 2016). From these figures, it is clear that the number of Chinese tourists coming to Australia is likely to continue to grow, and hence Australian wine producers cannot ignore this huge opportunity to create awareness about their products as well as generate sales by attracting these tourists to visit their wineries through organised package tour agencies and other local players.

Performance of Australian wine in China

Notwithstanding the rapid rise in production and consumption of wine, it still accounts for less than 5% of total domestic alcohol consumption in China, which is reflected in a very low per capita consumption, about 1 litre per annum (Qing et al., 2015). While this may present a tremendous opportunity in the long term because even a slight increase in this per capita consumption would accelerate the total demand for wine in China, it also poses a big challenge for Australian wine producers because it restricts the size of its target market in China. Specifically, it is clear that most of the current wine consumers in China are limited to the urban areas and people in the rest of the country are probably not ready to buy or drink wine. Moreover, Australian wines have a stronger presence in the popular segment that has a significantly lower per capita consumption of wine compared to the premium segment in which Australian wines are relatively weaker. Hence, a low per capita wine consumption presents a strong challenge to Australian wines in their quest to grow their presence in China.

In addition to the low per capita consumption, there is very low brand awareness for foreign wines among consumers outside the major Chinese cities and these consumers tend to be relatively more price sensitive. Therefore, Australian wines along with other new players (e.g. United States, New Zealand and South Africa) face a strong challenge in competing with the relatively cheaper and more popular local Chinese wine brands in these smaller Chinese cities. This situation is further compounded by the dumping of cheaper stocks of imported wines from other wine producing countries such as Chile, Portugal and Spain (Corsi et al., 2016).

Most of the sales of wine in China are highly concentrated in the top 11 cities (Shanghai, Guangzhou, Beijing, Shenzhen, Nanjing, Chengdu, Wuhan, Chongqing, Tianjin, Xi'an and Hangzhou), which account for 76% of imported wine sales and 60% of domestic wine sales by volume (Bouzdine-Chameeva et al., 2014). Hence, most foreign wine producers compete very aggressively with each other for market share. Unfortunately, Australian wine producers lack the heritage and premium attached to the traditional wines from countries like France and Italy, and hence, they are forced to compete with the rest of the field for a share of store shelves, wine cellars and dinner tables.

Wine distribution in China is dominated by big wholesalers, hotels, restaurants and bars (80%) although the retail sector and online players are also growing. Unfortunately, most Australian wine producers are relatively small compared to the big wine producers in Europe and elsewhere in the world; hence, they have to rely mostly on the support from wholesalers to push their stocks in the Chinese market. These wholesalers expect significantly higher margins from the relatively less popular Australian wine producers because they need to make extra efforts to sell their products. All this puts pressure on the bottom lines of Australian wine producers and makes it not very profitable for them to sell their wine in China.

The demand for wine in China is highly seasonal, with about 60% of the sales during the two main holidays (Chinese New Year and Mid-Autumn Festival), mostly for gifting purposes, which is an integral part of Chinese culture, as high-lighted earlier. Hence, foreign players need to be ready to bring in and move huge inventories of wine during these couple of weeks every year, which is not pos-sible for most of them, as they lack support from local supply chains in terms of transportation and warehousing. Australian wine producers suffer in particular in this regard due to their relatively smaller volumes because it is not quite viable for them to invest in developing their own supply chains in China and they need to rely more on the existing supply chains compared to some of the bigger players from Europe.

One of the reasons for the low per capita consumption of wine is that it is still considered mainly an aspirational drink to be consumed on special occasions rather than an integral part of the everyday dining experience. This is a chal-lenge not unique to Australian wines but it may partly explain why despite high country-of-origin recall (66%), Australian wines still have very low penetration (4%) and repeat demand (16%) in China (Corsi et al., 2016).

Finally, despite many efforts to promote wine tourism among the rapidly grow-ing numbers of Chinese tourists in Australia, there is still a huge lack of awareness and popularity of wine tourism among outbound Chinese tourists (Corsi et al., 2016). This assumes an even more serious proportion for Australian wines because Australia is not considered as a traditional wine producer by Chinese consumers. Clearly, Australian wine producers and Wine Australia need to work with Chinese tourism authorities and tour operators to promote the various wine regions in Australia (e.g. Barossa Valley, McLaren Vale, Margaret River, Hunter Valley, Yarra Valley etc.) as interesting destinations among the outbound Chinese tourists. For this purpose, they may even offer special package tours to Chinese tourists, com-bining visits to major Australian cities (e.g. Sydney or Melbourne) with free visits to wineries near these cities.

Opportunities for Australian wine in China

Despite several serious challenges faced by Australian wine exporters in China, the potential for growth in the demand for their products can be anticipated to con-tinue to be high in the coming years. As explained earlier, even a slight increase in per capita consumption would result in a dramatic growth in the total demand for wine in China, and this would help Australian wine producers as much as those from the other emerging wine producing countries, such as New Zealand and South Africa. There are many sources for this growth in per capita consumption.

First, more and more consumers are likely to switch from other alcoholic drinks such as beer and traditional Chinese rice wine, because drinking wine is increasingly being associated with being more educated, wealthier and with having higher class and style, especially among younger Chinese consumers (Corsi et al., 2016). Moreover, as the popularity and demand for wine increases, it is likely to spill over to Tier 2 cities and upcountry towns, which would further fuel the per capita consumption of wine. Australian wine producers can possibly be in a very good position to gain a major share out of this market growth from this younger crop of Chinese consumers because their brands are typically associated with fun, enjoyment and good health, which are values much sought after by younger educated Chinese consumers (Corsi et al., 2016).

The Chinese wine market is not as homogenous as some would believe and it has two very distinct value segments; a highly brand-conscious premium segment in major cities and a relatively more price-sensitive mass segment outside the major cities. Once again, Australian wine brands can probably capitalise on the huge opportunities in smaller cities because of their relatively lower prices coupled with their good quality and positive brand associations. Moreover, with the growing importance of offline and online retailers, Australian wine producers may find it possible to expand their distribution reach in this mass segment without relying too heavily on the traditional supply chains and wholesale networks.

Australian wines have enjoyed relatively higher country-of-origin recall compared to their competitors from other emerging wine producing countries, such as New Zealand, South Africa and even the United States (Corsi et al., 2016). Australian wines also enjoy quite a positive image in terms of quality and flavour in the minds of Chinese consumers (Corsi et al., 2016). Australian wine producers need to leverage these high country-of-origin recall and positive product images, to further penetrate the Chinese market, both in terms of increasing depth in its existing strongholds as well as its width by reaching newer markets.

Finally, the China Australia Free Trade Agreement (ChAFTA), which came into force on 20 December 2015 building on Australia's strong and successful commercial relationship with China, secures and improves access to the huge Chinese market across a range of key businesses, including goods, services and investments (DFAT, 2016). In fact, China is already Australia's largest trading partner and according to the Australian government, trade and investment with China is key to Australia's future prosperity (DFAT, 2016).

According to the figures released by the Department of Foreign Affairs and Trade, Australian Government, China bought A$85.9 billion of Australian exports, more than a quarter of Australia's total exports to the world in 2015–16. China is also the top overseas market for Australian agriculture, resources and services. In fact, Chinese investment in Australia has also grown steadily in recent years, reaching almost A$75 billion by the end of 2015. China is Australia's largest export market for agriculture, forestry and fishery products, worth around A$10 billion in 2015–16, which has doubled in the last five years, from A$5 billion in 2010–11.

In this context, China's wine import market provides a tremendous opportunity for Australian wine producers, because it has grown dramatically, more than doubling in size from 2009–10 to be worth about A$3.2 billion in 2015–16. In view of the excellent commercial relations between China and Australia, it is not surprising to see that China is already Australia's second-largest export market for wine, worth A$415 million in 2015–16. However, we must also note that Australian wines compete with those from New Zealand and Chile, both of which have preferential wine access under their FTAs with China. In fact, China's wine imports from Chile have increased almost seventeen-fold in size since its FTA with China entered into force in 2006 (Corsi et al., 2016). Under ChAFTA, tariffs of 14 to 20 per cent on Australian wine imports are being eliminated by 1 January 2019 and tariffs of up to 65 per cent on other alcoholic beverages and spirits are being eliminated by 1 January 2019 (Corsi et al., 2016). All these changes are expected to be of great benefit to Australian wine producers in the next four to five years.

Conclusion and recommendations

From the above discussion, it is clear that the Chinese market presents a huge opportunity for Australian wine producers and exporters but there are also several major challenges faced by them in entering and consolidating their presence in the vastly complex Chinese market. In this regard, Wine Australia and Austrade play important roles by continuing trade education via trade events, trade journals and promotions, as well as helping Australian wine exporters increase their penetration in retail and online channels to reach Tier 2 cities and upcountry markets. Efforts to promote Australian wine at airports and duty free shops and leveraging the relatively weaker Australian dollar at present to boost exports will also help.

Australian wine producers will also benefit from the recently signed China Australia Free Trade Agreement (ChAFTA) as more tariffs are eliminated over the next few years. Efforts to strengthen the positioning of Australian wine as 'fun', 'enjoyable' and 'healthy' will also help expand the wine market in China, by making drinking wine an integral part of everyday dining by Chinese consumers and moving away from the occasional formal wine consumption and gifting during special occasions. Finally, pairing Australian wines with popular Chinese cuisine would also go a long way in cementing their place in the minds of Chinese consumers.

References

Barclays (2012). China's rich top Asia in collectibles ownership. London, UK: Barclays PLC, 1–2.

Becken, S. & Scott, N. (2016). How Australia can capitalise on Chinese tourism. *The Conversation* [Available Online], July 8, 2016.

Bouzdine-Chameeva, T., Zhang, W. & Pesme, J.-O. (2014). The evolution of wine emerging markets: the case of China. *Asian Journal of Management Research*, 4(4): 683–698.

Corsi, A.M., Cohen, J. & Lockshin, L. (eds.) (2016). *China Wine Barometer: A Look Into the Future*. Adelaide, SA, Australia: Ehrenberg-Bass Institute for Marketing Science & Wine Australia, 1–67.

CTRI (2016). China outbound tourism in 2015 [Available Online], https://www.travel chinaguide.com/tourism/2015statistics/outbound.htm

DFAT (2016). China Australia Free Trade Agreement – outcomes at a glance [Available Online, http://dfat.gov.au/trade/agreements/chafta/fact-sheets/Pages/key-outcomes.aspx

Galbreath, J., Gao, G., Geneste, L., Georgiou, K., Hynes, N. & Weber, P. (2016). *WA Wine Exports: Building an Economic Future with China*. Perth, WA, Australia: Bankwest Curtin Economics Centre, 1–128.

Hussain, M., Simeon, R. & Sayeed, L. (2015). Chinese consumers' involvement in wine consumption and their willingness to visit wineries in California. *Atlantic Marketing Association* [Available Online],http://digitalcommons.kennesaw.edu/ama_proceedings/2015/Track3/3/

Masset, P., Weisskopf, J.-P., Faye, B. & Fur, E.L. (2015). Red obsession: ascent of fine wine in China. *European Financial Management Association (EFMA) Meeting*, Amsterdam, Netherlands: 1–38.

OIV (2016). State of the vitiviniculture world market [Available Online], http://www.oiv.int/public/medias/4587/oiv-noteconjmars2016-en.pdf

Qing, P., Xi, A. & Hu, W. (2015). Self-consumption, gifting and Chinese wine consumers. *Canadian Journal of Agricultural Economics/Revue canadienne d'agroeconomie*, 63(4): 601–620.

WTCF (2015). Market research report on Chinese outbound tourist (city) consumption (2014–2015) [Available Online], http://en.wtcf.org.cn/pdf/worldtourismcities10eng.pdf

13 Responses to Chinese tourists' interest in wine and food

An Italian perspective

Michael Volgger and Harald Pechlaner

Introduction

This chapter explores the link between Italian wine and food offers and consumer behaviour of Chinese tourists to Italy from three inter-linked perspectives. The first is an analysis of consumer behaviour of Chinese tourists regarding wine and food while on vacation in Italy. The second is a perspective of current destination responses, which focuses on how Italian tourism supply actors are currently responding to Chinese tourists' interests for wine and food. The third is the forward-looking advancement of destination responses, which investigates how, in future, Italian tourism supply actors might improve their presentation of the Italian wine and food offer to Chinese tourists.

When dealing with these three topics, this chapter adopts a perspective of the sociology of consumption. The underpinning idea is that consumption is not just a bare transaction of material goods or services, or an act to satisfy individual, or even physiological needs, but is deeply embedded into webs of meaning that make up for desires and attach sense to goods (Croll, 2006a). Understood in this manner, consumption has wide social implications ranging from signaling of social status to stratification and identity construction (Bocock, 1993).

Early economists and sociologists, such as Tocqueville (1835/1840), Marx (1859), Durkheim (1893) and Weber (1922) emphasise the symbolic value of purchased goods. Many authors belonging to this stream of thought consider a formalistic reduction of economic behaviour to the mere rational calculations of individualistic utility to be overly simplistic. They rather tend to embrace the notion of culturally embedded economic behaviour (Polanyi, 1944), which suggests an indissoluble intertwining between any type of transaction and the surrounding network of meanings and social relationships. In consequence, any attempt to conceptually separate economy from culture or society is misconceived.

For instance, one could consider situations when 'product utility derives not from personal consumption in a literal sense but from the value placed on the purchase by other individuals or social groups whose opinions are of significant importance to the buyer' (Mason, 1981, p. vii). Consumption that conforms to this description has been called 'conspicuous consumption' (Veblen, 1899), which is not motivated by the need to satisfy physiological needs or by immediate economic cost-benefit-rationality but is rather primarily driven by the objective to impress peers and thus to acquire social prestige and status. Therefore, within this type of consumption, the value in use of the products consumed vanishes in comparison to the value derived from the reaction of others to the showcased purchasing power (Mason, 1981). As this value is relative to one's peers' endowment, that is to what others have and to what they can demonstratively consume, such goods have been termed 'positional goods' (Hirsch, 1977). In this vein, this chapter explores the interplay between Italian wine and food offers and Chinese tourists' demand by interpreting consumer behaviour within its wider context of meaning.

Consumption in contemporary China

We explicitly acknowledge the extraordinary diversity within China and thus are aware of potential risks of simplification and, in addition, are willing to avoid an exaggerated essentialism. Nevertheless, it might be a useful attempt to work out noticeable features in contemporary consumption in China. For instance, scholars argue that economic interactions in China need to be conceived at the interface of three overlapping socio-economic logics: the socialist state-driven, redistribute economy, the capitalist economy and what has been termed 'the gift economy' or 'ritual economy' (Latham, 2006; Stafford, 2006). Commentators also refer to a 'new materialism' that allegedly filled the gap left by the mortification of meaning-providing Maoism (Tang, 1996; Latham, 2006). Today, it seems that identity in everyday life in China is at least partly defined by consumption (Liu, 1997).

This shift towards the 'need for greed' (Sedláček, 2011) has unfolded itself in a highly expedited manner. Since the advent of the reform government in 1978, China has literally seen a consumer revolution: '[C]onsumption symbolised the new freedom after years of controlled scarcity' (Croll, 2006b, p. 23). Consumption of more varied food was among the earliest harbingers of the new freedom and warmly welcomed, especially in urban China (Croll, 2006a). Western fast food had a non-negligible role in this process. Altogether, this illustrates that the sudden and vigorous turn to consumer goods in China must at least partly be conceived as a compensating behaviour with respect to past deprivations.

Deng Xiaoping, the key figure of the first reform generation, started to advocate that 'to get rich was glorious' (Croll, 2006b, p. 23). This notion increasingly reflected itself among many Chinese in a strong desire for visiting shopping malls, gazing at luxury products and shopping for leisure (Croll, 2006b). More importantly, goods developed into signifiers of difference. In the 1990s, increasing preoccupations of maintaining authenticity and 'Chineseness' accompanied this triumph of Western-born consumerism (ibid).

Direct, inter-personal relations (*guanxi*) traditionally have played a fundamental role in Chinese societies (Sun, 1983), but blossomed in centrality especially over the reform years and got intrinsically linked up with the exchange of goods (Yan Yunxiang, 1996; Croll, 2006b). Gift giving, or what Stafford (2006, p. 53) calls the 'ritual economy', occupies a central role in everyday life: '20 per cent of household net income in one [. . .] village is used for gift exchange.'

To sum up, Chinese consumption patterns are strongly embedded in a system of vertical and horizontal inter-personal ties. Within this system, ostentatious and demonstrative consumption can support the signaling of status and the acquiring of prestige. Wealth is probably the strongest signal of status in contemporary China (Blok, 2002). Thus, being able to afford exchange through gift giving can support maintenance and extension of one's network, which has become so vital in the reform years (Croll, 2006a). Foreign brands and certain eating and drinking habits have become recognised means to indicate and reaffirm social status. This might be linked to a recently observed increase in demand for premium food, imported food and processed food within China, while significant differences between rural and urban areas as well as between poor and rich segments of China's society should not be neglected (Zhou et al., 2014).

In an overview of Chinese wine consumption, Liu and Murphy (2007) argue that regular alcohol consumption in private contexts in China is focused on traditional Chinese spirits (produced from grains such as rice) and, occasionally, cheap wines. This is empirically confirmed by a relatively limited per capita consumption of wine and happens despite China having a long history of (elite) wine consumption and domestic wine production. However, especially well-educated and wealthy Chinese specifically use expensive foreign red wine as a means to foster social status in the *mianzi* (i.e. face value) and *guanxi* systems on particular occasions such as important business meetings, socially relevant get-togethers, feasts or holidays (Tang, 2006; Liu & Murphy, 2007). Liu and Murphy (2007) point out that in China general knowledge on wine is limited, which leads to price and country of origin (e.g. France) being key indicators for the symbolic value of wines.

Not less, travel abroad, which remains a relatively rare practice in China, is a strong symbolic consumption act that signals living a 'good life' (ibid). Although having a long history of travelling within the country, today's growing Chinese outbound tourism can only be understood in a context of increasing consumerism enforced by a mimicking of Western lifestyles and tourism behaviour (Zhang, 1989; Shu, 1995; Arlt, 2006). In this context of contemporary China, outbound travel is recognised to deliver sense and meaning, and to impress peers (Blok, 2002).

Chinese tourists in Italy

With 135 million outbound travellers in 2016 (including same day visitors), the 'latecomer' China has recently grown to world's leading country in outbound tourism (UNWTO, 2017). After a first wave of tourism development focused on domestic movements and trips to Hong Kong and Macau, Chinese outbound tourism took off in the mid 1990s (Arlt, 2006). It needs to be noted that short-haul trips to Hong Kong, Macau and Taiwan account for approximately 70% of the Chinese outbound tourism, with about 20% travelling to other Asian countries and 10% travelling to long-haul destinations (Dai et al., 2017).

In 2015, 12 million Chinese tourists visited Europe (European Commission, 2016). Outside the Asia-Pacific region, Europe is the number one destination for long-haul Chinese outbound tourists (ibid). Within Europe, Italy and France are the most visited destinations, accounting for around 23.0% of Chinese tourism arrivals in the EU each year (ibid). Among the nights that all non-EU tourists spent in Italy in 2014, 6.1% were from China (EUROSTAT, 2015). Most remarkably, the number of Chinese visitors in Italy almost tripled in the ten years between 2003 and 2013 (ISTAT, 2014).

Chinese outbound tourists to Italy tend to concentrate in a few world-famous cities, such as Venice and Verona in the Veneto region (27%), Milan in the Lombardy region (26%), Florence in the Tuscany region (26%) and the capital Rome in the region of Lazio (18%) (Osservatore Nazionale del Turismo, 2011). While the focus of Chinese tourists' visits in these cities are cultural attraction points (all of them offer UNESCO sites) and shopping opportunities (Guo, 2014), all listed cities are located nearby or at the centre of well-established wine growing regions. According to experts, food and wine experiences are decisively gaining in importance for Chinese visitors in Italy; at the same time, their full integration into Chinese travel patterns remains limited by an under-promotion of non-urban parts of Italy to Chinese visitors (Dall'Ara, 2016). In addition, weaknesses on the destination image level emerge: only 20% of recently interviewed Chinese indicate Italy as the number one wine producing country (well beyond France) (Dall'Ara & Grötsch, 2014). In contrast, another study indicates that Chinese rank Italian food ahead of French food as the most preferred among Western cooking styles (ibid).

Wine and tourism in Italy

A detailed account of the extensive Italian food and wine sector is beyond the scope of this chapter. Therefore, this chapter limits itself to exemplify the rich endowment of Italy with food and wine offers by discussing a few peculiarities of its wine production and pointing out some weaknesses in integrating wine with tourism products. Although Getz and Brown (2006) illustrate that wine tourism products are embedded in a context where culinary experiences are most important, due to space limitations, the food sector will not be touched upon.

Having a centuries-old tradition in wine growing even before the Romans, Italy is among the most important producers, exporters and consumers of wine worldwide. In 2015, with an estimated production quantity of 48.9 million hl, Italy has become the biggest wine production country worldwide (OIV, 2015a,b). The country ranks fourth regarding the 'area under vine' (surface), second in wine export (both in volumes and values) and third in wine consumption (OIV, 2015a,b). Italy differs significantly from its main competitor France in its characteristics in wine growing. For instance, in contrast to France, wine is produced across all Italian regions. This leads to a regionalised character of wine production, which is reflected in a high variety of cultivated grapes and terroirs. Approximately 350 registered indigenous grape varieties are cultivated in more than one million Italian vineyards (Filiputti, 2016).

This wine-related set up is translated into a variety of tourism products ranging from more than 100 wine roads, a variety of farm holidays, authentic restaurants ('*osterie*'), slow tourism and slow food offers, food and wine events such as the world-famous exposition 'Vinitaly', wine tasting and visits to wineries (*'cantine aperte'*, since 1993) as well as a number of high-profile publications (e.g. Gambero Rosso), which play an important role in canonisation and quality assurance (Croce & Perri, 2010; Mason & Paggiaro, 2012; Lemmi & Siena Tangheroni, 2015).

As far as China is concerned, Italy is (only) the fifth most important wine-export country for bottled wine in volume and value (I numeri del vino, 2016). This indicates that, while Italy is a major player in international wine production and has shown strong recent growth figures in the export of sparkling wines (e.g. Prosecco), it is currently not the most desired country of origin for wine in China. As an indication for this assertion, the keyword 'Chianti', one of the internationally most well-known Italian wines, receives about 1.2 million search results on the Chinese search platform www.baidu.com (in May 2017), whereas the entries relating to French 'Bordeaux' receives 5.6 million.

This refers to a major weakness of international wine-related tourism in Italy. Estimations speak of about 3% to a maximum of 8% of all tourist arrivals in Italy being somehow wine-related (Festa et al., 2015). Although having a strong resource endowment and product base, following Festa et al. (2015), Italy could do even better in wine-related tourism if it managed to increase the development of shared wine tourism strategies and marketing approaches, if collaboration among wine producers was improved and if – most importantly – it enhanced its overseas prominence of smaller, high quality wine growing areas. The diversity and peculiarity of Italian grape varieties turns out to create confusion on the international wine markets and on the wine-related tourism markets as well. Thus, the fragmented and complex nature of Italian wine production, at least in a short-term perspective, seems to limit success on international markets that are driven by standardisation and recognisability. However, the authors of this chapter are convinced that, if managed and communicated well, in the long run this endowment is likely to become a unique selling proposition in wine-related tourism.

Wine- and food-related tourism decision-making

From a consumer perspective, scholars speak of 'wine tourism' when wine offers such as visits to vineyards, wine tasting or the attractiveness of wine growing landscapes are major motivation factors in tourists' choice making and behaviour (Hall et al., 2000; Byrd et al., 2016). It is obvious that there is a range of different types of wine tourists. Tourism researchers repeatedly suggested a segmentation approach that refers to the degree of interest in and knowledge about wine. It posits a continuum ranging from 'wine lovers' (and 'wine connoisseurs') to 'wine novices' (sometimes termed 'curious tourists'), whereas the middle category of 'wine interested' is probably the most prevailing (Charters & Ali-Knight, 2002). Similarly shaded categories can be applied to the relationship between food and tourism (Hall, 2003).

This chapter will take a model on wine tourism involvement developed by Sparks (2007) as its point of reference, and, given the above, expand its application to food-related tourism. This model is chosen because it adopts a broad perspective on wine-related tourism decision-making and does not only focus on cellar door behaviour. In addition, Sparks' model fits with the above outlined principles of the sociology of consumption, in particular with the consideration of peer groups in travel behaviour and decision-making. Sparks (2007) argues that consumers are more likely to develop an intention to participate in wine-related tourism if they

(1) value the experience (*wine tourism experience*),
(2) expect the peer group to approve the behaviour (*normative influence*),
(3) believe they have the resources (*perceived control*), and
(4) have previous experiences with wine and tourism (*familiarity*).

According to Sparks, factor one, the wine tourism experience, consists of three elements, namely the wider destination experience, the core wine experience and personal development. Other research has confirmed that winery service quality (which resembles what Sparks refers to as the 'core wine experience') and complementary products (which is similar to the above 'destination experience') are major features that may positively trigger the perceived wine tourism experience (Carlsen & Boksberger, 2015; Quintal et al., 2015; Tsai & Wang, 2017). The third element, termed 'personal development' by Sparks (2007), is demand-orientated and emphasises that wine tourists, and in particular the segment of wine lovers, have a profound interest in education about wine (Charters & Ali-Knight, 2002).

Sparks' factor 'normative influence' highlights the influence of peer groups in deciding whether to engage in wine-related tourism or not. Although under-researched, the influence of reference groups on travel behaviour is recognised in tourism research (Woodside & Ronkainen, 1980; Bieger & Laesser, 2004). For instance, regarding Chinese tourists, Hsu et al. (2006) demonstrate that opinions voiced from family, relatives and friends are more influential on travel decision-making than a travel agency's advice.

Sparks' evidence supports the claim that 'perceived control' of tourists over the wine-related trip influences behaviour. This variable relates to subjective feelings regarding financial and time resources and regarding having the appropriate knowledge to undertake the trip. Lai et al. (2013) find that time may be an especially relevant constraint for Chinese outbound tourists.

Finally, Sparks' category 'familiarity' indicates that past tourism experiences located at the interface with wine influence current and future behaviour. The relevant role of past experiences has been shown repeatedly (Hall et al., 2000; Sparks, 2007), and highlights the importance of quality in wine-related tourism.

Research design and method

The purpose of this study is to investigate consumer behaviour of Chinese tourists regarding wine and food while on vacation in Italy and to link that up with current and potential destination responses. As little research has been conducted on that particular topic, the research design was exploratory and followed a qualitative approach. We conducted five semi-structured interviews with suppliers or intermediaries at the interface between Chinese outbound tourism to Italy and Italians' wine- and food-related offerings. Interviewees' selection was guided by principles of knowledgeability and diversity of views (Eisenhardt & Graebner, 2007). In detail, we interviewed destination managers, politicians, tour guides and tour operators. Interviewees were selected via snowball sampling starting with two key stakeholders, one being a previous CEO from the Italian National Tourist Office (ENIT), the other one the mayor of one of the most visited tourism places in the midst of the Tuscany wine regions, the UNESCO World Heritage declared city of San Gimignano. The collected data was analysed with GABEK, which allows for a semantic keyword-based analysis of qualitative interviews. It transforms statements into semantic networks or networks of concepts, which in the final output resemble mind maps (Zelger, 2000; Pechlaner & Volgger, 2012).

The result section uses the output of causal net graphs, which are based on a coding of suggested cause-and-effect relationships included in the statements of interviewees. Let us imagine, for instance, an interviewee suggests that '[t]he Chinese wine tourist exists. They are businessmen who come to Italy to make business.' To capture its key message, this statement might be coded by using the following keywords: Chinese tourist, wine, business, travel motivation and Italy. The list of keywords should, as a set, roughly represent the statement. After all interviews have been coded, GABEK depicts these statements as networks of keywords in at least two varieties: as association graphs and as causal net graphs. Whereas association graphs link up all keywords that appear together at least in one interview statement, causal net graphs are the product of an additional causal coding. Here, perceived cause-and-effect relationships as alleged by interviewees are coded and then represented by directed relationships (i.e. arrows) in the network graphs.

For instance, in the above example statement, one might interpret the reported link between 'business' (tourism) and 'wine' as a suggested causal link and represent the relationship between these keywords with an arrow pointing from 'business' to 'wine'. The following results section draws on causal net-graphs and contemporaneously links keywords to (1) inductively produced higher order concepts and (2) to the theoretical categories derived from Sparks' model (2007). The inductively produced concepts are given in bold and the appropriate theoretical concepts are indicated in brackets.

Results and discussion

The following selected webs of constructs and meanings emerge from the interviews. They are arranged with regards to whether they indicate interpretations of the current situation or whether they provide suggestions to improve the future offer at the interface between Italian wine- and food-related tourism supply and Chinese outbound tourists' interests.

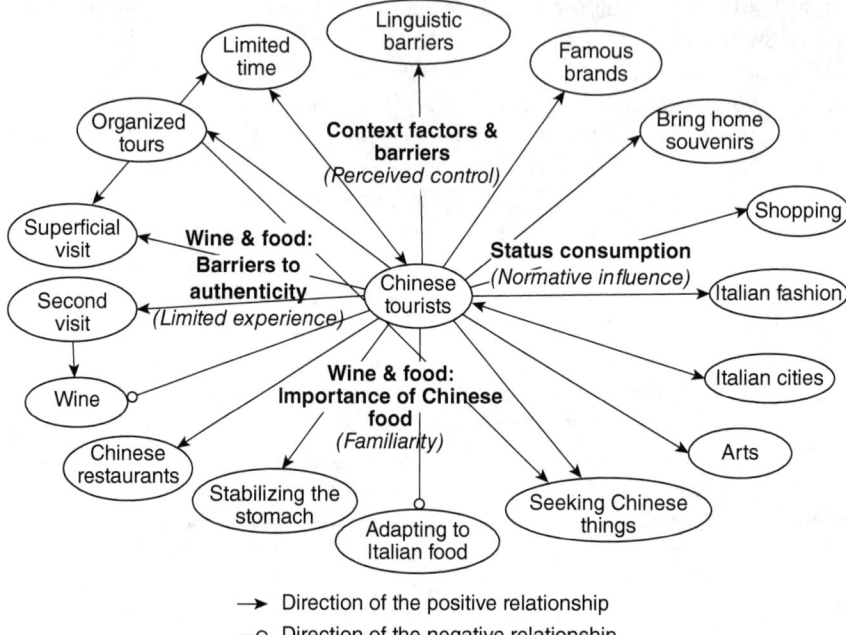

Direction of the positive relationship
Direction of the negative relationship

Figure 13.1 GABEK-causal net graph indicating aspects of interviewees' understanding
of the current situation regarding Chinese tourists in Italy

Note: All shown relationships are based on interviews. Coded interview keywords appear in ellipses;
inductively produced concepts are given in bold and the appropriate theoretical concepts are indicated
in brackets. All shown relationships are supported by at least one interview statement.

Current situation

Figure 13.1 illustrates major associations of interviewees with Chinese tourists
in Italy. First, respondents argue that the current situation of Chinese visitation to
Italy needs to be understood in the context of their limited available time, of existing linguistic barriers and the prevailing trip form of organised tours. Referring
to the apposite theoretical category suggested by Sparks (2007), this indicates
limited perceived control over the situation.

Second, referring to another but related theoretical category, according to
interviewees, *wine and food experiences* of Chinese visitors in Italy remain
limited by a travel behaviour that does not facilitate access to authentic experiences. Interviewees spoke of an impression of 'superficiality', which might be
linked to scarcity in available time compared to the planned travel itinerary,
or to a high awareness of value for money (i.e. efficiency) as argued by Arlt
(2006). In contrast, Chinese second time visitors seem to have a higher likelihood to get in contact with authentic wine- and food-related experiences in
Italy, as they might value it more:

This middle-class Chinese tourist tends to behave a bit 'superficially' regarding Italian food and wine. Often, they cannot afford to go to better restaurants, and end up at a restaurant proposed by the tour operator [. . .]. In this way, the Chinese tourist is not able to have the experience of slow Italian food. (Interviewee 2)

The problem is that tourists who do the traditional tours prepared by tour operators that organise tours for small time budgets, do not meet touring guides that try to explain things more in detail. (Interviewee 1)

Third, according to interviewees, what drives Chinese tourists' interests into visiting Italy is the cultural heritage in major Italian cities, and Italy's notoriety for fashion and shopping, which is reified in internationally renowned brands. Interviewees feel that the aim of bringing home (valuable) souvenirs is a major factor underpinning these interests. Thus, a primary motivation of Chinese tourists in Italy appears linked to what has been discussed above as ostentatious consumption and to what Sparks (2007) calls *normative influence* because of its underlying objective to impress reference groups:

The middle-class Chinese tourist comes to Italy because some cities in Italy are regarded as fundamental stages to elevate the status of the middle-class Chinese. So, a Chinese who has the economic capacity to afford a trip abroad visits Italy exactly because of its cities that are known worldwide for their symbolic value. (Interviewee 2)

Finally, when it comes to consuming Italian food and wine while being on vacation in Italy, according to the interviewees, access to familiar food remains important for Chinese tourists. Supported by their tour guides, they seem to be proactively seeking Chinese food and restaurants:

In Italy, Chinese tourists need to deal with tastes they are not accustomed to, e.g. the aged cheese, that they dislike. But they love fish dishes; often they order spaghetti with seafood. (Interviewee 5)

This interest in Chinese food abroad is often justified in biological terms with the need to 'stabilize the stomach'. Full adaptation to authentic Italian eating and drinking habits appears to remain challenging for visiting Chinese, which underscores the relevance of the *familiarity* category to wine- and food-related travel behaviour. Opposing a too individualistic interpretation of this behaviour as 'uncertainty avoidance', Arlt (2006) offers an alternative reading by arguing that such familiarity-seeking food habits abroad can be seen as a signal to context-boundedness of Chinese if not a contribution to the above-mentioned 'Chineseness'.

Figure 13.2 summarises the reported current situation at the interface between wine- and food-related tourism supply in Italy and the respective Chinese demand. Chinese tourists in Italy seem to be motivated by acquiring status symbols while on vacation in Italy. However, authentic but internationally less known

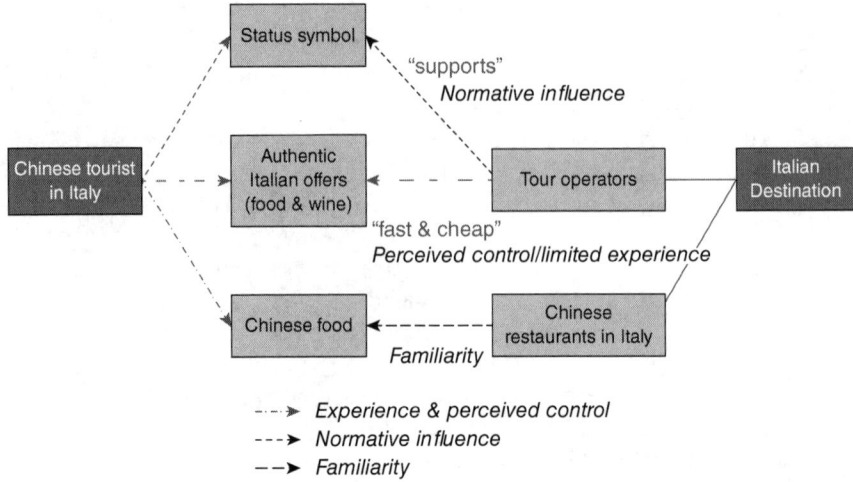

Figure 13.2 Supply–demand interaction in the current situation

Italian food and wine offers do not necessarily fall under this category of socially recognisable status symbols. Chinese visitors to Italy seem to seek typical Italian food and wine offers, but these often need to fulfill the 'fast and cheap' efficiency criterion. All these aspects underscore the relevance of the theoretical categories of *normative influence, perceived control* and *tourism experience* to describing and understanding the behaviour of many Chinese visitors to Italy in a wine- and food-related context. On the supply side, due to the organised nature of many of these trips, the tour operator plays a crucial role in ensuring that the status-seeking objectives are fulfilled and the interest into authentic Italian food and wine offers is balanced with available time and budgets. Moreover, eating Chinese food seems to remain important for Chinese visitors while on vacation (*familiarity*), making Chinese restaurants a relevant supply factor for Chinese tourists in Italy.

Beyond these general considerations, the GABEK association graph reported in Figure 13.3 illustrates interviewees' views about specifically wine-interested Chinese tourists to Italy. Indeed, when questioned about the sheer existence of such tourists, interviewees confirm that wine-interested tourists from China occur. They are not reported to be an exceptionally big group but increasing in numbers. Interviewees see this niche to be located mainly within the luxury segment, i.e. where *perceived control* over the trip tends to be relatively high. For instance, one interviewee argued:

> The Chinese wine tourist exists, there is interest. Chinese use wine less for everyday use, as we [Italians] do, but rather as a gift for friends. They come to visit the wineries for wine tasting. They also appreciate the process and the culture behind it. (Interviewee 5)

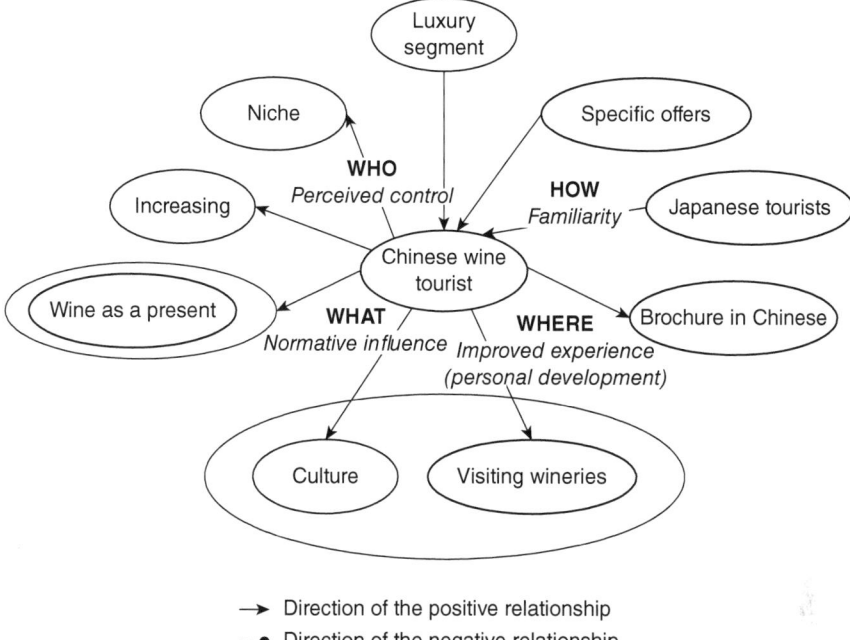

Figure 13.3 GABEK-causal net graph indicating associations of interviewees regarding wine-interested Chinese visitors to Italy

Note: All shown relationships are based on interviews. Coded interview keywords appear in ellipses; inductively produced concepts are given in bold and the appropriate theoretical concepts are indicated in brackets. All shown relationships are supported by at least one interview statement.

This statement indicates at least two aspects: It strengthens the already suggested hypothesis that wine-related tourism behaviour of Chinese visitors in Italy must in many instances be understood in relation to gift-giving activities. This enforces the relevance of Sparks' (2007) *normative influence* category. Moreover, the statement highlights that Sparks' *wine experience* category is no less crucial: it emphasises that visits to wineries are means to experience authentic Italian culture and Italian winemaking.

Interviewees argue that mitigating the newness of some offers and facilitating access is highly appreciated by wine-interested Chinese visitors. Practical and functional adaptation in the sense of specific offers or brochures in Chinese may help to reduce un-*familiarity* and thus to boost Chinese tourists' wine experience. According to interviewees, due to the cultural distance, wine experiences can be improved by a mitigation of cultural distance. However, it is not the first time that Italian wine-related tourism suppliers are dealing with relatively big cultural distances. Interviewees refer to consistent waves of Japanese tourists visiting the country, which led to the development of targeted information material. Such association of Chinese travellers to Japanese is often observed for tourism service providers in Western contexts. However, while parallels in cultural distance might exist, it cannot be emphasised clearly enough that culturally imprinted behavioural patterns differ highly between the two countries (Arlt, 2006).

Potential for improving offers and fostering demand

As indicated in Figure 13.4, interviewees recognise that Italian wine and food is currently not among the most prominent travel motives for Chinese tourists that choose Italy as their travel destination. In comparison to other source markets, such as Germany, Scandinavia or the United States, it seems that the food and wine aspect has less importance relative to other features of the vast Italian tourism offer; however, it plays a role for Chinese visitors as well.

To increase the prominence of Italian wine and food as a major travel motive to visit Italy (which has the advantage that it is also highly suitable to repeat visits), interviewees provided suggestions that can be linked to the theoretical categories of *perceived control, improved food and wine experience* and *normative influence.*

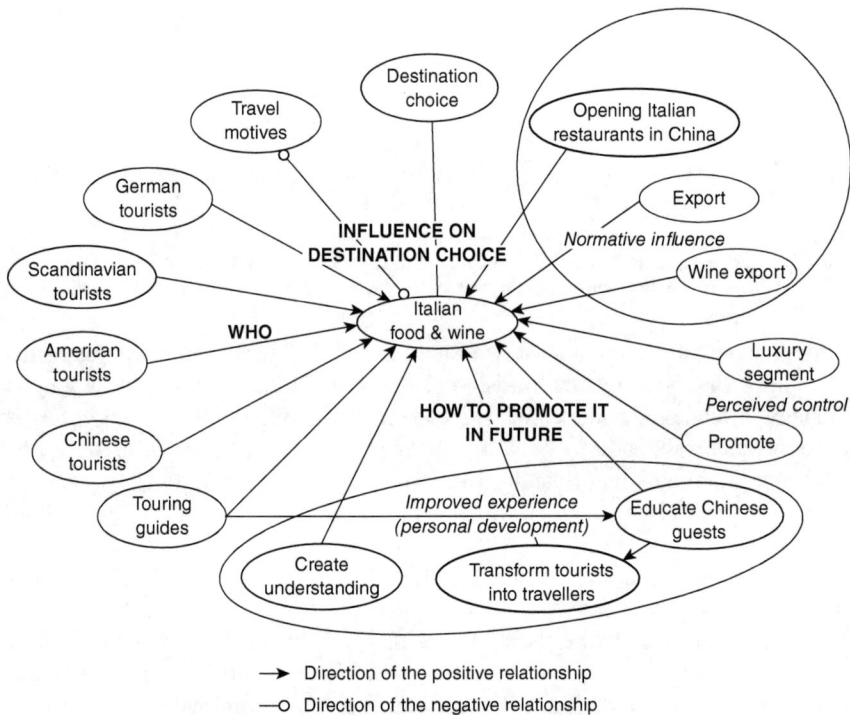

Figure 13.4 GABEK-causal net graph indicating selected ideas of interviewees to improve the food- and wine-offer for Chinese tourists in Italy

Note: All shown relationships are based on interviews. Coded interview keywords appear in ellipses; inductively produced concepts are given in bold and the appropriate theoretical concepts are indicated in brackets. All shown relationships are supported by at least one interview statement.

According to interviewees, the *perceived control* factor is linked to targeting particular groups such as the luxury segment, which have more time and resources available to experience authentic wine and food offers in Italy. In addition, tour guides seem to be crucial gatekeepers for the selection of wine and food experiences for group tourists as well as regarding their (cultural) translation:

> One thing is fashion, the second thing will be food and wine. [. . . T]our operators increasingly put taste among their offers [. . .], taste and fashion. It is like giving lessons on the [Italian] way of wearing certain clothes, to creating a certain style. This style will then be imitated. In the field of wine and food you must do this work as well. The tasting is a good starting point. (Interviewee 3)

In this vein, tour guides are key players in improving the *food and wine experience* of Chinese visitors. According to interviewees, tour guides and all service providers involved need to try to ensure that, to speak in terms of communication theory, the message is sent in such a manner that it can reach the recipient. In other words, interviewees hold that a style of communication and presentation conscious of the inter-cultural situation is required:

> It is necessary to proactively present the quality by making the Chinese visitors try the products together with a local and a Chinese-speaking person that can really convey the product quality. (Interviewee 2)

Moreover, interviewees maintain that to boost interest in wine and food offers, tour guides ought to support the transformation of more sightseeing-directed and ticking-off-orientated visitation patterns into experience seeking patterns:

> The potential is enormous, but it is necessary to 'educate' the tourist to make he/she aware of the quality of things he/she encounters and it is necessary to support him/her in these encounters. (Interviewee 3)

In everyday parlance, this difference is sometimes referred to as a contrast in connotations between the concepts of 'tourist' and 'traveller' (Theobald, 2005). In this context, however, it is interesting to note that this contrast in semantic connotations between tourists and travellers is markedly less pronounced in China and Chinese 'tourists' need to carry less negative semantic ballast (Huang & Chen, 2016).

Finally, according to interviewees, Italy might also play its cards related to the theoretical construct of *normative influence*. One suggestion relates to facilitating the gift-giving mechanism in a very practical manner by removing export barriers for tourists:

> A useful thing could be to give Chinese tourists the possibility to buy here and then to ship the product to China. At home, if Chinese people go to the market they do not fully trust the products they find because they could be fake. Unfortunately, they cannot buy too much while they are here on

vacation. So, it is necessary to find a mechanism that makes it easier to send the purchased products to China. (Interviewee 5)

Other suggested instruments are not confined within the Italian territory or to tourism in a narrow sense of the word, but are related to wine export and the creation of desire and symbolic value within the source market context, i.e. prior to tourism decision-making. As the argument goes, Italy should promote its wine export to China to increase familiarity and prominence of Italian wines as recognised status symbols. Status symbols are always socially constructed and as such subject to processes of creation and replacement. Thus, it seems that a long-term investment into the promotion of Italian wine as a status symbol in China might be helpful to improve demand for related tourism offers in Italy. According to the interviewed tourism service providers, the same holds for Italian food, where Italian restaurants in China may play a central role:

> I am in favour of promoting Italian wine and food directly in China. I think it is not enough to say 'Made in Italy', which is not an absolute synonym for quality in the eyes of Chinese people. (Interviewee 2)

Figure 13.5 summarises the instruments that interviewees proposed to further increase demand and to improve suitable supply around wine- and food-related tourism offers for Chinese visitors to Italy. From the interviews it emerges that it

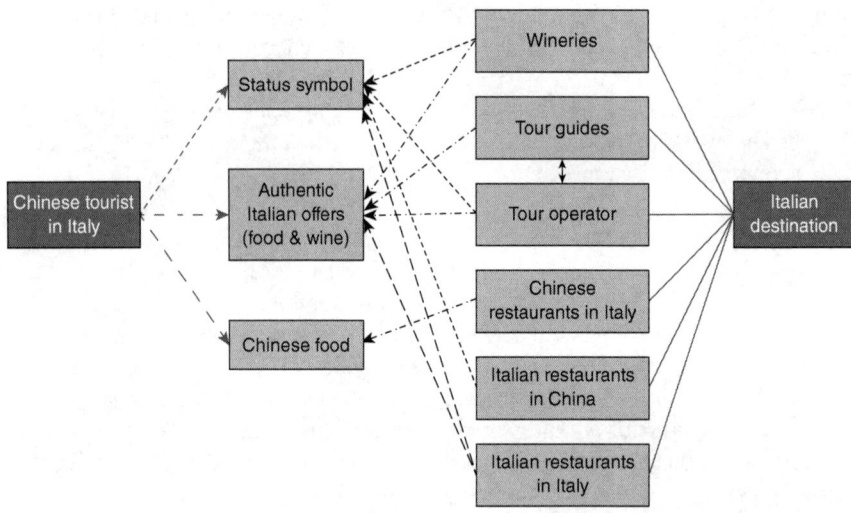

Figure 13.5 Potential for improved supply–demand interactions in the future

is probably necessary to employ instruments that are located both in the destination and in the source country and that are not limited to tourism in a narrow sense. Interviewees suggest combining different buttons on Sparks' claviature (2007):

1) Service providers as manifold as the Italian wine industry, tour operators and Italian restaurants in China seem to have a crucial role in preparing the ground in a medium-term outlook by making sure that Italian wine is positioned more strongly as a status symbol – i.e. as a *brand* (Arlt, 2006) – in China;
2) Wineries, tour operators and tour guides also have a pronounced responsibility in guaranteeing memorable wine and food tourism experiences in Italy, where the gradual introduction of Chinese outbound tourists to authentic and traditional wine and food offers is of paramount importance. The consideration of the authenticity dimension seems crucial because Chinese visitors strive for typical offers as argued previously (Arlt, 2004);
3) Finally, Italian restaurants, conscious of the inter-cultural distance, shall make an effort to adjust their offerings to Chinese tourists, without renouncing authenticity. Such a play of mitigating unfamiliar taste needs to keep a balance between paying respect to the visitor and its 'Chineseness-affirming behaviour' (Arlt, 2006) on the one hand and over-adjusting or over-dressing authentic offers on the other hand. Showcasing local specificities in a manner that is perceived respectful to China and Chinese appears to be a promising strategy.

Conclusion and implications

This chapter explored wine- and food-related consumption of Chinese tourists in Italy and linked the demand perspective to current and potentially improved destination responses. The chapter employed a lens rooted in the sociology of consumption regarding aspects of conspicuous, status-orientated consumption patterns. The literature review suggested that such forms of consumption continue to be relatively widespread in contemporary China, where both the capitalist economy and 'the gift economy' play significant roles (Croll, 2006b; Latham, 2006; Stafford, 2006).

The empirical part of this exploratory study indicated that wine and food motives are still relatively contained when it comes to major pull factors for Chinese to visit Italy. Moreover, the study illustrated that the reasons for this situation can be caught in a parsimonious manner by applying a model on wine tourism involvement developed by Sparks (2007), thus confirming the empirical applicability and suitability of the model in question. As Sparks' model suggests, (1) the value of the wine tourism experience, (2) the belief of having resources to afford the experience, (3) the feeling to be familiar with the specific wine and tourism context and (4) the expected effects on the peer group are critical factors to interpret the limited but increasing interest into current wine and food offers in Italy among Chinese visitors.

Considering the adopted theoretical lens rooted in the sociology of consumption, this study leads to the following hypothesis: To understand current consumption patterns of Chinese visitors to Italy regarding wine- and food-related offers and to achieve future improvements at the interface between demand and supply, it is necessary to consider this consumption not only as being driven by individualistic experience seeking (i.e. confined to Sparks' first category) but also as being positioned within the context of the Chinese gift economy. In brief, it is hypothesised that Italian wine needs to strengthen its symbolic value as a (recognised) status symbol in China in order to promote wine- and food-related tourism of Chinese visitors towards Italy. The motivations behind such visits seem to have considerable parallels with what Veblen (1899) coined 'conspicuous consumption' and Dai et al. (2017) named 'irrational consumption'.

This hypothesised finding implies that getting the socially embedded impression-making mechanism right is a key requirement for promoting wine- and food-related visitation of Chinese to Italy in a medium-term outlook. Inbound/outbound tour operators are asked to help by highlighting Italian wine as a status symbol and to gradually introduce authentic wine and food experiences into their travel packages. Italian wineries and restaurants shall adapt their offers in a manner that facilitates adoption by Chinese visitors, e.g. by ensuring that the experience communication is cross-culturally appropriate.

However, in this perspective, it is not enough to employ short-term marketing measures, but it is vital to improve the symbolic value attached to products. This does not necessarily mean that economic classes and their respective cultural symbols need to be reified through food and wine communication (by making reference to 'luxury' etc.) (de Jong & Varley, 2017), but it emphasises the need of proactively assembling and consciously modelling the symbolic value of culinary offers. Hardly any single tourism organisation, single inbound or outbound tour operator or any tourism service provider is currently positioned to implement a long-term strategy towards influencing consumer behaviour in distant markets. Hence, this implies a requirement for strengthening collaborative agency, strategic outlook, cross-sectoral understanding, organisational set-ups and umbrella branding approaches to improve consistency as well as comprehensiveness of the storytelling. A narrow conception of tourism destination management seems to be more inappropriate than ever. This is worsened by a tourism destination marketing that misinterprets calls for an ever more individualised quest for experiences. It risks conceiving tourism behaviour in a manner that does not account for the profound social implications and multi-faceted cultural embeddedness of consumptive behaviour. This is a major problem not only but especially with Chinese outbound tourists.

Acknowledgement: The authors would like to thank Greta Erschbamer for her support in data coding.

References

Arlt, W.G. (2004). Chinesischer Outbound Tourismus in Deutschland. Entwicklung – Perspektiven – Herausforderungen. In: Maschke, J. (ed.), *Jahrbuch für Fremdenverkehr 2004* (pp. 7–34). München: DWIF.

Arlt, W.G. (2006). *China's Outbound Tourism*. Abingdon: Routledge.

Bieger, T. & Laesser, C. (2004). Information sources for travel decisions: toward a source process model. *Journal of Travel Research*, *42*(4), 357–71.

Blok, A. (2002). China social anthropology study. Scandinavia Tourism Research Report, Scandinavian Tourist Board.

Bocock, R. (1993). *Consumption*. London, New York: Routledge.

Byrd, E.T., Canziani, B., Hsieh, Y C.J., Debbage, K. & Sonmez, S. (2016). Wine tourism: motivating visitors through core and supplementary services. *Tourism Management*, *52*, 19–29.

Carlsen, J. & Boksberger, P. (2015). Enhancing consumer value in wine tourism. *Journal of Hospitality & Tourism Research*, *39*(1), 132–144.

Charters, S. & Ali-Knight, J. (2002). Who is the wine tourist?. *Tourism Management*, *23*(3), 311–319.

Croce, E. & Perri, G. (2010). Il turismo enogastronomico: progettare, gestire, vivere. *L'Integrazione Tra Cibo, Viaggio, Territorio* (2nd ed.). Milan: Franco Angeli.

Croll, E.J. (2006a). *China's New Consumers: Social Development and Domestic Demand*. Abingdon, New York: Routledge.

Croll, E J. (2006b). Conjuring goods, identities and cultures. In: Latham, K., Thompson S. & Klein, J. (eds.), *Consuming China: Approaches to Cultural Change in Contemporary China* (pp. 22–41). Abingdon, New York: Routledge.

Dai, B., Jiang, Y., Yang, L. & Ma, Y. (2017). China's outbound tourism: stages, policies and choices. *Tourism Management*, *58*, 253–258.

Dall'Ara, G. & Grötsch, K. (eds.) (2014). Il libro bianco sul turismo Cinese in Italia: un'analisi del fenomeno del turismo Cinese in Italia. Chinese Friendly Italy.

Dall'Ara, G. (2016). Introduzione al mercato turistico Cinese. Chinese Friendly Italy.

de Jong, A. & Varley, P. (2017). Food tourism policy: deconstructing boundaries of taste and class. *Tourism Management*, *60*, 212–222.

De Tocqueville, A. (1835/1840). *De la Démocratie en Amérique*. Pagnerre: Paris.

Durkheim, É. (1893). *De la Division du Travail Social: Étude sur l'Organisation des Sociétés Supérieures*. Paris: Félix Alcan.

Eisenhardt, K.M. & Graebner, M.E. (2007). Theory building from cases: opportunities and challenges. *Academy of Management Journal*, *50*(1), 25–32.

European Commission (2016). *Tourism in Focus: the Chinese Outbound Travel Market*. Brussels: European Commission, Directorate-General for Internal Market, Industry, Entrepreneurship and SMEs.

EUROSTAT (2015). US and Russia account for a third of all non-EU tourism nights in the EU. *News release 165/2015* – 25 September 2015.

Festa, G., Farace, S., Rossi, M., Festa, A. & Vitale, P. (2015). *XII Rapporto sul Turismo del Vino in Italia: Caratteristiche Attuali e Dinamiche Evolutive del Turismo del Vino in Italia*. Milano: Città del Vino – Associazione Nazionale.

Filiputti, W. (2016). *The Modern History of Italian Wine*. Geneva: Skira.

Getz, D. & Brown, G. (2006). Critical success factors for wine tourism regions: a demand analysis. *Tourism Management*, *27*, 146–158.

Guo, M. (2014). *Il Turismo Cinese in Italia*. Thesis, Università La Sapienza.

Hall, C.M., Johnson, G., Cambourne, B., Macionis, N., Mitchell, R. & Sharples, L. (2000). Wine tourism: an introduction. In: Hall, C.M., Sharples, L., Cambourne, B. & Macionis, N. (eds.), *Wine Tourism Around the World: Development, Management and Markets* (pp. 1–23). Oxford: Butterworth-Heinemann.

Hall, C.M. (2003). Preface. In: Hall, M. (ed.), *Wine, Food and Tourism Marketing* (pp. xiii–xiv). Binghamton: The Haworth Press.

Hirsch, F. (1977). *The Social Limits to Growth*. London: Routledge & Kegan Paul.

Hsu, C.H., Kang, S.K. & Lam, T. (2006). Reference group influences among Chinese travellers. *Journal of Travel Research*, *44*(4), 474–484.

Huang, S. & Chen, G. (2016). *Tourism Research in China: Themes and Issues*. Bristol, Buffalo, Toronto: Channel View Publications.

ISTAT (2014). Anno 2013: Capacità degli esercizi ricettivi e movimento dei clienti. 10 December 2014. Retrieved from: https://www.istat.it/it/archivio/141531 (Accessed 1 January 2015).

I numeri del vino (2016). Cina – importazioni di vino 2015. Retrieved from: http://www. inumeridelvino.it/2016/06/cina-importazioni-di-vino-2015.html (Accessed 7 October 2016).

Lai, C., Li, X.R. & Harrill, R. (2013). Chinese outbound tourists' perceived constraints to visiting the United States. *Tourism Management*, *37*, 136–146.

Latham, K. (2006). Introduction: Consumption and cultural change in contemporary China. In: Latham, K., Thompson S. & Klein, J. (eds.), *Consuming China: Approaches to Cultural Change in Contemporary China* (pp. 1–21). Abingdon, New York: Routledge.

Lemmi, E. & Siena Tangheroni, M. (2015). Food and wine tourism as a pull factor for Tuscany. *AlmaTourism: Journal of Tourism, Culture and Territorial Development*, *6*(11), 36–53.

Liu, F. & Murphy, J. (2007). A qualitative study of Chinese wine consumption and purchasing: implications for Australian wines. *International Journal of Wine Business Research*, *19*(2), 98–113.

Liu, K. (1997). Popular culture and the culture of the masses in contemporary China. *Boundary 2*, *24*(3), 99–122.

Mason, R.S. (1981). *Conspicuous Consumption: A Study of Exceptional Consumer Behaviour*. Farnborough: Gower.

Mason, M.C. & Paggiaro, A. (2012). Investigating the role of festivalscape in culinary tourism: the case of food and wine events. *Tourism Management*, *33*, 1329–1336.

Marx, K. (1859). *Zur Kritik der Politischen Ökonomie*. Berlin: Franz Duncker (W. Besser's Verlagshandlung).

OIV (2015a). *2015 Global Economic Vitiviniculture Data*. Organisation Internationale de la Vigne e du Vin.

OIV (2015b). *World Vitiviniculture Situation*. 38th World Congress of Vine and Wine, 6 July 2015, Mainz (Germany).

Osservatore Nazionale del Turismo (2011). Schede mercato: Cina. Retrieved from http:// www.ontit.it (Accessed 07/01/2015).

Pechlaner, H. & Volgger, M. (2012). How to promote cooperation in the hospitality industry: generating practitioner-relevant knowledge using the GABEK qualitative research strategy. *International Journal of Contemporary Hospitality Management*, *24*(6), 925–945.

Polanyi, K. (1944). *The Great Transformation*. New York: Farrar & Rinehart.

Quintal, V.A., Thomas, B. & Phau, I. (2015). Incorporating the winescape into the theory of planned behaviour: examining 'New World'wineries. *Tourism Management*, *46*, 596–609.

Sedláček, T. (2011[2009]). *Economics of Good and Evil: The Quest for Economic Meaning from Gilgamesh to Wall Street*. New York: Oxford University Press.

Shu, T. (1995). The establishment and development of tourism in China. In: Umesao, T., Befu, H. & Ishimori, S. (eds.) *Japanese Civilisation in the Modern World: IX Tourism* (pp. 155–167). Osaka: National Museum of Ethnology.

Sparks, B. (2007). Planning a wine tourism vacation? Factors that help to predict tourist behavioural intentions. *Tourism Management*, *28*(5), 1180–1192.

Stafford, C. (2000). Deception, corruption and the Chinese ritual economy. In: Latham, K., Thompson S. & Klein, J. (eds.) *Consuming China: Approaches to Cultural Change in Contemporary China* (pp. 42–55). Abingdon, New York: Routledge.

Sun Longji (1983). *The Deep Structure of Chinese Culture.* Hong Kong: Yishan.

Tang, W.L. (2006). Make sure you understand consumers. *Journal of Chinese Famous Brand,* 6 March 2006.

Tang, X. (1996). New urban culture and the anxiety of everyday life in contemporary China. In: Tang, X. & Snyder, S. (eds.) *In Pursuit of Contemporary East Asian Culture* (pp. 107–108). Boulder: Westview Press.

Theobald, W.F. (2005). The meaning, scope and measurement of travel and tourism. In: Theobald, W.F. (ed.) *Global Tourism* (pp. 5–24). Amsterdam: Elsevier.

Tsai, C.T.S. & Wang, Y.C. (2017). Experiential value in branding food tourism. *Journal of Destination Marketing & Management,* 6(1), 56–65.

UNWTO (2017). World tourism barometer – Volume 15. 31/03/2017. Retrieved from: http://cf.cdn.unwto.org/sites/all/files/pdf/unwto_barom17_02_mar_excerpt_.pdf (Accessed 15 May 2017).

Veblen, T.B. (1899). *The Theory of the Leisure Class: An Economic Study of Institutions.* New York: MacMillan.

Weber, M. (1922). *Wirtschaft und Gesellschaft.* Tübingen: Mohr.

Woodside, A.G. & Ronkainen, I.A. (1980). Vacation travel planning segments. *Annals of Tourism Research,* 7, 385–94.

Yan Yunxiang (1996). *The Flow of Gifts: Reciprocity and Social Networks in a Chinese Village.* Stanford: Stanford University Press.

Zelger, J. (2000). Twelve steps of GABEK WinRelan: A procedure for qualitative opinion research, knowledge organisation and system development. In: Buber, R. & Zelger, J. (eds.) *GABEK II. Zur Qualitativen Forschung* (pp. 205–220). Innsbruck, Wien, München: Studien-Verlag.

Zhang, G. (1989). Ten years of Chinese tourism: profile and assessment. *Tourism Management,* 10(1), 51–62.

Zhou, Z.Y., Liu, H. & Cao, L. (2014). *Food Consumption in China: The Revolution Continues.* Cheltenham: Edward Elgar Publishing.

14 Wine tourists' perspectives of 'New World' winescapes

Australia, USA and China

Vanessa Ann Quintal, Ben Thomas,
Yu-An Huang and Ian Phau

Introduction

'New world' wine economies such as Australia, the USA and China are producing and consuming more wines, reinforcing their status as major players in the global wine industry. According to the International Organisation of Vine and Wine (2016), wine production in 2016 is estimated to be 22.5mhl (+2%) in the USA, 12.5mhl (+5%) in Australia and 11.5mhl (status quo) in China. This respectively ranks the USA, Australia and China as the fourth, fifth and sixth largest wine producers globally. Moreover, the Wine Institute (2016) observed that in 2014, wine consumption in these countries was significant, with the USA accounting for 32mhl (+3.2%), China for 16mhl (−9.6%) and Australia for 5mhl (+19.2%). This respectively puts the USA, China and Australia in the first, fifth and tenth place as the largest wine consumers globally. 'New world' wines embody an entrepreneurial spirit, accentuating less on traditional and more on contemporary practices of winemaking (Vinepair, 2016). In these 'New World' wine economies, growth in consumer demand has impacted on wine distribution and sales. Conventional retail outlets such as supermarkets, as well as online retailers, now include wines in their portfolio. There is also increased popularity of wine conferences, festivals, fairs and wine-tasting events that showcase 'New World' wines (Liu et al., 2014).

Wine tourism and the opportunity to experience 'good food and wine' while on vacation (Department of Foreign Affairs & Trade, 2016) is a noticeable trend in Australia, the USA and China (Qiu et al., 2013). At a winery, its cellar door and surrounds, referred to as the winescape (Thomas, Quintal & Phau, 2016), provides a platform for the wine producer to create a favourable winery experience and instigate purchase from the wine tourist. This is of particular importance for small to moderate wine producers in Australia and the USA (IBISWorld – Australia, 2015; IBISWorld – USA, 2016) where access and direct sales to wine tourists through the cellar door is a growing channel (Bruwer et al., 2015). Thus, a positive winery experience is vital to develop brand awareness and involvement (O'Neill & Charters, 2000), customer-based brand equity (Nella & Christou, 2014), post-visit purchase behaviour (Quintal, Thomas & Phau, 2015) and ongoing sales (Bruwer et al., 2015). For these reasons, it is imperative to consider the perspectives of wine tourists in Australia, the USA and China to establish what winescape attributes are important in shaping their attitude and behavioural intention toward wineries in these 'New World' wine economies.

Literature review

The notion of the winescape appears to be introduced by Myerscough-Walker (1968) when the author observed how visitors engage with inns, hotels, restaurants and clubs. Some 30 years later, Peters (1997, p. 124) referred to the winescape as a cultural/viticultural landscape with 'a winsome combination of vineyards, wineries and supporting activities necessary for modern production.' Following this, reference to the winescape has been made in some 35 wine-related articles (e.g. Bruwer et al., 2015; Carmichael, 2005; Getz & Brown, 2006b; Johnson & Bruwer, 2007; Sparks, 2007; Thomas, Quintal & Phau, 2016).

Although some 50 years have elapsed, the literature remains fragmented due to varying conceptual and operational definitions of the winescape. A commonly accepted conceptualisation of the winescape is lacking as the literature remains divided by two distinct approaches (Byrd et al., 2016). On the one hand, the conventional approach adopts a macro perspective, conceptualising the winescape as a wine region (e.g. Bruwer & Lesschaeve, 2012; Getz & Brown, 2006b; Sparks, 2007). However, the variability and coalescence of other activities that co-occur with wine activities within a wine region make conceptualising and operationalising all these dimensions an onerous task (Quintal, Thomas & Phau, 2015). On the other hand, the micro perspective conceptualises the winescape as winery-specific (e.g. Griffin & Loersch, 2006; O'Neill & Charters, 2000). This sets delimitations for conceptualising and operationalising the winescape and permits specific attributes to be assessed for their impacts within a specific winery's identified landscape (Quintal, Thomas & Phau, 2015).

To exacerbate the first gap in winescape conceptualisation, a combination of demand-related and supply-related attributes are utilised in existing winescape-related scales (e.g. Getz & Brown, 2006b; Sparks, 2007). Internal need motives of the wine tourist drive demand-related attributes (Mitchell & Hall, 2004), whereas the winescape itself drives supply-related attributes (Getz & Brown, 2006a). These definitions highlight that demand-related and supply-related attributes are distinct from each other and that introducing both into a scale can confound what is actually being measured. To illustrate, both demand-related (e.g. 'opportunities for me to escape from routine/stress of daily life') and supply-related (e.g. 'beautiful surroundings') items are introduced in Sparks' (2007) wine destination experience. This presents problems with conceptualising and specifying the construct domain (Churchill, 1979; DeVellis, 2003).

Finally, the literature is limited in its empirical support and theoretical underpinning of a winescape framework. Within the extent of 35 winescape publications in the literature, there is lack of theory in explaining how the winescape operates on wine tourist behaviour. Notable exceptions are a few contemporary studies (e.g. Bruwer & Lesschaeve, 2012; Quintal, Thomas & Phau, 2015) which have drawn on servicescape theory (Bitner, 1992), multiattribute theory (Cohen, Fishbein & Ahtola, 1972), destination choice (push-pull) theory (Crompton, 1979; Dann, 1981) and the theory of reasoned action (Fishbein, 1967) to underpin conceptualisations of the winescape and the decision-making framework within which it operates.

The current study addresses these three research gaps by setting two key research objectives. It validates conceptualisation and operationalisation of the winescape attributes across five winery-specific studies in Australia, the USA and Taiwan, China, all 'New World' wine economies. Then, it introduces the winescape attributes to an empirical framework, namely, the theory of reasoned action, to examine how specific winescape attributes directly predict wine tourist attitude and indirectly predict behavioural intention in these wine economies.

Conceptual model and hypotheses

Fishbein (1967) introduced the theory of reasoned action (TRA) to psychology to describe a person's decision-making behaviour. Key to the theory is behavioural intention to execute a given action. According to Ajzen and Fishbein (1980), behavioural intention is conceptualised as a signal for willingness to perform a particular action and is influenced by two conceptually independent predictors, namely, attitude and subjective norms. Attitude is conceptualised as a person's positive or negative evaluation of performing the particular behaviour. Subjective norms are conceptualised as a person's perception of society's approval or disapproval over their performance of a particular behaviour. In subsequent studies, Ajzen (2001) observed that behavioural intention is useful in predicting actual behaviour. Actual behaviour is conceptualised as an observable response in a particular context which relates to a given outcome (Ajzen & Fishbein, 1980).

A diversity of disciplines has utilised the TRA across varied contexts. These include psychology (e.g. Doswell et al., 2011), health (e.g. Hennessy et al., 2014), consumer behaviour (e.g. Ruefenacht et al., 2015), education (e.g. Hussein & Mourad, 2014), hospitality (e.g. Okumus & Bilgihan, 2014), tourism (e.g. Chang, Backman & Huang, 2014) and sustainable economics (e.g. Gilinsky et al., 2015).

In the current study, the 'pull' supply-related attributes of the winescape are operationalised as: (1) setting; (2) atmospherics; (3) wine quality; (4) wine value; (5) wine complementary product; (6) signage; and (7) wine service staff (Thomas, Quintal & Phau, 2016). As shown in Figure 14.1, these winescape attributes are introduced to the TRA to assess whether the winescape attributes directly predict attitude and indirectly predict behavioural intentions to revisit and recommend the winery.

Winescape setting, which includes the natural landscape, surrounds and vineyards that create ambience, impacts on wine tourist attitude (Thomas, Quintal & Phau, 2016). Carmichael (2005) as well as Griffin and Loersch (2006) called attention to the landscape as a key driver of wine tourists' favourable assessments of a wine region. Similarly, the TRA-related study conducted by Sparks (2007) observed the influence that beautiful surroundings have on wine tourists' approving attitude towards taking a wine holiday. Consequently:

H1a: Winescape setting will have a positive effect on wine tourist attitude towards the winery.

According to Thomas, Quintal and Phau (2016), winescape atmospherics, which includes the winery's design, building materials, furnishings, heritage artefacts and displays that forge identity, stimulates wine tourist attitude. Pan, Su and Chiang (2008), who examined the interior and exterior architecture of buildings, reported that design affects a positive relationship with wine tourist attitude. It has been pointed out that heritage artefacts and displays (Dodd, 1995, 2000; Hall et al., 2002) hold some sway on wine tourist attitude. Consequently:

H1b: Winescape atmospherics will have a positive effect on wine tourist attitude towards the winery.

As conceptualised by Thomas, Quintal and Phau (2016), winescape wine quality, which refers to the superiority of the wines produced at the winery and delivered at its cellar door, affects wine tourist attitude. Dodd (2000) as well as O'Neill, Palmer and Charters (2002) noted that good wine quality and wine taste shape affirming evaluations of the overall winery experience. Consequently:

H1c Winescape wine quality will have a positive effect on wine tourist attitude towards the winery.

Winescape wine value, which relates to the worth of the wines produced at the winery and offered at its cellar door, holds influence on wine tourist attitude (Thomas, Quintal & Phau, 2016). Affordably priced wines at a cellar door have a positive effect on wine tourist attitude (Bruwer et al., 2015; Dodd & Gustafson, 1997) and in creating an enjoyable experience (Liu et al., 2014; Roberts & Sparks, 2006). Consequently:

H1d: Winescape wine value will have a positive effect on wine tourist attitude towards the winery.

According to Thomas, Quintal and Phau (2016), winescape complementary product, which refers to ancillary services that include dining facilities, accommodation, local produce and crafts, art, attractions, shopping, entertainment and activities that support the primary winery experience, shapes wine tourist attitude. Dining (Carmichael, 2005), fine cuisine (Carlsen & Dowling, 2001), accommodation (Hall et al., 2002), visits to local producers (Byrd et al., 2016) to sample their cheese and olive oil offerings (Roberts & Sparks, 2006), art and handicraft

Winescape Attributes   **Theory of Reasoned Action**

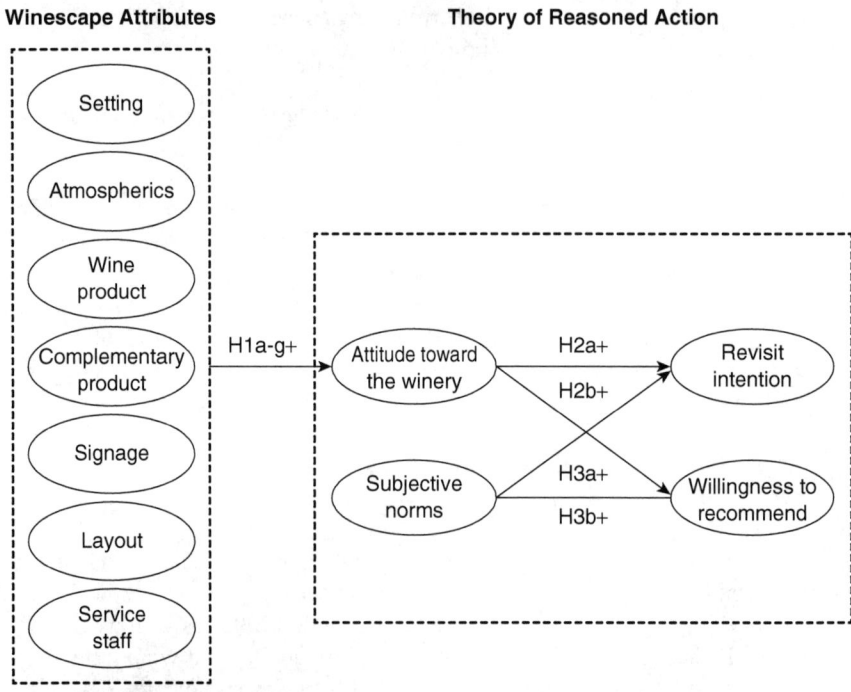

Figure 14.1 Conceptual model

activities (e.g. Cohen & Ben-Nun, 2009), shopping (Carmichael, 2005) and entertainment (Quadri-Felitti & Fiore, 2013) all affect approving evaluations of the winery's offering. Consequently:

> *H1e: Winescape complementary product will have a positive effect on wine tourist attitude towards the winery.*

As conceptualised by Thomas, Quintal and Phau (2016), winescape signage, which includes the use of promotion materials to explain the wine tourists' role in the winery, direct them around the winery and communicate the winery's desired brand image, impacts wine tourist attitude. Adequate and well-located directional signage, as well as informative tasting notes, are highlighted as influencers on assenting wine tourist attitude towards wineries (Griffin & Loersch, 2006). Consequently:

H1f: Winescape signage will have a positive effect on wine tourist attitude towards the winery.

Winescape service staff, which relates to the winery staff's knowledge and personal skills that enhance the quality of their interaction with the wine tourist, is commonly proposed to stimulate wine tourist attitude (Thomas, Quintal & Phau, 2016). Knowledgeable cellar door staff (Dodd, 2000) with personal skills, namely, friendliness, understanding and attentiveness, contribute to positive wine tourist evaluations of the cellar door (Griffin & Loersch, 2006; O'Neill, Palmer & Charters et al., 2002). Consequently:

H1g: Winescape service staff will have a positive effect on wine tourist attitude towards the winery.

Fishbein's (1967) TRA premises that attitude affects behavioural intention such as willingness to revisit. Lee and Chang (2012) as well as Quadri-Felitti and Fiore (2013) concluded that favourable evaluations by wine tourists in Taiwan, China and the USA respectively reinforce their intention to revisit a winery. Consequently:

H2a: Favourable attitude towards the winery will have a positive effect on wine tourists' willingness to revisit the winery.

Drawing on the TRA, attitude influences behavioural intention such as willingness to recommend. Gill, Byslma and Ouschan (2007) as well as Quadri-Felitti and Fiore (2013) examined wine tourists in Australia and the USA respectively and reiterated that their approving evaluations of a winery generate their willingness to suggest the winery and its products to their friends and relatives. Consequently:

H2b: Favourable attitude towards the winery will have a positive effect on wine tourists' willingness to recommend the winery.

According to theoretical underpinnings from the TRA and social identity theory (Tajfel, 1979; Trepte, 2006), subjective norms shape behavioural intention such as willingness to revisit. TRA-related studies conducted by Sparks (2007) as well as Quintal, Thomas and Phau (2015) accentuated the significant role that subjective norms play on intention to take a wine holiday for potential wine tourists. Consequently:

H3a: Higher subjective norms will have a positive effect on wine tourists' willingness to revisit the winery.

Finally, citing the TRA and social identity theory, social norms instigate on behavioural intention such as willingness to recommend. Social approval affects wine tourists' intention to suggest the winery to friends and relatives (Quintal, Thomas & Phau, 2015; Sparks, 2007). Consequently:

H3b: Higher subjective norms will have a positive effect on wine tourists' willingness to recommend the winery.

Methodology

Research sites

Five winery-specific studies were conducted to fulfil the two identified research objectives. The micro winery-specific approach was utilised so that the supply-related winescape attributes could be studied for their effects in a specific winery. Study 1's winery was sited at the Swan Valley, Western Australia (N = 301); Study 2's winery at the Yarra Valley, Victoria, Australia (N = 250); Study 3's winery at the Barossa Valley, South Australia (N = 231); Study 4's winery at the Napa Valley, California, USA (N = 345); and Study 5's winery at Houli, Taichung, Taiwan, China (N = 350). Each wine region accounts for the production of a sizeable volume of quality wines.

Survey instrument and measures

The study's instrument comprised a self-administered survey. Thomas, Quintal and Phau's (2016) 20-item winescape scale, which represented seven attributes, namely, setting, atmospherics, wine quality, wine value, complementary product, signage and service staff, was adopted due to its reliability (≥0.70) and validity. The theory of reasoned action measures were chosen from existing scales (see Bagozzi, Dholakia & Basuroy, 2003; Quintal, Lee & Soutar, 2010) for their reliability (≥0.70) and relevance to the wine tourism context. All constructs utilised a seven-point Likert-style scale, ranging from 1 for '*strongly disagree*' to 7 for '*strongly agree*'. An exception to this was the attitude construct, which utilised a semantic bi-polar scale.

Sampling method

For the Australian studies, a convenience sampling approach intercepted respondents as they exited a specific winery at the Swan Valley, Yarra Valley and Barossa Valley, to elicit their response to the winescape. As an incentive for respondents' participation, a prize draw with the opportunity to win a case of selected wines was used. For the American and Chinese studies, samples were randomly drawn from online consumer panels who had visited a specific winery at the Napa Valley and the Houli, Taichung respectively. These respondents had given their prior consent to participate and were paid (in points) for their time.

Results

Reliability and validity

Confirmatory factors analysis utilising one-factor congeneric models with AMOS 22 assessed the winescape scale. Across the five samples, the seven winescape attributes had acceptable goodness-of-fit indices ($\chi^2/df \leq 3.0$; RMSEA≤ 0.08;

CFI≥0.90; TLI≥0.90; IFI≥0.90) (Hair, Babin & Anderson, 2010). Standardised parameter estimates scores derived from confirmatory factor analysis calculated construct reliability and average variance extracted. Construct reliabilities for the seven winescape attributes were acceptable (≥0.70) (Hair, Babin & Anderson, 2010), suggesting they had stable dimensions. Average variance extracted scores of the seven winescape attributes across the five samples were high (≥0.50), demonstrating convergent validity (Fornell & Larker, 1981). Correlations between the seven winescape attributes were low (≤0.80), implying discriminant validity (Bagozzi & Heatherton, 1994). Further, across the five samples, average variance extracted scores for the seven winescape attributes exceeded the squared correlations between any two attributes, also suggesting discriminant validity (Fornell & Larker, 1981).

Hypotheses testing

The research model's hypothesised relationships were assessed using path analysis with AMOS 22 and the results are shown in Table 14.1. The goodness-of-fit criteria was implemented for each of the five samples to test model fit (χ^2/df≤3.0; RMSEA≤0.08; CFI≥0.90; TLI≥0.90; IFI≥0.90). Although the CFI and TLI and IFI did not exceed the critical level of 0.90 in the Chinese model, it was deemed acceptable because it met stipulated thresholds for the normed chi-square and RMSEA.

Winescape setting had a significant and positive effect on wine tourist attitude towards the winery in Study 4 ($\beta = 0.26$, p = 0.01), partly supporting *H1a*. Winescape atmospherics produced a significant and positive effect on wine tourist attitude towards the winery in Study 5 ($\beta = 0.12$, p = 0.05), partly supporting *H1b*. Given that the current study conceptualises atmospherics as the heritage aspects of the built environment and viticulture is relatively young in Australia and the USA (Vinepair, 2016), it is possible respondents do not perceive any heritage aspects in their winescapes; hence the non-significant effect.

Winescape wine quality had a significant and positive effect on wine tourist attitude towards the winery in Study 4 ($\beta = 0.16$, p = 0.05), partly supporting *H1c*. Since each study's sampling frame was drawn from a general population of wine tourists with diverse wine knowledge in Australia and Taiwan, China, this makes their evaluations about wine quality complex and difficult to assess (Liu et al., 2014). Winescape wine value did not produce a significant effect on wine tourist attitude towards the winery, which did not support *H1d*. It is likely that Australia, the USA and Taiwan, China may be reputed for their value-for-money wines (Liu et al., 2014) and this is taken for granted by wine tourists.

Winescape complementary product had a significant and positive effect on wine tourist attitude towards the winery in Study 1 ($\beta = 0.22$, p = 0.01); Study 2 ($\beta = 0.28$, p = 0.05); and Study 3 ($\beta = 0.25$, p = 0.001), partly supporting *H1e*. Interestingly, this observation only pertains to Australia. Conversely, winescape complementary product had no significant effect on wine tourist attitude in the USA and Taiwan, China. Winescape signage produced a significant and positive

Table 14.1 Standardised path coefficients and model fit

Hypotheses	Standardised regression paths				
	Study 1 Swan Valley (N = 301)	Study 2 Yarra Valley (N = 250)	Study 3 Barossa Valley (N = 231)	Study 4 Napa Valley (N = 345)	Study 5 Houli Taichung (N = 350)
H1a: SET → ATT	0.11	0.01	0.06	0.26**	0.10
H1b: AT → ATT	0.07	0.09	0.07	0.04	0.12*
H1c: WQ → ATT	0.08	0.13	0.09	0.16*	−0.06
H1d: WV → ATT	0.05	−0.05	0.11	0.10	0.10
H1e: CP → ATT	0.22**	0.28*	0.25***	−0.14	0.05
H1f: SI → ATT	−0.05	0.19	0.05	−0.01	0.24**
H1g: SS → ATT	0.38***	0.33***	0.28***	0.49***	0.12*
H2a: ATT → RI	0.35***	0.51***	0.35***	0.45***	0.89***
H2b: ATT →WTR	0.35***	0.33***	0.39***	0.60***	0.07
H3a: SN → RI	0.11*	−0.02	0.28***	0.31***	−0.09
H3b: SN → WTR	0.23***	0.09	0.31***	0.30***	0.46***
Model fit statistics					
χ^2	1172.62	1046.82	968.13	1147.80	2436.40
df	511	511	511	511	511
p-value	0.001	0.001	0.001	0.001	0.001
RMSEA	0.07	0.07	0.06s	0.06	0.08
CFI	0.91	0.90	0.94	0.93	0.86
TLI	0.90	0.89	0.93	0.92	0.84
IFI	0.92	0.90	0.94	0.93	0.86

Note: χ^2 = chi-square, df = degrees of freedom, RMSEA = root mean square error of approximation, GFI = goodness of fit indices, CFI = comparative fit indices, NFI = normative fit indices

* p ≤ 0.05, ** p ≤ 0.01, *** p ≤ 0.0011s

effect on wine tourist attitude towards the winery in Study 5 (β = 0.24, p = 0.01), partly supporting *H1f*. Since a winery experience mainly instigates the search for hedonic value (Quadri-Felitti & Fiore, 2013), signage may serve a temporary utilitarian function (Newman, 2007) before it is dispensed with; hence the non-significant effect.

Winescape service staff had a significant and positive effect on wine tourist attitude towards the winery in Study 1 (β = 0.38, p = 0.001); Study 2 (β = 0.33, p = 0.001); Study 3 (β = 0.28, p = 0.001); Study 4 (β = 0.49, p = 0.001); and Study 5 (β = 0.12, p = 0.05), supporting *H1g*.

Wine tourist attitude towards the winery produced a significant and positive effect on willingness to revisit the winery in Study 1 (β = 0.35, p = 0.001); Study 2 (β = 0.51, p = 0.001); Study 3 (β = 0.35 p = 0.001); Study 4 (β = 0.45, p = 0.001); and Study 5 (β = 0.89, p = 0.001), supporting *H2a*. Similarly, wine tourist attitude towards the winery had a significant and positive effect on willingness to recommend the winery in Study 1 (β = 0.35, p = 0.001); Study 2 (β = 0.33, p = 0.001); Study 3 (β = 0.39, p = 0.001); and Study 4 (β = 0.60, p = 0.001), partly supporting *H2b*.

Subjective norms produced a significant and positive effect on willingness to revisit the winery in Study 1 (β = 0.11, p = 0.05); Study 3 (β = 0.28, p = 0.001); and Study 4 (β = 0.31, p = 0.001), partly supporting *H3a*. Likewise, subjective norms had a significant and positive effect on willingness to recommend the winery in Study 1 (β = 0.23, p = 0.001); Study 3 (β = 0.31 p = 0.001); Study 4 (β = 0.30, p = 0.001); and Study 5 (β = 0.46, p = 0.001), supporting *H3b*.

Discussion and conclusion

The current study identified two research objectives. The first research objective was to validate the winescape attributes in the context of 'New World' wine economies. The seven winescape attributes demonstrated reliability and validity at specified wineries across five wine regions in Australia, USA and Taiwan, China. Theoretically, this validates applicability of the 'pull' supply-related attributes in 'New World' winescapes. The second research objective was to introduce the winescape attributes to the theory of reasoned action (TRA) model to determine how specific winescape attributes directly predict wine tourist attitude and indirectly predict behavioural intention. Theoretically, the extended TRA model offers an empirical framework that explains wine tourist behaviour in 'New World' winescapes.

Clearly, wine tourists identify service staff as the key winescape attribute in shaping their attitude towards the winescape. All five studies across Australia, the USA and Taiwan, China corroborate the critical role of service staff in creating a positive winery experience. Interestingly, wine tourists appear to single out the *personal* skills of service staff such as being friendly, courteous and paying attention to them, rather than the *professional* skills of service staff such as being knowledgeable about the winery's products and history.

From a managerial perspective, the critical input of staff in the winescape needs to be acknowledged. It is imperative that the recruitment, training and retention of qualified staff remain a priority. Wine regions and managers of wineries would do well to create opportunities for the self-improvement of motivated service staff and local communities. To illustrate, the Napa Valley Wine Academy conducts wine study courses, both online and face-to-face, in several locations including the Napa Valley, Denver, Santa Barbara, Cleveland, Nashville, Las Vegas, Orlando, Miami and Tampa. These courses, ranging from a day to eight weeks, are designed for a diverse range of stakeholders who include professionals, connoisseurs and novices (Napa Valley Wine Academy, 2016). It is in the interest of the Napa Valley and other wine regions to advance wine knowledge and wine appreciation in individuals since this creates a core of stakeholders committed to progressing the social, environmental and economic sustainability of the wine region.

Across all three wine regions in Australia, wine tourists highlight the importance of complementary products in forming their attitude towards the winescape. It is apparent that wine tourists expect to participate in wine-tasting as well as engaging in entertainment, galleries, art and sampling of local produce. This suggests that Australia as a 'New World' wine economy may be developing its own unique reputation for offering a *holistic experience* that encompasses a diversity of activities.

Managerially, it is opportune for all Australian wine regions to develop an innovative strategy that cross-sells complementary products along the wine route. To illustrate, the Margaret River wine region recently introduced a free mobile app which enables wine tourists to navigate their way to cellar doors, breweries, restaurants, galleries, art, shops, beaches, surf spots, sightseeing areas, bike trails and hikes (Margaret River Region, 2016). Plans to offer the app in key languages, particularly Mandarin, are already underway. This corroborates Tourism Australia's objective to better target, inspire and convert visitors in key markets, particularly China, through innovative marketing methods that incorporate new digital platforms (Department of Foreign Affairs & Trade, 2016).

Wine tourists in Taiwan, China identify atmospherics as a significant winescape attribute in shaping their attitude towards the winescape. Evidently, such tourists seek out the heritage aspects of the winery's built environment. Since the Houli District has its beginnings as a rural township of Taichung County, wine tourists expect to observe and learn about the winery's traditional and progressive advancements. This observation suggests that Taiwan, China should capitalise on its unique positioning as a 'New World' wine economy that offers a *nexus* between heritage and contemporary charm. To illustrate, the 'Domaine Shu Sheng Golden Muscat Fortified Wine' was awarded the gold medal at the prestigious 2014 Vinalies Internationales wine competition in Paris. The golden muscat grapes, which were introduced to Houli, Taichung by the Japanese in 1942, harnessed mature, internationally competitive technology to make premium award-winning wine using domestic ingredients (China Post, 2014).

From a marketing perspective, international awards and success stories need to be packaged within the brand to develop a sound positioning strategy for a winery and wine region. Atmospherics created from interactive display areas can showcase a winery's traditions and advancements (Qiu et al., 2013). This is highly pertinent to Chinese wine tourists who are 'not culturally exposed or accustomed to wine in their daily lives' and have low knowledge of wine (Qiu et al., 2013, p. 1128). There is significant potential for 'New World' wine economies to educate such tourists about wine culture and knowledge through interactive display areas and tasting rooms.

The current study identified delimitations in its micro approach to the winescape and its focus on three 'New World' wine economies. The micro approach was justified for being winery-specific and constrained to determine the effects the winescape attributes had in a specific environment. Although the research model demonstrates generalisability across wineries in Australia, the USA and Taiwan, China, a forthcoming research agenda identifies a macro approach to the winescape which is region-specific. This will replicate and validate the research model across various wine regions. It is also pertinent to mount future studies in other 'New World' wine economies such as Argentina, Chile and South Africa to discover and accentuate parallels of the winery experience in these contemporary wine economies.

Clearly, two winescape attributes feature prominently in shaping wine tourist attitude towards the winery, namely, service staff and the complementary product. Consequently, it may be feasible to explore these two winescape attributes individually and in greater detail. Determining what professional skills of service staff resonate most with wine tourists may help to enhance staff-visitor interactions in the winescape. Identifying the diverse complementary activities that augment a winery visit can make the difference between a mediocre or a memorable winery experience for tourists.

In summary, the current study set out to validate seven winescape attributes in the context of three 'New World' wine economies. The seven winescape attributes demonstrated reliability and validity across Australia, the USA and Taiwan, China. The winescape attributes were introduced to the theory of reasoned action (TRA) model. Generally, service staff and the complementary product directly predicted wine tourist attitude, which in turn, predicted behavioural intention in these wine economies.

References

Ajzen, I. (2001). Nature and operation of attitudes. *Annual Review of Psychology*, 52, 27–58.

Ajzen, I. & Fishbein, M. (1980). *Understanding Attitudes and Predicting Social Behaviour*. Engle-wood-Cliffs, New Jersey: Prentice-Hall.

Bagozzi, R., Dholakia, U. & Basuroy, S. (2003). How effortful decisions get enacted: the motivating role of decision processes, desires,and anticipated emotions. *Journal of Behavioural Decision-Making*, 16(4), 273–295.

Bagozzi, R.P. & Heatherton, T.F. (1994). A general approach to representing multifaceted personality constructs: application to state self-esteem. *Structural Equation Modelling*, 1(1), 35–67.

Bruwer, J. & Lesschaeve, I. (2012). Wine tourists' destination region brand image perception and antecedents: conceptualisation of a winescape framework. *Journal of Travel and Tourism Marketing*, 29(7), 611–628.

Bruwer, J., Lockshin, L., Saliba, A. & Hirche, M. (2015). Cellar door: trial-purchase-repurchase of the brand: how does a cellar door visit impact future sales? *Wine and Viticulture Journal*, 30(1), 56–59.

Byrd, E.T., Canziani, B., Hsieh, Y-C.J., Debbage, K. & Sonmez, S. (2016). Wine tourism: motivating visitors through core and supplementary services. *Tourism Management*, 52, 19–29.

Carlsen, J. & Dowling, R. (2001). Regional wine tourism: a plan of development for Western Australia. *Tourism Recreation Research*, 26(2), 45–59.

Carmichael, B. (2005). Understanding the wine tourism experience for winery visitors in the Niagara Region, Ontario, Canada. *Tourism Geographies*, 7(2), 185–204.

Chang, L-L., Backman, K.F. & Huang, Y.C. (2014). Creative tourism: a preliminary examination of creative tourists' motivation, experience, perceived value and revisit intention. *International Journal of Culture, Tourism and Hospitality Research*, 8(4), 401–419.

China Post (2014). Taiwanese wine wins gold medal at international competition in Paris. 8 March 2014. Retrieved on 11 November 2016. http://www.chinapost.com.tw/taiwan-business/2014/03/08/402332/Taiwanese-wine.htm

Churchill, G.A. (1979). A paradigm for developing better measures of marketing constructs. *Journal of Marketing*, 16(1), 64–73.

Cohen, E. & Ben-Nun, L. (2009). The important dimensions of wine tourism experience from potential visitors' perception. *Tourism and Hospitality Research*, 9(1), 20–31.

Cohen, J.B., Fishbein, M. & Ahtola, O.T. (1972). The nature and uses of expectancy- value models in consumer attitude research. *Journal of Marketing Research*, 9, 456–460.

Crompton, J.L. (1979). Motivations for pleasure vacation. *Annals of Tourism Research*, 6, 408–424.

Dann, G.M.S. (1981). Tourist motivation: an appraisal. *Annals of Tourism Research*, 8, 187–219.

Department of Foreign Affairs and Trade (2016). Innovative tourism marketing helps drive growth in Australia's tourism industry. 14 April 2016. Retrieved on 11 November 2016. http://ministers.dfat.gov.au/richardcolbeck/releases/Pages/2016/rc_mr_160414.aspx

De Vellis, R.F. (2003). *Scale Development: Theory and Applications.* Thousand Oaks, California: SAGE Publications.

Dodd, T.H. (1995). Opportunities and pitfalls of tourism in a developing wine industry. *International Journal of Wine Marketing*, 7(1), 5–16.

Dodd, T.H. (2000). Influences on cellar door sales and determinants of wine tourism success: results from Texas wineries. In Hall, C.M., Sharpies, L., Cambourne, B., Macionis, N. (eds.), *Wine Tourism Around the World: Development, Management and Markets.* Oxford: Elsevier Science (pp. 136–149).

Dodd, T.H. & Gustafson, W.A. (1997). Product, environmental and service attributes that influence consumer attitudes and purchases at wineries. *Journal of Food Products Marketing*, 4(3), 41–59.

Doswell, W.M., Braxter, B.J., Cha E-S. & Kim, K.H. (2011). Testing the theory of reasoned action in explaining sexual behaviour among African American young teen girls. *Journal of Pediatric Nursing*, 26, 45–54.

Fishbein, M. (1967). *Readings in Attitude Theory and Measurement.* New York: John Wiley.

Fornell, C. & Larker, D.F. (1981). Structural equation models with unobservable variables and measurement error: algebra and statistics. *Journal of Marketing Research*, 18(3), 382–388.

Getz, D. & Brown, G. (2006[a]). Critical success factors for wine regions: a demand analysis. *Tourism Management*, 27(1), 146–158.

Getz, D. & Brown, G. (2006[b]). Benchmarking wine tourism development: the case of the Okanagan Valley, British Columbia, Canada. *International Journal of Wine Business Research*, 18(2), 78–97.

Gilinsky, A., Newton, S.K., Atkin, T.S., Santini, C., Cavicchi, A., Casas A.R. & Huertas, R. (2015). Perceived efficacy of sustainability strategies in the US, Italian and Spanish wine industries: a comparative study. *International Journal of Wine Business Research*, 27(3), 164–181.

Gill, D., Byslma, B. & Ouschan, R. (2007). Customer perceived value in a cellar door visit: the impact on behavioural intentions. *International Journal of Wine Business Research*, 19(4), 257–275.

Griffin, T. & Loersch, A. (2006). The determinants of quality experiences in an emerging wine region. In Carlsen, J., Charters, S. (eds.), *Global Wine Tourism: Research, Management and Marketing.* Wallingford: CAB International (pp. 153–60).

Hair, J.F., Babin, B.J. & Anderson, R.E. (2010). *Multivariate Data Analysis: A Global Perspective*. New Jersey: Pearson Education.

Hall, C.M., Sharples, L., Cambourne, B. & Macionis, N. (2002). In Hall, C.M., Sharples, L., Cambourne, B., Macionis, N. (eds.), *Wine Tourism Around the World: Development, Management and Markets*. Auckland, New Zealand: Butterworth-Heinemann.

Hennessy, M., Bleakley, A., Mallya, G. & Romer, D. (2014). The effect of household smoking bans on household smoking. *American Journal of Public Health*, 104(4), 721–727.

Hussein, R.M.S. & Mourad, M. (2014). The adoption of technological innovations in a B2B context: an empirical study on the higher education industry in Egypt. *Journal of Business & Industrial Marketing*, 29(6), 525–545.

IBISWorld (2015). Global wine manufacturing. IBISWorld industry report. Sydney, NSW.

IBISWorld (2016). Wineries in the US. IBISWorld industry report 31213. Sydney, NSW.

International Organisation of Vine and Wine (2016). Statistical report on world vitiviniculture. Paris, France.

Johnson, R. & Bruwer, J. (2007). Regional brand image and perceived wine quality: the consumer perspective. *International Journal of Wine Business Research*, 19(4), 276–297.

Lee, T.H. & Chang, Y.S. (2012). The influence of experiential marketing and activity involvement on the loyalty intentions of wine tourists in Taiwan. *Leisure Studies*, 31(1), 103–121.

Liu, H.B., McCarthy, B., Chen, T., Guo, S. & Son, X. (2014). The Chinese wine market: a market segmentation study. *Asia Pacific Journal of Marketing and Logistics*, 26(3), 450–471.

Margaret River Region (2016). Retrieved on 11 November 2016. https://www.margaretriver.com/app/

Mitchell, R. & Hall, C.M. (2004). The post-visit consumer behaviour of New Zealand winery visitors. *Journal of Wine Research*, 15(1), 39–49.

Myerscough-Walker, R. (ed.) (1968). *Innscape [and] Winescape, Sussex: A Graded Guide to Inns, Hotels, Restaurants and Clubs*. Chichester, England: Myerscough Maps.

Napa Valley Wine Academy (2016). Retrieved on 11 November 2016. http://napavalleywineacademy.com/

Nella, A. & Christou, E. (2014). Linking service quality at the cellar door with brand equity building. *Journal of Hospitality Marketing and Management*, 23(7), 699–721.

Newman, A.J. (2007). Uncovering dimensionality in the servicescape: towards legibility. *Services Industry Journal*, 27(1), 15–28.

Okumus, B. & Bilgihan, A. (2014). Proposing a model to test smartphone users' intention to use smart applications when ordering food in restaurants. *Journal of Hospitality and Tourism Technology*, 5(1), 31–49.

O'Neill, M. & Charters, S. (2000). Service quality at the cellar door: implications for Western Australia's developing wine tourism industry. *Managing Service Quality*, 10, 112–123.

O'Neill, M., Palmer, A. & Charters, S. (2002). Wine production as a service experience: the effects of service quality on wine sales. *Journal of Services Marketing*, 16(4), 342–362.

Pan, F., Su, S. & Chiang, C. (2008). Dual attractiveness of winery: atmospheric cues on purchasing. *International Journal of Wine Business Research*, 20(2), 95–100.

Peters, G.L. (1997). *American Winescapes: The Cultural Landscapes of America's Wine Country*. USA: Westview Press.

Qiu, H.Z., Yuan, J.J., Ye, B.H. & Hung, K. (2013). Wine tourism phenomena in China: an emerging market. *International Journal of Contemporary Hospitality Management*, 25(7), 1115–1134.

Quadri-Felitti, D.L. & Fiore, A.M. (2013). Destination loyalty: effects of wine tourists' experiences, memories and satisfaction on intention. *Tourism and Hospitality Research*, 3(1), 47–62.

Quintal, V.A., Lee, J.A. & Soutar, G.N. (2010). Risk, uncertainty and the theory of planned behaviour: A tourism example. *Tourism Management*, 31(6), 797–805.

Quintal, V.A., Thomas, B. & Phau, I. (2015). Incorporating the winescape into the theory of planned behaviour: examining 'New World' wineries. *Tourism Management*, 46, 596–609.

Roberts, L. & Sparks, B. (2006). Enhancing the wine tourism experience: the customers' viewpoint. In Carlsen, J., Charters, S. (eds.), *Global Wine Tourism: Research, Management and Marketing*. Wallingford: CAB International (pp. 47–55).

Ruefenacht, M., Schlager, T., Maas, P. & Puustinen, P. (2015). Drivers of long-term savings behaviour from the consumers' perspective. *International Journal of Bank Marketing*, 33(7), 922–943.

Sparks, B. (2007). Planning a wine tourism vacation? Factors that help to predict tourist behavioural intentions. *Tourism Management*, 28(5), 1180–1192.

Tajfel, H. (1979). Individuals and groups in social psychology. *British Journal of Social and Clinical Psychology*, 18(2), 183–190.

Thomas, B., Quintal, V.A. & Phau, I. (2016). Wine tourist engagement with the winescape: scale development and validation. *Journal of Hospitality and Tourism Research*, 1–36.

Trepte, S. (2006). Social identity theory. In Bryant, J., Varderer, P. (eds.), *Psychology of Entertainment*. New York: Routledge (pp. 255–271).

Vinepair (2016). Retrieved on 17 November 2016. http://vinepair.com/wine-101/guide-old-world-vs-new-world-wines/

Wine Institute (2016). Trade data and analysis: world wine consumption 2011–2014. Retrieved on 17 November 2016. http://www.wineinstitute.org/resources/statistics

15 The potential of wine tourism to enhance Chinese holidaymakers' experiences in New Zealand

Insights from Chinese ITOs

Joanna Fountain

Introduction

The changing Chinese visitor market

Chinese visitors to New Zealand have grown from being an emerging market to the second largest – and fastest growing – source of visitors to the country over the past decade. In the year ending September 2016, New Zealand hosted 405,504 Chinese visitors, including 311,232 holidaymakers, an increase of 25.4% over the previous year (Tourism New Zealand, 2016a). Until relatively recently most Chinese holidaymakers to New Zealand were very short staying visitors on group or 'shopping tours'. In the last five years, however, there has been a substantial shift in the characteristics of Chinese holidaymakers and the types of trips they are making to the country. In particular, there has been a significant growth in Free Independent Tourists (FITs) and premium group tours who generally stay longer and spend more. For example, figures from 2015 show that while the majority of Chinese holidaymakers are on group tours (142,040), a third of holidaymakers are now FITs (73,000); most of the growth in the Chinese market is coming from the FIT market, which increased by 60% between 2014 to 2015 (Tourism New Zealand, 2016b). In 2016, for the first time since New Zealand acquired Approved Destination Status (ADS), ADS tour groups constituted less than half of all Chinese visitors in the country (46%) and FITs more than a quarter (29%) (Teo, 2016).

There are many explanations for the shift in characteristics in this market. In part, this trend may be viewed as a consequence of changing travel laws in China, which saw increased restrictions put on the shorter 'shopping tours' and resulted in a substantial decline in ADS group tours in 2013/2014 (Teo, 2016). Chinese holidaymakers are becoming more experienced travellers also, with greater confidence to make their own travel plans based on previous travel experiences (Arlt, 2013; Li, 2015). Younger tourists, in particular, are much more likely to speak some English, making it easier for them to plan and book their own trips (King & Gardiner, 2015; Prayag et al., 2015). At the same time, there have been concerted efforts by the New Zealand government and tourism industry over recent years to attract a greater proportion of longer staying, higher-spending Chinese visitors to New Zealand (Teo, 2016; Tourism New Zealand, 2016a), with considerable focus being placed on improving the quality of their experiences in the country (Expert Advisory Group, 2012). These efforts have included the launch in 2013 of the Premier Kiwi Partnership programme (PKP), which offers quality assured accreditation to Chinese tour operators offering longer-stay, higher quality tour itineraries of New Zealand (Teo, 2016). This programme incentivises tourist products that offer a range of memorable holiday experiences. In this context, wine tourism may offer opportunities for such experiences.

An overview of wine tourism in New Zealand

Wine tourism has been a feature of the wine and tourism industry in New Zealand for the last two decades (Baird & Hall, 2014). The oft-cited definition by Hall et al. (2000, p.3) describes wine tourism as 'the visitation to vineyards, wineries, wine festivals and wine shows for which grape wine tasting and/or experiencing the attributes of the grape wine region are the prime motivating factors for visitors.' This definition is often extended with the observations of Getz (2000) that wine tourism is not only a form of tourist behaviour, but also a destination strategy, whereby places can develop and market wine-related attractions and imagery, and a marketing opportunity for wineries to sell their product, build brand recognition and brand loyalty. This is an important rationale for wine tourism involvement by New Zealand wineries, with cellar door sales offering an important distribution channel, particularly for smaller operators (Baird & Hall, 2014).

In a New Zealand context, wine tourism is most closely associated with visiting a winery cellar door or tasting room. In a national study conducted in 2010, 68% of the wineries surveyed operated a cellar door; a figure similar to that reported in 2003 (Baird & Hall, 2014). While the term 'cellar door' might conjure up images of visits to wine cellars deep underground, most New Zealand wineries welcome guests to purpose-built premises, often situated amongst, or within sight, of the vineyards. Depending on the size of the company, these premises may be visually and architecturally impressive, offering panoramic views over vineyards. Smaller wineries may rely on a more modest room attached to an existing building, or may utilise more rustic and small-scale purpose-built facilities. Studies of wine tourism globally have identified a range of motivations and activities undertaken by wine tourists, but primarily focused around tasting and buying wine, enjoying a day out, and socialising with friends and family (Hall & Mitchell, 2008). Depending on the levels of wine interest or wine knowledge of the visitor, learning about wine may be an important motive for visiting a winery (Charters & Ali-Knight, 2002). The immediate setting and surrounds of a winery and the wider regional setting – described as the 'winescape' (Peters, 1997) – has also been shown to positively influence wine tourism experiences (Bruwer & Lesschaeve, 2012; Quintal et al., 2015; Sparks, 2007). Experience of winescape may be enhanced by the presence of a café or restaurant on site, enabling visitors to linger longer in this setting.

Wine tourism experiences are available in all of the wine regions of New Zealand, however the level of development and degree of involvement by wineries varies by region. Some of the most successful and well-known wine tourism destinations in New Zealand, such as Central Otago and Waiheke Island, are not the largest wine regions. A combination of features and attractions enhances the wine tourism experiences offered in these regions, so that visitors can appreciate the natural and cultural environment and setting. For example, Waiheke Island offers a relatively convenient getaway only 30 minutes' ferry ride from New Zealand's largest city, Auckland. While producing less than 1% of New Zealand's wines, a focus on high quality wines, and exceptional food and hospitality experiences makes it an important tourist and day-trip destination for wine tourists (Baragwanath & Lewis, 2014). Central Otago is another region with relatively small-scale wine production but which benefits from the stunning setting and an interesting array of wine tourism opportunities (Fountain & Thompson, 2017).

Chinese tourists and wine tourism in New Zealand: potential synergies?

To date, very limited academic research has been conducted on the engagement of Chinese tourists in wine tourism during their trip to New Zealand, and given the rapid development and shifts in the Chinese market (for wine and tourism), industry data on this issue is outdated. For example, data from 2008 which indicates only two per cent of Chinese tourists visited a winery while in New Zealand is of limited value (Ministry of Tourism, 2009). Similarly, evidence suggesting that between 2009 and 2013 approximately 2,000 Chinese holidaymakers visited a winery in New Zealand per year says little about the nature of these experiences (Tourism New Zealand, 2014). Academic research on the topic of Chinese wine tourists is beginning to emerge, both globally and in a New Zealand context. For example, there is recent research exploring domestic wine tourism opportunities and motivations in China (Howson et al., 2013, 2014; Ye et al., 2014; Zhang et al., 2013) and some aspects of Chinese wine tourism in New Zealand (e.g. Deng, 2013; Huang, 2014). Most recently, a 2015 study of 123 short-stay (<30 days) Chinese holidaymakers to New Zealand found that a third of respondents (35.7%) had visited, or would visit, a winery during their trip (Fountain, 2016). This represents a much higher engagement in wine tourism in New Zealand than has been reported in industry data. It should be noted, however, that these respondents were not typical of the typical Chinese holidaymaker in New Zealand. The 'average' Chinese holidaymaker stays in the country eight days and is likely to be on a group tour (Teo, 2016; Tourism New Zealand, 2016a); by contrast, these respondents were staying on average 12.2 days and only a quarter were travelling entirely on a group tour (Fountain et al., 2014).

The fact that FIT holidaymakers represents the fastest growing segment of the Chinese market in New Zealand (Tourism New Zealand, 2016a) may suggest that interest in wine tourism will grow as this market share increases. Furthermore, the expanding wine culture in China (Balestrini & Gamble, 2006; Bouzdine-Chameeva et al., 2013; Camillo, 2012; Liu & Murphy, 2007; Yu et al., 2009) may mean more of these tourists will arrive with some interest, and knowledge, of wine. Furthermore, many of the characteristics of the wine tourism experience align with the interests of Chinese tourists to New Zealand. This includes the chance to enjoy landscape and scenery, experience a clean and unpolluted environment and meet local friendly people (Insights Team TNZ, 2013; Tourism New Zealand, 2016a). The importance to the Chinese market of gift buying during their travels and the desire to purchase high quality local produce in New Zealand (Expert Advisory Group, 2012) suggest some potential synergies with wine tourism, particularly given the role of foreign wine in gift-giving in China (Camillo, 2012; Guinard, 2005; Li et al., 2011; Noppe, 2012). Furthermore, there is evidence that Chinese visitors are impressed by the good quality wine in New Zealand and are interested in participating in 'food and wine events/shows' during their visit (Tourism New Zealand, 2012).

The aim of this research is to explore the current interest in wine tourism amongst Chinese holidaymakers to New Zealand. It seeks to explore what barriers exist to greater participation by this market in wine tourism and to identify strategies to increase participation by Chinese holidaymakers in this tourist activity. This chapter presents the first research that has explored the potential of wine tourism to enhance the Chinese tourism experience in New Zealand from the perspective of Chinese Inbound Tour Operators (ITOs); stakeholders who are uniquely placed to provide an insider's perspective of Chinese holidaymakers' perceptions and experiences of wine tourism experiences; perceptions which have implications for wine tourism providers if they are to target the expanding Chinese market.

Methodology

This chapter is based on a series of in-depth semi-structured interviews with Chinese Inbound Tour Operators (ITOs) based in New Zealand. The eleven Chinese ITOs who at the time of study were registered as part of the Premier Kiwi Partnership (PKP) programme were contacted by email and asked if they would be willing to participate in the study. The option was given to complete an interview by telephone or face-to-face in English, or in Mandarin with the support of an interpreter. Six of these ITOs agreed to be interviewed and five interviews were completed (four face-to-face and one by telephone). The face-to-face interviews were conducted at the workplace of the respondents, who were marketing officers or owners of the company. The face-to-face interviews took on average 35 minutes, while the phone interview was approximately 20 minutes long. With the permission of the respondents, all face-to-face interviews were audio-recorded to be transcribed later, while notes were taken during the telephone interview. These notes and transcriptions were then analysed to identify themes.

The companies represented by these respondents had been operating in the tourism market with Chinese clients for between five and more than thirty years, although not all that time in New Zealand. Most of the respondents had been working in the tourism industry in New Zealand for five years or more. While the number of interviews is relatively small for a study of this type, the insights and opinions of respondents were remarkably consistent, so it is felt that data saturation was reached, and no follow up interviews were sought from other ITOs.

Findings

Is there demand for wine tourism amongst Chinese holidaymakers?

When asked about the potential of wine tourism for their clients, the Chinese ITOs were positive, but realistic, about the immediate possibilities in this market. There was a clear differentiation in their responses between the interests and demands of their ADS group tourists and their FIT or premium group tour clientele. For example, one ITO manager, whose tour groups were generally shorter staying (4–5 days) than those of the other ITOs represented, suggested 'we don't have a lot of visitors interested in winery tours. I think at this stage the level of their interest is still very low.' He explained that there is growing demand in China for grape wine (as opposed to Chinese wine) but that this trend had not yet translated into demand for wine tourism in New Zealand; a demand his company had tested:

> *We did a few winery itineraries and sent them to overseas agents but we didn't receive any feedback or any tourists specialising in wineries, like ten days in New Zealand only for wineries. Maybe some of them would like ten days in New Zealand but maybe spend only two or three hours wine tasting.*
>
> (ITO 1)

The other ITOs similarly felt that the ADS group tours they catered for generally had limited interest in wine tourism experiences. As well as the fact that the majority of Chinese people do not regularly consume wine, there were two other reasons given for the lack of interest in wine tourism amongst group tourists. The first, and primary, reason related to priorities; one respondent summarised the situation highlighted by them all:

> *For the normal holidaymaker they may want to see more sightseeing or programmes, rather than just going to a winery.*
>
> (ITO 2)

It was acknowledged that many of these groups are on relatively short stays and want to fit in as many activities and sights as possible, as was explained:

> *Chinese people, they are a bit different when they go for a holiday . . . [They] want to see as many things within a day, even though they feel very tired. Some other people may just want to relax – holiday means holiday – but Chinese people just want to see as much as they can. Even though they can't remember the names when they go back they think it is worth it.*
>
> (ITO 2)

In this situation, if a winery is included in an itinerary, particularly if it is one that takes some time and effort to get to, it would limit other options:

> *In the itinerary if you go to Waiheke it will take the whole day but sometimes you want to take the whole day to Rotorua or either Taupo, so it is a bit tricky. If they only have a limited time, to spend a whole day on one activity is a bit much.*
>
> (ITO 4)

Some of the ITOs also mentioned that the limited facilities and small size of many wineries meant logistically it was difficult to find places willing to accept groups of 30 or more tourists. The nature of the itineraries also meant that the largest wine regions in New Zealand were generally excluded from itinerary consideration:

> *[The winery needs to be] convenient to the major destinations. Like Marlborough and Hawkes Bay may be the two biggest wine regions but the thing is they are a bit far away from the major cities, like Christchurch or Wellington or even Auckland.*
>
> (ITO 1)

As well as these logistical issues, respondents highlighted the fact that expectations and knowledge about New Zealand meant that wine tourism was not something Chinese visitors expected or looked for during a trip. As one respondent stated:

> *The natural setting, sitting on the bus, watching the landscape, sheep, cows, that's what they love . . . I think the main priority when they come to New Zealand is still the natural scenery, that's the number one, and all the experiences of different cultures. Very small numbers of clients want to visit wineries*
>
> (ITO 1)

The experience expected in New Zealand was contrasted with expectations of a trip to France:

> *Actually it depends on location because most of the people from China come to New Zealand for the beautiful natural scenery, and for example, if we change the [location], going to France, they definitely would want to go to winery, yeah.*
>
> (ITO 2)

As well as expectations around 'typical' New Zealand tourist experiences limiting demand for wine tourism experiences, the ITOs acknowledged that most of the ADS group visitors lacked awareness about New Zealand wine more generally. As one operator explained:

They don't really know until they have a relative or friends in New Zealand that do come and visit and tell them that New Zealand [has] good wine, otherwise they don't really know New Zealand wine.

(ITO 2)

Despite the relative lack of interest and knowledge amongst their clients, these ITOs acknowledged that winery visits were often a part of the itineraries they prepared, with most groups visiting a winery in either the Auckland region or in Central Otago. As one respondent explained:

The people who asked [for a winery visit] are really rare. But if you put a winery on the itinerary, they will accept it.

(ITO 4)

This viewpoint was supported by another operator:

I think the reason they visit wineries is partly because they are suggested by the overseas travel agents. Like . . . do something different from the traditional itineraries. Maybe the travel agent says 'ah, why not go to winery and have a tour, wine tasting' . . . I don't think a lot of tourists request tours of wineries.

(ITO 1)

As alluded to in the previous comment, there was some feeling that for ADS group tours, the inclusion of a winery experience offered something unique or 'special' to the itinerary; something of a status symbol, which some operators were trying to exploit:

We have some agents, they really try to promote the winery products . . . . They want to make the itinerary special, and they want to think it is a kind of unique product, to make them like high-end clients . . . For normal people like us, we don't really drink wine. We drink wine but we don't know the history or the culture or whatever . . . normally rich people will buy wine.

(ITO 3)

Where winery visits were included for ADS group tours, usually this included just a wine tasting, but some included a winery tour or lunch. Again, time was a consideration and a limiting factor in the development of programmes including wine, as one operator explained:

We . . . planned to have lunch at the winery but actually it is not really practical, so now we change, like we have lunch on the beach, BBQ lunch on the beach at Murawai, and then we go to have a visit to the winery . . . No time for a tour, just overlook the vineyard, and have some wine tasting, just to experience the wine tour, but . . . we don't have too much time for this.

(ITO 4)

There was a very clear sense from the operators that ADS group tourists would not be interested in visiting more than one winery – it was something to 'tick off' rather than explore in many different regions. This was explained in relation to limited time, but also reflected a limited understanding of wine culture, and the differences between regions in terms of wine and terroir:

> *They don't really understand what the difference [is between wines/wineries]. They just think it is just wine; you just taste different wine. They may not really know about the culture or the background of the winery . . . so for tourists they may feel OK, just wine, I just taste it, whether good or not, whether I buy it or not.*
>
> (ITO 3)

When asked what their ADS group clientele enjoyed most about their winery experiences, generally this was framed in terms of having a new experience, or enjoying the scenery and surroundings:

> *I think first of all they enjoy the scenery, as I experienced a winery tour in Waiheke, I feel it is a very good, beautiful area and really elegant environment to enjoy, and very relaxing, and you can see very far away to the sea coast and about the vineyards. And you enjoy wine, so it is a very good experience.*
>
> (ITO 2)

Buying wine was not really a priority for ADS group tourists, but if wine was purchased it was primarily purchased as a gift for friends and colleagues in China, rather than for personal consumption. The Chinese ITOs reported that there were two key issues restricting wine purchases; one logistical and the other relating to the reputation of the wine. An issue that faces all international tourists is the customs limitation on the importation of wine. Added to this, these group travellers were also hesitant to buy wine early on in their trip to New Zealand if it meant transporting it around with them on a tour bus:

> *If the [winery] visit is close to the end of the trip then the people buy some wine; if it is at the beginning of the trip they would maybe hesitate because they have to carry the box all through the trip.*
>
> (ITO 1)

Another issue is that because New Zealand wine is not very well known in China it may not be seen as prestigious, compared to French wines. For this reason, New Zealand wine may be viewed as not suitable for gifting to a high status recipient, rather it might be viewed as a 'normal gift for their friends or colleagues' (ITO 1). As another respondent suggested:

If they really want to buy for friends, or either for colleague or boss, they
will [buy] quite expensive wines . . . . If they buy for themselves they will just
choose the average price, normal product . . .

(ITO 3)

In this situation, the challenge is knowing if the recipient of the gift will understand and appreciate the value, or expense, of a New Zealand wine if the product
is not well known in China.

Though the interest in wine tourism amongst ADS group visitors is relatively
limited at the current time, most of the Chinese ITOs interviewed were enthusiastic about the potential and growing appeal of wine tourism experiences amongst
the premium group tourists or FIT visitors whose itineraries they plan. This is
partly because these visitors are generally on longer itineraries, so have more
time to explore a range of tourism experiences in different parts of the country,
but also because they are more likely to be wine consumers already, who are more
knowledgeable, and interested, in New Zealand wine:

New Zealand wine is actually getting popular in China because of their
limited quantity, and getting expensive as well. Some of them who really
know the wine, they are getting interested in New Zealand wine because it
depends on whether the quality is good . . . [So they are] interested in buy-
ing wine, that is the purpose, and they want to taste and want to know the
wine knowledge.

(ITO 2)

Those ITOs that dealt with more of these premium tour parties, made up of family
groups travelling together or business and government officials, reported having
many requests for winery visits, with some groups wanting to visit many wineries during their stay. These respondents acknowledged that they were not clear
whether this interest was entirely leisure-based, or whether it had some business
purpose, particularly where the delegation was made up of government officials
or business people. As one ITO acknowledged:

Business people, even though it is holiday time, they are still looking for a
business opportunity. That's the reason why the want to try the winery.

(ITO 2)

Another operator outlined a special tour her company had arranged for a group
of 80 VIPs:

We arranged a very successful meeting on Waiheke Island; there's lots of
vineyards. The itinerary is: about two o'clock they get on board for the
Fullers ferry and they take about one hour to get to the island, and then they
had a two-hour tour around the island, to look at the scenery, something like

that, then they go to the winery, and they had a tour, and somebody give them an introduction of the winery and how they make the wine and information about the wine, and also they visited the cave, and they taste, and then they had like an event, like all the people have dinner together at the winery, that was very successful.

(IT0 4)

The general sense from the Chinese ITOs was that while these premium and VIP groups were at times looking for business opportunities, they also have a genuine interest in tasting and learning about wine and wine culture more generally. One respondent outlined her own first experiences of wine in New Zealand to explain the potential of wine tourism to advance wine knowledge for not only business and VIP travel, but potentially for more general tourists also:

Before I come to New Zealand I do not know much about wine knowledge. When I first come to New Zealand . . . we went only half an hour driving go to winery and then the first time I know 'Oh wine has a lot of stories, different stories', that was quite interesting to me . . . When we [were] tasting wine, they give a lot of . . . How they are made, because before I don't have this knowledge. It make me so interested, ah that is so different.

And you think your high end visitors are interested in this too?

Yes definitely, some of them are already. If they are really high end they do visit all over the world, they already maybe have a little knowledge, but not really knowledge, but if you give . . . this kind of knowledge they are really, really interested. Even the group [tourists] because . . . it is kind of an innovation for Chinese people drinking wine.

(ITO 3)

What would increase Chinese holidaymakers' demand for wine tourism?

The Chinese ITOs interviewed were asked what might increase interest in wine tourism amongst their clientele. Most of the operators felt that demand for wine tourism would increase over time as wine consumption in China continues to become more widespread and as the characteristics of Chinese visitors trend towards more FITs and premium group tourists with longer periods of time in New Zealand. This growth would be gradual, however:

The traveller is more and more mature. Before that, people more rely on the travel company but nowadays more and more people prefer to develop their own holidays. Especially for the young people, the new generation, they can speak very good English, so for them they don't have as much barriers, language barriers . . . also in China it is very popular to travel FIT, it looks like you are more professional, more capable, something like that.

(ITO 4)

One thing respondents agreed on was that greater awareness and exposure to New Zealand wine in China, through friends, marketing or greater purchasing opportunities, would result in more Chinese tourists seeking out wine tourism opportunities in New Zealand:

> *I think it will promote well in China, but it will really take time, because peo-ple really don't know about it, or even if they know, they may want to spend much time to do sightseeing. Because tourism product is the thing that you come to see, it's a kind of experience. [So it will be promoted through] word of mouth, and also the promotion of the winery and wine in the China market. You can promote the winery as a product in the itinerary, or you promote the wine, so as the wine is so famous you want to see the winery, so it can be both things.*

> (ITO 3)

In this regard, a couple of respondents explicitly mentioned a recent visit by the New Zealand Prime Minister to China as a good starting point:

> *John Key [is] doing good things. He goes to China for a visit and did men-tion about [wine] and it makes people think, 'oh there is something differ-ent, something good' . . . and then they start asking, 'oh New Zealand have wine, do they have something similar like south of France? [Do they] have a winery or what it look like?' and we start contracting. So I think this is an improvement and a lot of potential, honestly.*

> (ITO 2)

How can the wine tourism experience for Chinese holidaymakers be improved?

While ITOs were generally satisfied with the wine tourism opportunities for their clients, there was a sense that there were some modifications that winer-ies and cellar door managers could make to improve the attractiveness of the offering to this market. A central issue related to the language barrier many Chinese visitors currently face, and the limited interpretation or translation offered at most wineries at the present time. Respondents spoke of the lack of information available in Chinese about the wineries and their wines on their websites, which saw them resorting to translating material themselves to pass on to offshore partners and tour guides. The wineries themselves also often lacked even basic signage or information in Chinese. Given the fact that a number of the ITOs felt that Chinese visitors, particularly premium visitors, saw their winery visit as an opportunity to learn about wine, the fact that most wineries did not have Chinese-speaking staff to explain the wines or tell stories about the wineries, or Chinese language audio-guides to fulfil the same role, was a real limitation:

> *Most of the Chinese tourists who come here their English is not that good, so they still expect at least you have something translated to them, they under-stand what is happening, or story, make more interesting. Otherwise they don't know what you talk about.*
>
> (ITO 2)

While all their tour groups were accompanied by tour guides, the operators stressed that wineries should not rely on these guides to translate for them:

> *Most of [the guides] don't have this winery knowledge, so we expect winery staff they do explain in Chinese, or have Chinese headset.*
>
> (ITO 2)

Another issue related to the focus of the tour guide, who might be managing a large group, only some of whom want to learn about wine:

> *The tour guide can do the translation but sometimes . . . some people just go other place to take photos, so the tour leader has to look after everyone, so sometimes it is a bit hard for the tour leader to just focus on the wine discussion. Because in a group of people, maybe 15 people when they come to a winery, it might be like: 'I'm going to go to the toilet, I'm going to go to this place'.*
>
> (ITO 3)

As well as these practical concerns, ITOs also suggested product develop-ments that could appeal to Chinese visitors, particularly centred on pairing food and wine in a tourism experience. Respondents mentioned that Chinese visitors enjoyed opportunities to taste fresh New Zealand food, and this is an element that to date has performed somewhat poorly in this market (Tourism New Zealand, 2016b). In this context, the idea of combining wine with fresh food experiences was repeated by the following two ITOs as options that would appeal to the high end market, with more time and money to spend on such experiences, but also to group tourists, for whom wine is still not a regular part of their lifestyles:

> *If we can put [food and wine experiences] in the itinerary that would be great for most people because always some people [think] New Zealand doesn't have much good food; experience good scenery, but not really experience for the food . . . But if we put something wine, people start [getting interested], people really love it, if we can put these things . . . Put it that way, if you are going to visit somewhere you want beautiful scenery definitely [but] you want some good food as well, because that is your holiday, right?*
>
> (ITO 2)

I think, my suggestion is that it is not just the wine tasting. You can combine some food, like lunch, or dinner, together . . . so it is more than wine. So it is about the experience of the very good quality New Zealand food and New Zealand wine . . . [For example] if you can arrange a very good winery dinner, it could be a very good option . . . like you enjoy a good wine, and you enjoy candles or something. It is a very good experience.

(ITO 4)

Discussion and conclusion

The previous discussion has provided an interesting glimpse into the Chinese visitor market, as perceived by a group of intermediaries critical to the development and execution of their New Zealand tour experience, Chinese ITOs. Their accounts reveal that while the market for wine tourism experiences amongst Chinese holidaymakers in New Zealand is still quite small, its appeal is growing, particularly amongst premium and FIT tourists. These respondents were of the belief that in time, as New Zealand wine becomes more well known and available in China, and as wine consumption spreads further into Chinese culture, this interest will only increase. There are ways in which the wine tourism experience can be improved for Chinese tourists, however, with suggestions for improvement generally revolving around overcoming language barriers and offering memorable experiences that combine food and wine.

From the perspective of the Chinese ITOs, the biggest improvement to New Zealand wine tourism offerings would involve removing language barriers for their clientele. This is particularly pertinent given a number of respondents highlighted the interest in learning about wine and wine culture of some of their clientele, which can only be hindered by communication barriers. There are a range of ways in which this could be facilitated, from hiring Chinese-speaking staff to work in cellar doors, to commissioning translators to work on promotional collateral and websites. Providing a total experience, including high quality food to be served alongside the wine, would also to add value to the Chinese winery experience. Some of these suggestions require considerable financial investment. Many wineries, particularly smaller operations, may question the value of this investment at the current time, given the relatively low level of interest of Chinese holidaymakers in buying wine at the cellar door, which is a primary rationale for involvement in wine tourism for many small wine producers (Baird & Hall, 2014).

The findings suggest Chinese holidaymakers appreciate the opportunity to relax and enjoy the setting or winescape of a winery; to a greater extent perhaps than other markets for wine tourism. At the same time, a factor limiting the interest of Chinese holidaymakers in wine tourism in New Zealand is a lack of awareness and exposure to the New Zealand wine industry. In this context, greater use of winescape imagery in the promotional campaigns of Tourism New Zealand and destination marketing organisations for the Chinese market may facilitate awareness, and interest, in including a winery experience on their New Zealand holiday.

References

Arlt, W.G. (2013). The second wave of Chinese outbound tourism. *Tourism Planning & Development*, 10(2), 126–133.

Baird T. & Hall, C.M. (2014). Between the vines: wine tourism in New Zealand. In P.J. Howland (ed.). *Social, Cultural and Economic Impacts of Wine in New Zealand* (pp. 191–208). New York: Routledge.

Balestrini, P. & Gamble, P. (2006). Country-of-origin effects on Chinese wine consumers. *British Food Journal*, 108(5), 396–412.

Baragwanath, L. & Lewis. N. (2014). Waiheke Island. In P.J. Howland (ed.). *Social, Cultural and Economic Impacts of Wine in New Zealand* (pp. 211–226). New York: Routledge.

Bouzdine-Chameeva, T., Pesme J-O. & Zhang, W. (2013). Chinese wine industry: current and future market trends. Paper presented at the 7th Conference of the American Association of Wine Economics, Stellenbosch, South Africa. 26–29 June.

Bruwer, J. & Lesschaeve, I. (2012). Wine tourists' destination region brand image perception and antecedents: conceptualisation of a winescape framework. *Journal of Travel & Tourism Marketing*, 29(7), 611–628.

Camillo, A.A. (2012). A strategic investigation of the determinants of wine consumption in China. *International Journal of Wine Business Research*, 24(1), 68–92.

Charters, S. & Ali-Knight, J. (2002). Who is the wine tourist? *Tourism Management*, 23(3), 311–319.

Deng, S. (2013). 'I prefer a dry red thanks': a consumer behavioural study of resident Auckland Chinese wine consumption and wine related tourism. Master of International Management Thesis, Auckland University of Technology, Auckland, New Zealand.

Expert Advisory Group (2012). *China Market Review*. Wellington, New Zealand.

Fountain, J. (2016). Just here for the scenery? Chinese holidaymakers and wine tourism in New Zealand. Paper presented at the 9th Academy of Wine Business Research Conference, Adelaide, Australia. 7–19 February.

Fountain, J. & Thompson, C. (2017). More than the mountains? Tourist perception of the winescape of Central Otago. Cauthe 2017: Time for big ideas? Rethinking the field for tomorrow. Paper presented at the 27th CAUTHE Annual Conference, University of Otago, Dunedin. 7–10 February.

Fountain, J., Wen, Y. & Menival, D. (2014). Chinese visitors' interest and engagement with wine tourism in New Zealand: a comparison of short-stay holidaymakers and students. Paper presented at the New Zealand Tourism and Hospitality Research Conference, , Hamilton, New Zealand. 9–12 December.

Getz, D. (2000). *Explore Wine Tourism: Management, Development & Destinations*. Cognizant Communication Corporation, New York.

Guinard, L. (2005). The Chinese taste for wine. *Wines and Vines*, Dec, pp. 42–44.

Hall, C.M. & Mitchell, R. (2008).Cellar door: direct sales, brand building and relationships. *In Wine Marketing: A Practical Guide*. (pp. 112–142). Oxford: Butterworth-Heinemann.

Hall, C. M., Sharples, L., Cambourne, B. & Macionis, N. (2009). *Wine Tourism Around the World*. London: Routledge.

Howson, C., Ly, P. & Begun, J. (2013). Developing a market for Chinese wine: tourism and education, *The Wine Economist*, 1 October. Accessed: http://wineeconomist.com/2013/10/01/china3/

Howson, C., Ly, P. & Begun, J. (2014). Chinese wine tourism: not just about wine. *Alternative Emerging Investor*, 20 August. Accessed: http://www.aeinvestor.com/story/chinese-wine-tourism-not-just-about-wine/

Huang, L. (2014). A study of characteristics of female Chinese tourists who participate in New Zealand wine tourism. Master of International Hospitality Management (MIHM) dissertation, Auckland University of Technology, Auckland, New Zealand.

Insights Team TNZ (2013). Market intelligence. Accessed: www.chinatoolkit.co.nz

King, B. & Gardiner, S. (2015). Chinese international students. An avant-garde of independent travellers? *International Journal of Tourism Research*, 17(2), 130–139.

Li, J.G., Jia, J.R., Taylor, D., Bruwer, J. & Li, E. (2011). The wine drinking behaviour of young adults: an exploratory study in China. *British Food Journal*, 113(10), 1305–1317.

Li, X.R. (ed.) (2016). *Chinese Outbound Tourism 2.0*. New Jersey: Apple Academic Press.

Liu, F. & Murphy, J. (2007). A qualitative study of Chinese wine consumption and purchasing: implications for Australian wines. *International Journal of Wine Business Research*, 19(2), 98–113.

Ministry of Tourism (2009). Tourism sector profile: wine tourism. Ministry of Tourism, Wellington. Accessed: https://www.med.govt.nz/about-us/pdf-library/tourism-publications/Wine%20Tourism%20Profile%20-456KB%20PDF.pdf

Noppe, R.P. (2012). Rise of the dragon: the Chinese wine market. Dissertation submitted to the Cape Wine Academy, Cape Town, South Africa.

Peters, G.L. (1997). American winescapes: the cultural landscapes of America's wine country. *Geographies of the Imagination (USA)*. Boulder, Colorado: Westview Press.

Prayag, G., Disegna, M., Cohen, S.A. & Yan, H. (2015). Segmenting markets by bagged clustering: Young Chinese travelers to Western Europe. *Journal of Travel Research*, 54(2), 234–250.

Quintal, V.A., Thomas, B. & Phau, I. (2015). Incorporating the winescape into the theory of planned behaviour: examining 'New World' wineries. *Tourism Management*, 46, 596–609.

Sparks, B. (2007). Planning a wine tourism vacation? Factors that help to predict tourist behavioural intentions. *Tourism Management*, 28(5), 1180–1192.

Teo, P. (2016). China ADS market trends. Paper presented at the China ADS Market Forum, Heritage Hotel, Auckland. Accessed: http://www.tourismnewzealand.com/.../china-ads-market-forum-jun-2016-paul-yeo-tnz.pdf

Tourism New Zealand (2012). Visitor experience monitor: Chinese market report 2011/2012. Tourism New Zealand, Wellington.

Tourism New Zealand (2014). Tourist special interest: wine tourism. Accessed: http://www.tourismnewzealand.com/media/1765/wine-tourism-profile.pdf

Tourism New Zealand (2016a). China insight session FY16 Q3. Accessed: http://www.tourismnewzealand.com/media/2426/china-insights-q3-fy16.pdf

Tourism New Zealand (2016b). Market trends. Accessed: http://www.tourismnewzealand.com/markets-stats/markets/china/market-trends/

Ye, B.H., Zhang, H.Q. & Yuan, J. (2014). Intentions to participate in wine tourism in an emerging market: theorisation and implications. *Journal of Hospitality and Tourism*

Research. Accessed: http://jht.sagepub.com/content/early/2014/03/05/1096348014525 637.abstract

Yu, Y., Sun, H., Goodman, S., Chen, S. & Ma, H. (2009). Chinese choices: a survey of wine consumers in Beijing. *International Journal of Wine Business Research*, 21(2), 155–68.

Zhang, Q.H., Yuan, J., Haobin Y.B. & Hung, K. (2013). Wine tourism phenomena in China: an emerging market. *International Journal of Contemporary Hospitality Management*, 25(7), 1115–1134.

16 At a crossroad

A study of Nyonya cuisine as intangible cultural heritage

Nazaruddin Haji Hamit

Introduction

The advent of trade between nations in the fifteenth and sixteenth century and Malacca being an imminent thriving port of call along the spice route in the Straits of Malacca have led to the sojourn of many traders from Europe and the Middle East. Likewise, traders from the Far East, such as China, made Malacca not only a port of call but a resting place while waiting for a change in wind direction before they could set sail back to their homeland (Tan, 1988).

The arrival of Chinese traders to the city and marrying of local women resulted in the emergence of a new community called the Baba and Nyonya community (Tan, 1988). The descendants of these early marriages are often referred to as the Peranakan or Straits Chinese, with the Malay term Peranakan being derived from the root word, '*anak*', meaning child. In these communities, the male is called

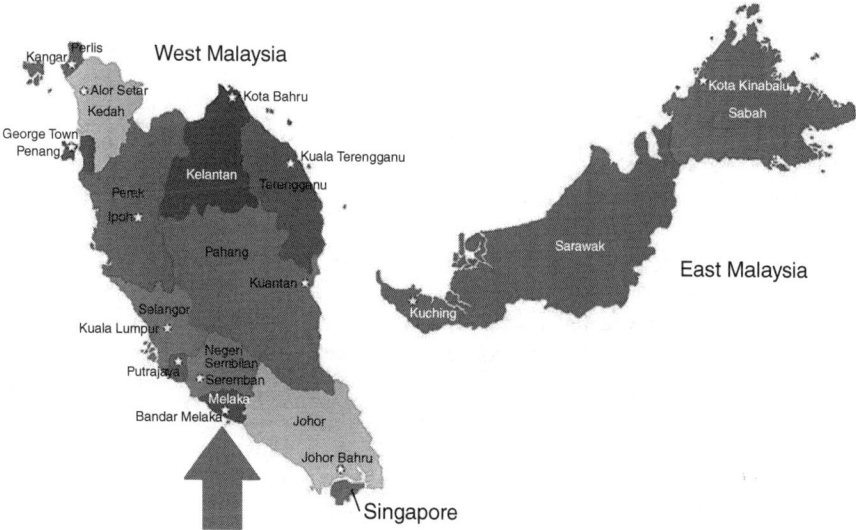

Figure 16.1 Location map of Malacca, Malaysia

Baba and the female Nyonya (Hutton, 2005). Today, the Peranakan community is found distributed throughout Malaysia and Singapore, with its strongholds in Malacca, Singapore and Penang.

The Peranakan immigrated into the bustling ports of Malacca, Penang and Singapore during the British colonial expansion in the nineteenth century and were successful as merchants and, later, as professionals within the community (Tan, 1988; 1993). They became extremely cosmopolitan and displayed their wealth and artistic sensibilities using items acquired from China, Southeast Asia and Europe. In due time, the Baba Nyonyas developed a distinct Malay-Chinese culture and saw themselves as distinct from any other migrant Chinese who came later to settle in the nineteenth and twentieth centuries. The Baba Nyonyas spoke a creolized Malay language enriched with Chinese loan words and syntax. The religion followed many traditional Chinese patterns, but over time the Peranakan Chinese developed distinct rituals. Later, with European colonial expansion, many converted to Christianity.

The Peranakan Chinese began as the go-between traders linking the local Malay population and China and later served as the liaisons between the colonial powers and the local populations (Tan, 1988; 1993). They acted as intermediaries for the Portuguese, Dutch and, later, the English. By the late nineteenth century many of the Babas had been educated in English schools and took upon themselves both the dress and culture of the English. As the Peranakan entered government bureaucracy and the professional classes they became further Anglicised and were even referred to as the King's Chinese. When the Japanese controlled British Malaya (which included Singapore), much of the Peranakan wealth and status was undermined. Furthermore, both Malaysian and Singaporean independence and development further diminished the special status of the Peranakan popula-tion. In recent years, there has been a great effort at reviving and sustaining the achievements of this unique community.

What is Nyonya cuisine?

Tan (2007) describes Nyonya cuisine as a product of the ingenuity of the Nyonya women who have produced food from a combination of their knowledge of Chinese and local indigenous cooking styles with the use of local ingredients. Lee (2008b) concurs that Nyonya cuisine is the original fusion of Chinese food using local Malay ingredients and Tan (2007) asserts that the use of numerous local ingredients not usually found in mainstream Chinese cooking has made Nyonya cuisine very distinct from the usual Chinese food. Key ingredients include coconut milk, galangal (a subtle, mustard-scented root similar to gin-ger), candlenuts as both a flavouring and thickening agent, laksa leaves, pandan leaves, belachan (shrimp paste), tamarind juice, lemongrass, torch ginger bud, jicama, fragrant kaffir lime leaves and rice or egg noodles. Another important ingredient is chinchalok, a pungent sour and salty shrimp-based condiment that is typically mixed with lime juice, chillies and shallots and eaten with rice, fried fish and other side dishes.

Nyonya cuisine, renowned for its delicious flavours, is a complex, serious and time-consuming affair, as traditionally the Nyonyas spent endless hours pounding *rempah* (spices) on a flat stone slab. The cuisine is tangy, aromatic, spicy and herbal and derived its varied influences from Chinese, Thai, Malay, Indonesian and European cuisine. Dishes from Malacca and Singapore exude a Malay and Indonesian influence, using more coconut milk, whilst their northern counterpart Penang has a stronger Thai influence and the food has a relatively sour taste by using tamarind.

Another observation made by Tan (2007) is that traditionally the Nyonya do not follow exact measurements when cooking. They adhere to the '*agak-agak*' principle, which translated from Malay literally means a 'guesstimate'. One has to learn by experience on how much of the ingredients to use and the thrill of producing delicious food in a non-standardised way (Tan, 2007) can create frustration amongst the younger generations who are more used to a cookbook format with exact measures and detailed methods.

In the past, Nyonya cuisine often required painstaking efforts and labourious hours in the kitchen (Hutton, 2005). There has been a growing concern expressed not only in popular media but also in conversations within the Baba Nyonya community of losing their food heritage, especially amongst the future generation. As Teo (2010) notes, 'the sound of pestle hitting mortar is a thing of the past in most Peranakan kitchens as it has been replaced by the electric blender, saving time and effort.' The author adds that the tradition of learning culinary skills from mothers and grandmothers is being lost with fewer Nyonya women choosing to do so under the pressure of modern lifestyles.

Lee (2008a) observes that the cultural practices of the Baba Nyonya community are at a crossroad. Factors contributing to this include globalisation, spatial dispersion and the ageing population of the 'living repositories' amongst the Baba Nyonya community. This is further aggravated by the disinterest of many of the younger generation towards their ancestral culture, with preference being given instead to instantaneous gratification of what modernity has to offer (Lee, 2009).

As Chai (2009, p. 1) commented:

> Early globalisation was what gave the Peranakan its birth, but are these very same forces threatening the survival of the sarong kebaya, dondang sayang and Baba Malay? Members of the community tell adaptability is essential for the revival of their unique legacy. The Babas and Nyonyas who came to the Straits Settlements centuries ago are no strangers to blowing storms. But with the winds of change blowing fast, there are fears their unique culture may be gone with the wind. Scholars, authors, bloggers and anthropologists have since characterised the Babas and Nyonyas as an endangered species. So, would this unique culture fade into oblivion?

The traditional Nyonya cuisine is one of the inherited cultural expressions that has built the identity of the Babas and Nyonyas in Malacca through the passage of time. Behind every recipe lies a hidden knowledge, for example about the perfect mixtures, doneness and ingredient qualities among others, which have been passed down from one generation to another, creating unique hallmarks in the food of this locality and a sense of belonging to this region and its community. Nevertheless, as with any other cultural expression that is part of an intangible cultural heritage, this knowledge is accumulated in the living memory of communities and is faced with the risk of being forgotten (Lee, 2008b).

Decline in home cooking, traditional food knowledge and practices

Knowledge is a critical component of the competencies and meanings that are associated with food practices. Braun and Beckie (2014) argue that the concept of traditional food knowledge is much more than a mere accumulation of information and facts, but it is a process lived out through experience and passed from generation to generation, continually being readapted, reformed and influenced. Fonte (2010) contends that traditional knowledge does not perform a specialised function in society, but rather embodies cultural values as an element integrated into a vast and complex set of beliefs and knowledge that is held collectively and transmitted both orally and through common practices, from generation to generation. It therefore represents the cumulative wisdom of many generations of people who have learned how to produce, prepare, store and teach the practices of food provisioning (Braun & Beckie, 2014). Braun and Beckie (2014) further add that it is this formal and informal sharing and education that ensures the knowledge is kept alive. Its scope extends beyond the technical skills required to procure and prepare food and includes the specific cultural meanings and historical context that has shaped the particular types of food prepared and consumed within that community. For this study, one aspect of traditional food knowledge that will be examined is food preparation in the form of cooking.

Anxiety over the 'impoverished state of domestic cooking' that is highlighted by the media and, more recently, by celebrity chefs garnered significant public interest and academic inquiry over the last decade (Lyon, Sydner, Fjellstrom, Janhonen-Abruquah & Schroder, 2011). In general, researchers have suggested that the erosion of skills held by previous generations is linked to the breakdown of traditional domestic divisions caused by women's labour market participation and the effects of technologies, leading to both de-skilling in the kitchen and distracting children from being in the kitchen to absorb tacit cooking skills (Meah & Watson, 2011).

In analysing cooking as practice, this study takes into account Meah and Watson's (2011) and Lyon et al.'s (2011) earlier findings in examining whether similar instances of de-skilling are taking place within the context of the Baba Nyonya community within the historic city of Malacca, Malaysia. The aim of this chapter is to examine and understand Nyonya heritage cuisine and how the ethnic community of the Babas and Nyonyas in the historic city of Malacca, Malaysia have sustained their cultural heritage through the passage of time. Specifically, this chapter will focus on examining how knowledge of food and gastronomic and culinary practices of traditional Nyonya cuisine is sustained in the face of globalisation and changing lifestyles.

Theoretical framework for this study: social practice theory

The underpinning theoretical framework of this study is social practice theory. Social practice theory is a group of theories that focus on social practices rather than the individual as the point of departure for social change and their patterns of consumption (Reckwitz, 2002a & 2002b). Social practice theory provides an insight into the inter-relations between actors and social structures. For social practice theorists, actors create social structures through the social practices that they sustain and reproduce while simultaneously these practices are embedded into social structures. Society is seen as constituted by social practices that are produced and reproduced over time and space (Ropke, 2009).

Practice can be defined as a shared understanding of what it means to carry out a particular activity that is a routinised type of behaviour (Shove & Pantzar, 2005). Shove, Pantzar and Watson (2012) present practices as the combination of three basic elements: (a) stuff (materials, technologies and tangible, physical entities), (b) images (domain of symbols and meanings), and (c) skills (competence, know-how and techniques) (Figure 16.2). Practices are created and used in everyday life when the links between these three elements are made, sustained and fossilised when these links are disintegrated.

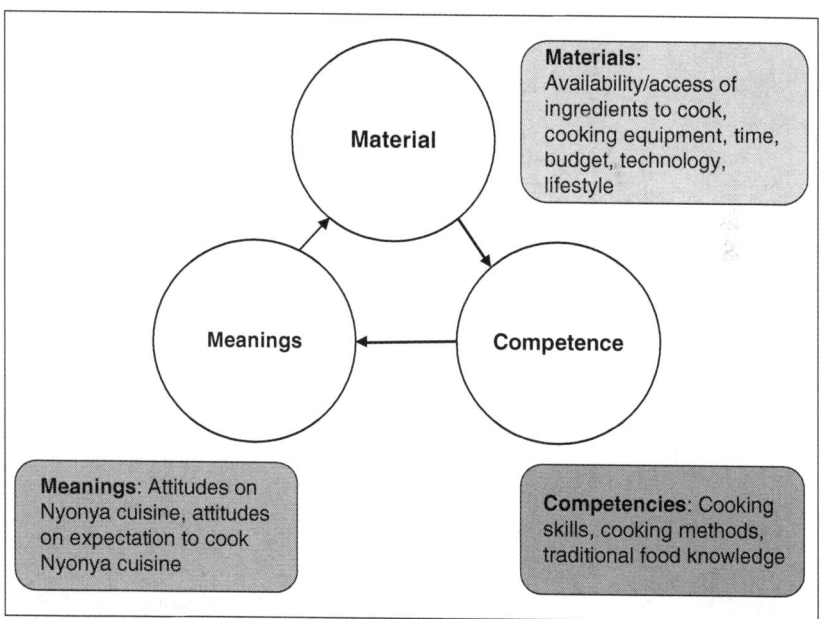

Figure 16.2 Adapted model of social practice theory from Shove, Pantzar & Watson (2012)

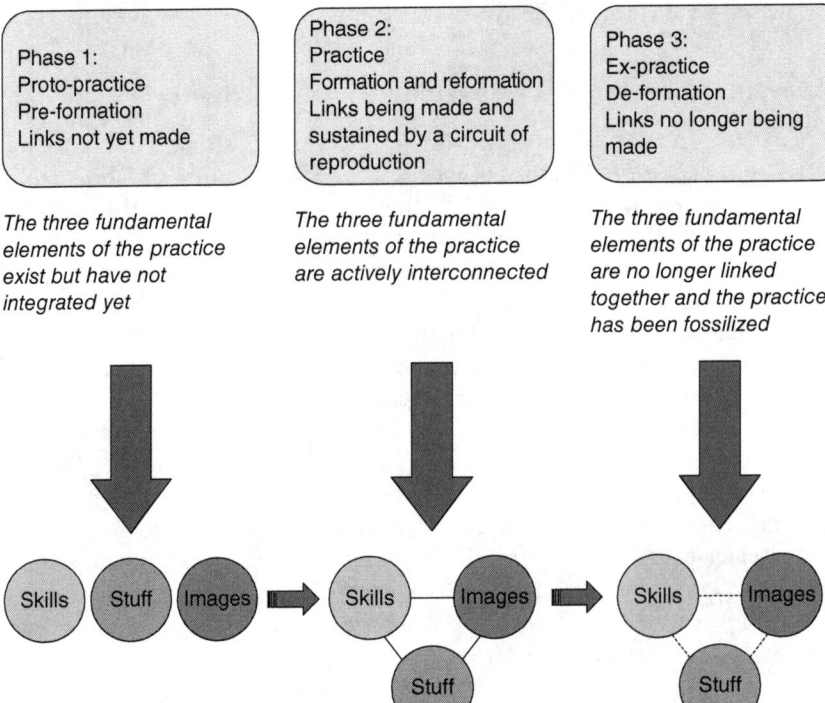

Figure 16.3 The practice of cooking cycle

Source: Adapted from the model of social practice theory from Shove, Pantzar & Watson (2012)

As the social-ecological landscape constantly changes, stable practices need constant reproduction from their carriers in order to persist. If this is not the case, Shove and Pantzar (2006) use the metaphor of 'fossilisation', drawn from the natural sciences to describe the process in which the soft parts of practice (skills and images) fade away leaving the material (stuff) behind. Thus, the fundamental elements of the practice have meaning and effect only when integrated into practice and only by 'doing' they are animated, sustained and reproduced. The separated and stranded elements though can return in new combinations and lead to the resurrection of the practice (Shove & Pantzar, 2006). The latter has been a key concept for this study (Figure 16.3).

Cooking up a storm, the '*agak-agak*' way

The following sections of this chapter present preliminary findings of a study that investigates the sustainability of Nyonya cuisine as an intangible cultural heritage through the passage of time in the historic city of Malacca, Malaysia. This study is designed to examine and document what constitutes Nyonya cuisine and the gastronomic culinary practices and traditions of the Baba Nyonya community in Malacca by applying social practice theory. Subsequently, the study attempts to explore the meanings, materials and competence in regards to the practice of cooking Nyonya cuisine drawing on the accounts of five interviewees (three women and two men) who participated in semi-structured interviews that were conducted between January and February 2016. The interviewees represent bearers of Baba Nyonya cultural heritage within the community. They include three restaurateurs (two females and one male), one retiree (male) and one food entrepreneur-cum-cookbook author/food writer (female), all in their 50s and 60s.

Observers arguably describe Baba Nyonya adhere to a rigid patriarchal tradition (Tan, 2007). Their womenfolk tended to spend their time within the confines of the house taking care of the family. This seems to be the agreement of all the interviewees who view their mothers or grandmothers as the caretaker of the family while the men were out at work. The kitchen or '*perut rumah*' (stomach) (Lee & Chen, 1998) of the house was where the Nyonya spent most of their time preparing dishes for the family. Having been deprived of an education, the Nyonyas focused on being a good daughter, wife and mother (Lee, 2008b), which gave rise to the exceptional cooking that was a fundamental yardstick to being a good woman (Tan, 2001; 2007). The kitchen was off limits to males, as one male interviewee (restaurateur) remembers being scolded by his grandmother for asking her about recipes as he was interested in learning how to cook.

When asked who had taught them cooking, the interviewees' replies were mixed. One female interviewee (restaurateur) mentioned that it was only at a later age (after the passing of her mother) that she learned cooking through her recollections of her mother's cooking. She attributed this late start to her being the youngest in the family, where all was taken care of in regards to her meals. In some ways, admittedly, she was rather pampered. Another female interviewee however did not have the luxury of having her mother to teach her. She recalls that her mom was working then and her times with her aunties during school holidays saw her pick up her culinary skills that turned into her vocation now as an entrepreneur and cookbook author. One male interviewee (restaurateur) recalls that it was only after the passing of his grandmother that he dared to ask his mom to teach him how to cook Nyonya cuisine. However, he mentioned that the experience he had was not really cooking as such but more of assisting his mom in the preparation of the Nyonya dishes and observing her in the cooking of those dishes. Eventually, when his mom was no longer fit enough to carry out cooking later in life he took over the kitchen which meant that he had to cater to up to 300 people for Chinese New Year celebrations. Surprisingly, although the kitchen domain is usually dominated by the womenfolk, one female interviewee (restaurateur) admits that she learnt a great deal of the finer points of taste in cooking Nyonya cuisine from her '*ah koong*' (grandfather), who happened to be a good cook, unlike her own father. She was blessed that her mom and grandmother encouraged her to cook in preparation for womanhood and marriage. Interestingly, another male interviewee claims that within his family he is a much better cook as far as Nyonya cuisine is concerned, whereas his spouse is much better at cooking Western dishes.

It was widely acknowledged by all interviewees that cooking Nyonya cuisine is labourious and time-consuming. The interviewees also emphasised that family matriarchs are well-known perfectionists and meticulous in every aspects of preparing, cooking and presenting Nyonya dishes. In the olden days, all Nyonya cuisine was prepared and cooked from scratch. For those interviewees who had the privilege of being encouraged to cook by their elders, this meant that frequent trips to the markets were inevitable to source fresh ingredients or to shops to buy spices for the dishes. All interviewees agreed that knowing what key ingredients look like is essential in the preparation and cooking of exquisite Nyonya dishes, as are the quality and freshness of the ingredients.

Having sourced the ingredients, an apprentice to the family matriarch needs to clean and prepare the fresh ingredients before they can be cooked into the various Nyonya dishes. Most interviewees claimed that Nyonya elders are fussy when it comes to the technique of cutting vegetables, poultry or seafood as this has an impact on food presentation. Whether it is cutting vegetables at an angle, slicing or dicing of onions or red chillies, cutting and de-boning of the poultry or seafood, all must be done immaculately.

One essential component in the cooking of Nyonya cuisine lies in the preparation of the '*rempah*' or spice mix. All of the interviewees noted its importance in their narrations. Each dish has its own '*rempah*' mix but as one male interviewee (restaurateur) notes, the fundamental spice mix is made up of seven ingredients namely galangal, lemongrass, turmeric, chillies, onions, shrimp paste and candlenut. This was also echoed by another male interviewee (retiree). Another female interviewee (restaurateur) also mentioned the freshness of the ingredients like chillies may have an impact on the taste of the dish, citing her personal preference of buying chillies grown in the nearby villages instead of the ones up in the highlands. According to the male restaurateur, other ingredients might be added to the basic '*rempah*' mix, such as coconut milk, to prepare Nyonya laksas. In the olden days, most of the '*rempah*' mixtures were pounded using the '*batu lesong*' or pestle and mortar, or they were grounded using a '*batu giling*' (rectangular shaped granite slab with a roller). All of the interviewees claimed they have used this kitchen equipment under careful and strict observation of their family matriarch. However, nowadays the interviewees admit that there is choice of not having to use that age-old method, instead opting for more efficient kitchen appliances such as an electric blender or food processor, although as one male interviewee (retiree) laments, the taste of the '*sambal*' is not as good as when made using the pestle and mortar.

As indicated in the preceding paragraph, Nyonya cooking is described as matriarchal in nature by the interviewees. Furthermore, family recipes rarely tend to be written down. Instead, each family's seemingly guarded recipes are passed down orally from one generation to the next and are expected to be mastered by the women in the family. Along with the recipes will be a list of commands (cooking instructions) handed down the generations that, arguably, should never be reinterpreted or adapted. Cooking under the watchful eye of the family matriarch meant that one is learning to cook from scratch in the hope of emulating the recipes to perfection. Cooking Nyonya dishes is no simple affair as it requires time, patience and skills, and it is challenging due to the lack of exact measurements. Most of the interviewees relate to the latter as the '*agak-agak*' way of cooking, as traditionally most Nyonyas do not follow exact measurements but estimates when cooking. Terms such as 'an inch of this', 'a handful of this' or a 'fistful of this' are often heard when the matriarchs talk about their recipes. Paraphernalias in the kitchen too were also used as units of measurement; a teaspoon of, tablespoon of, a '*sendok*' (ladle), a cup or a rice bowl. This can be overwhelming to follow as lamented by one interviewee (restaurateur) who was frustrated upon looking at his matriarch's handwritten recipe book and being unable to figure out which cup to use since the matriarch has passed on. He further commented that the matriarch's recipe book would contain unit measurements of ingredients using 'cents or a dollar of such and such' which baffles and frustrates him a lot. Further, '*agak-agak*' is also exercised in terms of the actual cooking time to check the doneness of the dish. So in this light, long hours of toiling in the family kitchen, assisting the family matriarch or cooking under her stern watchful eyes, in the eyes of the interviewees is key to the creation of excellent dishes based on intuition honed by time and experience.

Status quo of Nyonya cuisine in Malacca

From the above accounts of the men and women interviewed in this study it becomes apparent that they have obtained their food heritage/traditions either by observing, assisting or cooking the Nyonya dishes under the tutelage of their family matriarch. Their accounts of learning to cook Nyonya dishes seem to mirror Sutton's (2001) 'embodied experience'. According to the author food traditions are taught through a process of 'embodied apprenticeship' where culinary knowledge and skills are transmitted and learned through taking part in the physical performance of the cooking (Sutton, 2001, p. 126). For the Babas and Nyonya interviewed, learning to cook Nyonya cuisine took place by and large in an informal setting of observing their familial matriarch. Benny (2012) explains the imitative processes where cultural taste preferences and memories become embedded by observing, listening, smelling and tasting, which develops into a stock of knowledge that acts as a point of reference for future practice. Sutton (2001) emphasises that this knowledge is absorbed casually and without formal lessons whereby the body learns through habituated practice in a way that cannot be set down in more formal situations such as following instructions in recipe books. Sensory cues such as smell and taste are particularly important for indicating when food is correctly prepared according to custom and the cultural tastes of those who will be eating it (Choo, 2004).

The interviewed Baba Nyonya still identify with Nyonya cuisine as part of their Peranakan Chinese food tradition and identity but, as Lee (2008b) cautions, 'a culture is only alive for as long as it is practiced and observed.' Some of the interviewees lament the prospect of losing their food heritage especially among the younger generations. From those interviewed, it can be seen that a few interviewees learnt their cooking not from their mother but from other family matriarchs such as their aunties, grandmother or their father or grandfather. Moreover, when asked if those interviewed would pass down their knowledge to their children, it turned out not to be a question of whether they would but whether their children, who are often preoccupied with work, would find the time to pick up cooking from them. Only a handful has shown interest but only on occasions when it is easy to learn and replicate the recipes for cooking Nyonya cuisine. The seemingly laborious and long hours could possibly be a contributing factor adding to this disenchantment.

The younger generation of Babas and Nyonyas will miss out on the tradition of learning culinary skills from mothers and grandmothers as fewer chose to do so. Interviewees highlighted that future generations will not have the same level of cooking skills and the acquired taste that they have. A few interviewees even mentioned that the younger generations may not even know the basic ingredients let alone know how to cook. One interviewee suggests commercialism is slowly creeping in. The scarcity of time, seemingly prevalent amongst the younger generation of Baba Nyonya, has seen them resort to takeaway and convenience food, which is readily available in Malacca and all over Malaysia for that matter. Street foods are at every nook and corner, while fast food joints and restaurants are mushrooming due to growing demand.

Relevance of food in tourism

Growing evidence suggests that tourists are consciously seeking out food experiences. Abdul Karim and Chi (2010, p. 532) note 'there are many tourists who travel for reasons of seeking culinary experience. Tourism activity related to food has been labelled variously – food tourism, culinary tourism or gastronomy tourism.' These terms have been used interchangeably and carry the same meaning that food-lovers travel to a specific destination for the purpose of tasting, consuming and appreciating local cuisines. Whilst food/culinary/gastronomic tourism is a growing field of research, there is no agreement to define this type of tourism as observed by Sanchez-Canirez and Lopez Guzman (2012); see for example Smith and Xiao's (2008) argument on Long's (2004) articulation of the term culinary tourism and Du Rand and Heath's (2006) and Molz's (2007) discussion on Hall and Sharples' (2003) definition of food tourism. A more inclusive definition was offered by Ignatov and Smith (2006) who propose that culinary tourism can be defined as the purchasing or consuming of local food or the observing and researching of the process of food production by the tourist while travelling. This definition seems to suggest that food can be a primary motivator to travel to a particular destination or it can be an important travel activity at the destination.

Food, culture and identity

Dining or eating is not only an indelible part of travel but is also closely related to culture (Chuang, 2009; Tikannen, 2007; Hjalager & Richards, 2002). Cuisine has now become a cultural artefact (Bessiere & Tibere, 2013; Bessiere, 1998; 2001), cultural resource (Lopez-Guzman, Di-Clemente & Hernandez-Mogollon, 2014) and is central to cultural learning (Hegarty & O'Mahoney, 2001). Additionally, aspects of food rituals and customs may be linked to religion as they bind people to their faiths and belief systems (Mintz & Dubois, 2002), whilst other food customs may signify a society where cooking and eating transcends mere functionality (Harrington, 2005). Food tourism implies that local culture has interesting stories about their cuisines and that local or special knowledge and information that represent local culture and identities are being transferred (Smith & Xiao, 2008; Ignatov & Smith, 2006; Long, 2004). Accordingly, unique cuisines not only build popular travel destinations but also improve the culinary cultural image of a country, which makes the cuisine culture a major attraction within the travel destination (Horng & Tsai, 2012). In fact, cuisine has now been realised as an integral part of a destination's intangible heritage (Hassan, 2008).

Food image and destination branding

A country like Italy has successfully blended food into their Italian culture and thus connects it to the lifestyles of its people (Corigliano, 2002). Du Rands, Heath and Alberts (2003) concur that the cuisine of a country can showcase its cultural or national identity. Having been elevated to be an integral part of a destination heritage, food has now become an important element used to represent the destination image (Everett & Aitchison, 2008; Kivela & Crotts, 2005; Quan & Wang, 2004). Destinations can use food to represent their 'cultural experience, cultural identity and communication' (Frochot, 2003). As Jones and Jenkins (2002) add, beyond its functionality, food can positively present a destination's cultural element. Similarly, the study conducted by Abdul Karim and Chi (2010) echoes the notion that a favourable food image of a country has a significant positive relationship with visit intentions. It is an important marker of cultural distinctiveness (Dawson, 2012) and an important dimension of a destination's perceived image (Beerli & Martin, 2004). Food is a significant part of a national emblem and a key aspect of differentiation (Chuang, 2009). As such food has been used as a tool in destination branding and marketing (Seo, Kim, Oh & Yun, 2013; Okumus, Kock, Scantlebury & Okumus, 2013; Horng & Tsai, 2012b; Lin, Pearson & Cai, 2011; Karim & Chi, 2010; Harrington & Ottenbacher, 2010; Horng & Tsai, 2010; Everett & Aitchison, 2008; Okumus, Okumus & McKercher, 2007; Fox, 2007; Du Rand & Heath, 2006; Kivela & Crotts, 2005; Boyne, Hall & Williams, 2003).

Food as a conduit of destination to tourist experience

Quan and Wang (2004) note that food can convey unique experience and enjoyment to travellers. Food offers tourists a window into a destination's culture as it introduces new flavours and traditions (Field, 2002). Travellers can appreciate a destination's culture as they experience new foodways (Chang, Kivela & Mak, 2011; Fox, 2007). Studies by Bjork and Kauppinen-Räisänen (2014a, 2014b), Wijaya, King, Nguyen and Morrison (2013) and Everett (2008) affirm that food is an essential element contributing to the visitors' travel experience (Kivela &

Crotts, 2009), so much so that Quan and Wang (2004) aptly regard it as 'peak' experience that has a positive effect on tourists' holistic experiences of a destination and influencing their re-visit intentions (Kivela & Crotts, 2006).

Food and the motivation to travel

Quan and Wang (2004) argue that culinary tourists' primary motivation to travel is to seek new cuisines, a novelty-seeking behaviour that may contribute to their total memorable experience. To others, trying the unfamiliar could be a major trip accomplishment (Tikkanen, 2007). Tourists who attempt to explore unfamiliar local food could be motivated by physical, cultural, interpersonal and status and prestige reasons (Kim, Eves & Scarles, 2009). More recent work on this aspect includes Guan and Jones (2014), Kim, Eves and Scarles (2013), Mak, Lumbers, Eves and Chang (2013), Su (2013), Mak, Lumbers, Eves and Chang (2012), Mak, Lumbers and Eves (2012), Tsai and Lu (2012), Kim and Eves (2012), Sanchez-Canizares and Lopez-Guzman (2012), Kim, Kim and Goh (2011), Chang, Kivela and Mak (2011), Kim, Goh and Yuan (2010), Chang, Kivela and Mak (2010); and Kim, Eves and Scarles (2009).

The preceding discussion on food and tourism seems to suggest several key findings. First, it acknowledges the importance of food as a cultural marker of a community at a particular place or region. Second, food has also been recognised as a resource in tourism such that it has been used as a marketing and promotional tool by various destination-marketing organisations. Further, the literature has also deliberated on the various roles that food presents in tourists' experiences at a destination

Nevertheless, most of these studies pertaining to food in tourism were based on the consumers' perspectives of how they perceive food at a tourist destination. The literature seemingly fails to address the local community's views on this. Further, whilst acknowledging food as a cultural resource, the literature barely scratches the cultural aspect of food in tourism. If food is to be embraced like any other tourism resource, it certainly cannot escape the possible impact that tourism will have on it. We were once reminded by Reynold (1993, p. 48) that 'if a culture is to prove sustainable in the face of tourism, then traditional and ethnic food must be preserved along with other art forms.'

Although food/culinary/gastronomy studies have added a cultural perspective to sustainable tourism (Scarpato, 2002), there have been only few empirical studies into food-related tourism, particularly from a socio-cultural perspective (Boniface, 2003; Hjalager & Richards, 2002). Reynolds (1993, p. 53) asserts that if we are serious about sustainable tourism then the 'erosion of cultures and traditional skills must be investigated'. Du Rand and Heath (2006) echo a similar notion and suggest that the promotion of local and regional cuisine will lead to the preservation of culinary heritage as well as adding value to a destination's authenticity.

Whilst acknowledging the complexity of the relationship between heritage cuisine and tourism, Timothy and Ron (2013) also observe the changes in heritage cuisine. Metro-Roland (2013) notes the 'touristification' of the traditional Hungarian goulash to suit the palate of tourists. Similar instances were reported by Staiff and Bushell (2013) and Avieli (2013) in their study of local food in Laos and Vietnam. Such observations were also made by Teixeira and Robeiro's (2013) study of traditional food and foodways in Portugal, echoing an earlier study by Cohen and Avieli (2004) that described some form of negotiation in terms of ingredients, cooking style or method so as to conform to the typical mass tourist palate and sacrifice of their own culinary traditions.

Tourism has emerged from the backwaters of the Malaysian economy to become a major contributor in terms of foreign exchange. In 2016, Malaysia received 26.8 million tourists compared to 25.7 million tourists in 2015, with tourist receipts rising by 18.8%, contributing RM82.1 billion to the country's revenue against RM69.1 billion in 2015, which translates to an average per capita expenditure of RM3,068 (Tourism Malaysia, 2017). The top ten tourist generating markets to Malaysia in 2016 were Singapore (13.3 million), Indonesia (3.1 million), China (2.1), Thailand (1.8 million), Brunei (1.4 million), India (0.64 million), South Korea (0.44 million), the Philippines (0.42 million), Japan (0.41 million) and the United Kingdom (0.40 million). In the context of Malacca, it seems that of those 12.2 million international tourists who visited Malacca in 2015 1.1 million (or 31.27%) were tourists from Singapore, followed by 711,800 from China, Indonesia (484,738), Taiwan (121,799) and 100,462 visitors from Japan (Murali, 2016). Further, the author comments that the availability of a wide range of food seems to entice visitors to Malacca.

The federal, state and local governments have recognised food as one of the attractions that can lure tourists to visit Malaysia. The literature discussed by Seo, Kim, Oh and Yun (2013), Ryu and Jang (2006), Cohen and Avieli (2004) and Bessiere (1998) shows that food may act as a sense of place attraction for tourists wanting unusual or extraordinary gastronomic experience or food culture beyond their place of domicile. To this end, the National Heritage Department of Malaysia has enlisted 400 different dishes and ingredients of Malaysian cuisine as part of the country's heritage. This amalgam or culinary potpourri has enabled Malaysia to showcase its history and cultural heritage via its foodways (Omar S.R., Abdul Karim, Abu Bakar & Omar S.N., 2015; Khoo & Badarulzaman, 2014). Khoo and Badarulzaman (2014) also suggest that foodways support a strong sense of identity for individuals belonging to different religions and ethnicities and have the possibility to develop local heritage tourism, festivals and other celebrations.

Worden (2003) mentions that the heritage in Malacca is produced for international tourism. Some researchers argue that this might have gone too far and that the state has become 'a cultural theme park, with a particular past frozen and a city created largely for tourist consumption' (Lai & Ooi, 2015, p. 12). The slogan 'where it all began', together with *Melawat Melaka Bersejarah bererti melawat Malaysia*' ('visiting Malacca is equivalent to visiting Malaysia'), has been used in various official tourist promotion campaigns (Lai & Ooi, 2015; Goh, 2014). These slogans show how much significance is put into showcasing Malacca as a tourism attraction with a rich heritage background (Cartier, 1998).

The preceding discussions have shown that the linkages between food and tourism may provide the platform for economic development with the use of food experiences for branding and marketing destinations (Richards, 2012). The challenge nevertheless according to the author is in managing the shift towards intangible cultural heritage. The traditional Malaysian cuisine has historically been influenced by traders from neighbouring countries such as India, the Middle East, China and Indonesia (Zainal, Zali & Kassim, 2010). In the context of Nyonya cuisine, the potential of this food experience to tourists, in particular those from China, can provide an enormous opportunity to 'travel back in time', reminisce about the 'good old days' and recollect memories of the historical contributions of the early Chinese migrants to the development of Malacca. Further, Richard (2012) notes that food experiences for tourism may contribute to regional attractiveness, sustain the local environment and cultural heritage and strengthen local identities and sense of community. Only time can tell if food tourism can be a way to preserving Nyonya cuisine into the future. From the earlier discussions, it can be seen that Nyonya cuisine remains an important cultural marker for the Baba Nyonya community and certainly has a place in the realm of tourism in the context of Malacca and Malaysian tourism in general.

Conclusion

While Lee (2008a) may have been far sighted to seemingly predict that the cultural practices of the Baba Nyonya community are at a crossroad, nevertheless, the findings of this study have only scratched the surface of the phenomena under investigation. The preliminary interviews elucidated that there is a learning process experience by the elder generation from their parents or grandparents. Nevertheless, whilst most interviewees show interest in imparting their cooking knowledge and skills to their children, time is of the essence. The younger generations seem to be preoccupied with work and what modernity has to offer. Like any other Chinese community the Babas and Nyonyas face a similar dilemma; the decline of traditions, modernity and the growing number of mixed or inter-marriages. These factors arguably lead to changes in the Baba Nyonya culture and an uncertain future. The use of food experience in tourism may offer an avenue for the community to revive traditional cuisine such as Nyonya cuisine in the wake of demand by tourists who want to experience 'authentic' yet not too unfamiliar food during their holidays. While it is arguably a challenge to produce food that is palatable for foreign tourists, nonetheless, food tourism holds opportunities for Malacca as a destination to use Nyonya cuisine in their branding and marketing initiatives, thus differentiating it from any other destination.

References

Abdul Karim, S. & Chi, C.G. (2010). Culinary tourism as a destination attraction: An empirical examination of destinations' food image. *Journal of Hospitality Marketing and Management*, 19(6), pp. 531–555.

Avieli, N. (2013). What is 'local food'? Dynamic culinary heritage in the World Heritage Site of Hoi An, Vietnam. *Journal of Heritage Tourism*, 8(2–3), pp. 120–132.

Beerli, A. & Martin, J.D. (2004). Tourists' characteristics and the perceived image of tourist destinations: a quantitative analysis. A case study of Lanzarote, Spain. *Tourism Management*, 25(5), pp. 623–636.

Benny, H. (2012). When traditions become innovations and innovations become traditions in everyday food pedagogies. *Australian Journal of Adult Learning*, 52(3), November, pp. 595–615.

Bessiere, J. (1998). Local development and heritage: Traditional food and cuisine at tourists attractions in rural areas. *Sociologica Ruralis*, 38(1), pp. 21–34.

Bessiere, J. (2001). The role of gastronomy in tourism. In D. Hall and L. Roberts (eds.) *Rural Tourism and Recreation: Principles and Practices* (pp. 115–118). Wallingford: CABI.

Bessiere, J. (2013). 'Heritagisation', a challenge for tourism promotion and regional development: An example of food heritage. *Journal of Heritage Tourism*, http://dx.doi:10.1080/1743873X.2013.770861

Bessiere, J. & Tibere, L. (2013). Traditional food and tourism: French tourist experience and food in rural spaces. *Journal of Science and Food Agriculture*, 93, pp. 3420–3425.

Bjork, P. & Kauppinen-Raisanen, H. (2014a). Culinary-gastronomy tourism: A search for local food experience. *Nutrition and Food Science*, 44(4), pp. 294–309, http://dx.doi:10.1108/NFS12-2013.0142

Bjork, P. & Kauppinen-Raisanen, H. (2014b). Exploring the multi-dimensionality of travellers' culinary-gastronomic experiences. *Current Issues in Tourism*, http://dx.doi:10.1080/13683500.2013.868412

Boniface, P. (2003). *Tasting Tourism: Travelling for Food and Drink*. Burlington: Ashgate Publishing.

Boyne, S., Hall, D. & Williams, F. (2003). Policy, support and promotion for food related tourism initiatives: A marketing approach to regional development. *Journal of Travel and Tourism Marketing*, 14(3), pp. 21–34.

Braun, J. & Beckie, M.A. (2014). Against the odds: The survival of traditional food knowledge in a rural Albertan community. *Canadian Food Studies*, 1(1), pp. 54–71.

Braun, J.A.J. (2013). Pickles, beets and bread: Examining traditional food knowledge in a rural Albertan community. Master of Science in Rural Sociology, University of Alberta, Edmonton, Canada.

Chai, M.L. (2009, November 15). Can Peranakan culture survive? *New Sunday Times*, http://www.leesukim.net/pdf/NST_15nov09.pdf

Chang, J. & Tsieh, A.T. (2006). Leisure motives of eating out in night markets. *Journal of Business Research*, 59, pp. 1276–1278.

Chang, R.C.Y., Kivela, J. & Mak, A.H.N. (2010). Food preferences of Chinese tourists. *Annals of Tourism Research*, 37(4), pp. 989–1011.

Chang, R.C.Y., Kivela, J. & Mak, A.H.N. (2011). Attributes that influence the evaluation of travel dining experience: When east meets west. *Tourism Management*, 32 (2), pp. 307–316.

Chi, C.G-Q., Bee, L.C., Othman, M. & Abdul Karim, S. (2013). Investigating structural relationships between food image, food satisfaction, culinary quality and behavioural intentions: The case of Malaysia. *International Journal of Hospitality & Tourism Administration*, 14(2), pp. 99–120.

Choo, S. (2004). Eating Satay Babi: Sensory perception of transnational movement. *Journal of Intercultural Studies*, 25(3), pp. 203–213.

Chuang, H.T. (2009). The rise of culinary tourism and its transformation of food cultures: The national cuisine of Taiwan. *Copenhagen Journal of Asian Studies*, 27(2), pp. 84–108.

Cohen, E. & Avieli, N. (2004). Food in tourism. Attraction and impediment. *Annals of Tourism Research*, 31(4), pp. 755–778.

Corigliano, A. (2002). The route to quality: Italian gastronomy networks in operations. In A.M. Hjalager and G. Richards (eds.) *Tourism and Gastronomy* (pp. 166–185). London: Routledge.

Dawson, A.C. (2012). Food and spirits: religion, gender and identity in the 'African' cuisine of North Brazil. *African and Black Diaspora: An International Journal*, 5(2), pp. 243–263.

Du Rand, G. & Heath, E. (2006). Towards a framework for food tourism as element of destination marketing. *Current Issues in Tourism*, 9(3), pp. 206–234.

Du Rand, G., Heath, E. & Alberts, N. (2003). The role of local and regional food in destination marketing: A South African situation analysis. In C.M. Hall (ed.) *Wine, Food and Tourism Marketing* (pp. 77–96). New York: The Haworth Hospitality Press.

Everett, S. (2008). Beyond the visual gaze?: The pursuit of an embodied experience through food tourism. *Tourist Studies*, 8(3), pp. 337–358. http://.dx.doi: 10.1177/1468797608100594

Everett, S. & Aitchison, C. (2008). The role of food tourism in sustaining regional identity: A case study of Cornwall, South West England. *Journal of Sustainable Tourism*, 16(2), pp. 150–167.

Fields, K. (2002). Demand for the gastronomy tourism products: Motivational factors. In A.M. Hjalager and G. Richard (eds.) *Tourism and Gastronomy* (pp. 36–50). London: Routledge.

Fonte, M. (2010). Food relocalisation and knowledge dynamics for sustainability in rural areas. In M. Fonte & G. Papadopoulos *Naming Food After Places: Food Relocalisation and Knowledge Dynamics in Rural Development* (pp. 1–35). England: Ashgate Publishing

Fox, R. (2007). Reinventing the gastronomic identity of Croatian tourist destinations. *International Journal of Hospitality Management*, 26(3), pp. 546–559.

Frochot, I. (2003). An analysis of regional positioning and its associated food images in French tourism regional tourism brochures. In C.M. Hall (ed.) *Wine, Food and Tourism Marketing* (pp. 77–96). New York: The Haworth Hospitality Press.

Goh, D.P. (2014). Between history and heritage: Post-colonialism, globalisation and the remaking of Malacca, Penang and Singapore. *TRaNS: Trans-Regional and National Studies of Southeast Asia*, 2, pp. 79–101. doi:10.1017/trn.2013.17

Guan, J. & Jones, D.L. (2014). The contribution of local cuisine to destination attractiveness: An analysis involving Chinese tourists' heterogeneous preferences. *Asia Pacific Journal of Tourism Research*, pp. 1–19. http://dx.doi:10.1080/10941665.2014.889727

Hall, C.M. & Sharples, L. (2003). The consumption experience or the experience of consumption. An introduction to tourism of taste. In C.M. Hall et. al. *Food Tourism Around the World: Development, Management and Markets* (pp. 1–24). Oxford: Butterworth-Heinemann

Harrington, R.J. (2005). Defining gastronomic identity: The impact of environment and culture on prevailing components, texture and flavours in wine and food. *Journal of Culinary Science and Technology*, 4(2/3), pp. 129–152.

Harrington, R.J. & Ottenbacher, M.C. (2010). Culinary tourism: A case study of the gastronomic capital. *Journal of Culinary Science and Technology*, 8(1), pp. 14–32. http://dx.dpi.org/10.1080/15428052.2010.490765

Hassan, Y. (2008). *Local cuisines in the marketing of tourism destinations: The case of Kelantan.* Paper presented at ECER Regional Conference, Kelantan, Malaysia. http://www.iefpedia.com

Hegarty, J.A. & O'Mahoney, G.B. (2001). Gastronomy: A phenomenon of cultural expressionism and an aesthetic for living. *Hospitality Management*, 20, pp. 3–13.

Hjalager, A.M. & Richards, G. (eds.) (2002). *Tourism and Gastronomy*. London: Routledge.

Horng, J.S. & Tsai, C.T. (2010). Government websites for promoting East Asian culinary tourism: A cross-national analysis. *Tourism Management*, 31(1), pp. 74–85.

Horng, J.S. & Tsai, C.T. (2012a). Culinary tourism strategic development: An Asia-Pacific perspective. *International Journal of Tourism Research*, 14, pp. 20–55.

Horng, J.S. & Tsai, C.T. (2012b). Exploring marketing strategies for culinary tourism in Hong Kong and Singapore. *Asia Pacific Journal of Tourism Research*, 17(3), pp. 277–300. http://dx.doi: 10.1080/10941665.2011.625432

Ignatov, E. & Smith, S. (2006). Segmenting Canadian culinary tourists. *Current Issues in Tourism*, 9(3), pp. 235–255.

Jones, A. & Jenkins, I. (2002). A taste of Wales – Blass Ar Cymru: Institutional malaise in promoting Welsh food tourism products. In A.M. Hjalager and G. Richards (eds.). *Tourism and Gastronomy* (pp. 112–115). London: Routledge.

Khoo, S. & Badarulzaman, N. (2014). Factors determining George Town as a city of gastronomy, *Tourism Planning & Development*, 11(4), pp. 371–386.

Kim, Y.G. & Eves, A. (2012). Construction and validation on a scale to measure tourist motivation to consume local food. *Tourism Management*, 33, pp. 1458–1467.

Kim, Y.G., Eves, A. & Scarles, C. (2009). Building a model of local food consumption on trips and holidays: A grounded theory approach. *International Journal of Hospitality Management*, 33, pp. 423–431.

Kim, Y.G., Eves, A. & Scarles, C. (2013). Empirical verification of a conceptual model of local food consumption at tourist destinations. *International Journal of Hospitality Management*, 28, pp. 484–489.

Kim, Y.H., Goh, B.K. & Yuan, J. (2010). Development of a multi-dimensional scale for measuring food tourist motivations. *Journal of Quality Assuarance in Hospitality and Tourism*, 11(1), pp. 56–71.

Kim, Y.H., Kim, M. & Goh, B.K. (2011). An examination of food tourists' behaviour: Using the modified theory of reasoned action. *Tourism Management*, 32, pp. 1159–1165.

Kivela, J. & Crotts, J.C. (2005). Gastronomy tourism. *Journal of Culinary Science and Technology*, 4(2/3), pp. 39–35.

Kivela, J. & Crotts, J.C. (2006). Tourism and gastronomy: Gastronomy's influence on how tourists experience a destination. *Journal of Hospitality & Tourism Research*, 30(3), pp. 354–377.

Kivela, J. & Crotts, J.C. (2009). Understanding travellers' experiences of gastronomy through etymology and narration. *Journal of Hospitality & Tourism Research*, 33(2), pp. 161–192.

Kuake, B.M.J. (2010). Communicating the Baba Nyonya Identity. Published doctoral thesis, University Putra Malaysia.

Kwik, J. (2008a). Traditional food knowledge: A case study of an immigrant Canadian 'foodscape'. *Environments*, 36(1), pp. 59–74.

Kwik, J. (2008b). *Traditional Food Knowledge: Renewing Culture and Restoring Health*. Master of Environmental Studies, University of Waterloo, Ontario, Canada.

Lai, S. & Ooi, C.-S. (2015). *Experiences of Two UNESCO World Heritage Cities: National and Local Politics in Branding the Past*. Copenhagen: Center for Leisure and Culture Services Working Paper, Copenhagen Business School.

Lee, P. & Chen, J. (1998). *Rumah Baba: Life in a Peranakan House*. Singapore: National Heritage Board, Singapore History Museum.

Lee, S.K. (2008a). *In Search of Identity: Through the Eyes of Peranakan Women Writers.* Plenary paper presented at the Inaugural National Conference on Peranakan Chinese and Kwangtung Muslims, Penang.

Lee, S.K. (2008b). The Peranakan Baba Nyonya culture: resurgence or disappearance?, *Sari*, 26, pp. 161–170.

Lee, S.K. (2009). A Nyonya precedence. *Resonance*, 8, April–June 2009.

Lin, Y.C., Pearson, T.E. & Cai, L.A. (2011). Food as a form of destination identity: A tourism destination brand perspective. *Tourism and Hospitality Research*, 11, pp. 30–48.

Long, L.M (ed.) (2004). *Culinary Tourism*. Kentucky: The University Press of Kentucky.

Lopez-Guzman, T., Di-Clemente, E. & Hernandez-Mogollon, J.M. (2014). Culinary tourists in the area of Extremadura, Spain. *Wines, Economics and Policy*.

Lopez-Guzman, T. & Sanchez-Canizares, S. (2012). Culinary tourism in Cordoba. *British Food Journal*, 114(2), pp. 168–179.

Lyon, P., Sydner, Y.M., Fjellstrom, C., Janhonen-Abruquah, H. & Schroder, M. (2011). Continuity in the kitchen: how younger generations and older women compare in their food practices and use cooking skills. *International Journal of Consumer Studies*, 35, pp. 529–537.

Mak, A.H.N., Lumbers, M. & Eves, A. (2012). Globalisation and food consumption in tourism. *Annals of Tourism Research*, 39(1), pp. 171–196.

Mak, A.H.N., Lumbers, M., Eves, A. & Chang, C.Y. (2012). Factors influencing tourists' food consumption. *International Journal of Hospitality Management*, 31, pp. 928–936.

Mak, A.H.N., Lumbers, M., Eves, A. & Chang, C.Y. (2013). An application of the repertory grid method and generalised procrustes to investigate the motivational factors of tourist food concumption. *International Journal of Hospitality Management*, 35, pp. 327–338.

Meah, A. & Watson, M. (2011). Saints and slackers: Challenging discourses about the decline of domestic cooking. *Sociological Research Online*, 16(2), p. 6.

Metro-Roland, M. (2013). Goulash nationalism: the culinary identity of a nation. *Journal of Heritage Tourism*, 8(2/3), pp. 172–181.

Mintz, S.W. & Du Bois, C.M. (2002). The anthropology of food and eating. *Annual Review of Anthropology*, 31, pp. 99–119.

Molz, J.G. (2007). Eating difference: The cosmopolitan of culinary tourism. *Space and Culture*, 10(1), pp. 77–93.

Murali, R.S. (2016, January 8). Malacca draws the crowd. *The Star Online*. http://www.thestar.com.my/metro/community/2016/01/08/malacca-draws-thecrowd-high-volume-of-tourist-arrivals-recorded-last-year/

Okumus, F., Kock, G., Scantlebury, M.M.G. & Okumus, B. (2013). Using local cuisines in promoting small Caribbean island destinations. *Journal of Travel and Tourism Marketing*, 30(4), pp. 410–429. http://dx.doi:10.1080/10548408.2013.784161.

Okumus, B., Okumus, F. & McKercher, B. (2007). Incorporating local and international cuisines in the marketing of tourism destinations: The cases of Hong Kong and Turkey. *Tourism Management*, 28, pp. 253–261.

Omar, S.R., Abdul Karim, S., Abu Bakar, A.Z. & Omar, S.N. (2015). Safeguarding Malaysian Heritage Food (MHF): The impact of Malaysian food culture and tourists' food culture involvement on intentional loyalty. *Procedia – Social and Behavioural Sciences*, 172, pp. 611–618.

Quan, S. & Wang, N. (2004). Towards a structural model of tourist experience: An illustration from food experience in tourism. *Tourism Management*, 25, pp. 297–305.

Reckwitz, A. (2002a). The status of 'material' in theories of culture: From 'social structure' to 'artefacts'. *Journal of the Theory of Social Behaviour*, 32, pp. 195–217.

Reckwitz, A. (2002b). Towards a theory of social practice: A development in culturalist theorising. *European Journal of Social Theory*, 5, pp. 243–263.

Reynolds, P. (1993). Food and tourism: Towards an understanding of sustainable culture. *Journal of Sustainable Tourism*, 1(1), pp. 48–54.

Richards, G. (2012). An overview of food and tourism trends and policies. In *Food and the Tourism Experience* (pp. 13–46). The OECD-Korea Workshop, OECD Studies on Tourism, OECD Publishing.

Ropke, I. (2009). Theories of practice: New inspiration for ecological economic studies on consumption. *Ecological Economics*, 68, pp. 2490–2497.

Ryu, K. & Jang, S. (2006). Intention to experience local cuisine in a travel destination: The modified theory of reasonable action. *Journal of Hospitality & Tourism Research*, 30(4), pp. 507–516.

Sanchez-Canizares, S.M. & Lopez-Guzman, T. (2012). Gastronomy as a tourism resource: Profile of the culinary tourist. *Current Issues in Tourism*, 15(3), pp. 229–245.

Scarpato, R. (2002). Sustainable gastronomy as a tourist product. In A. Hjalager and G. Richards (eds.) *Tourism and Gastronomy* (pp. 132–151). London: Routledge.

Seo, S., Kim, O.Y., Oh, S. & Yun, N. (2013). Influence of informational and experiential familiarity on image of local foods. *International Journal of Hospitality Management*, 34, pp. 295–308.

Short, F. (2006). *Kitchen Secrets: The Meaning of Cooking in Everyday Life*. Oxford & New York: Berg.

Shove, E. & Pantzar, M. (2005). Consumers, producers and practices: Understanding the invention and reinvention of Nordic walking. *Journal of Consumer Culture* 5(1), pp. 43–64.

Shove, E. & Pantzar, M. (2006). Fossilisation. *Ethnologia Europaea: Journal of European Ethnology* 35(1/2), pp. 59–63.

Shove, E., Pantzar, M. & Watson, M. (2012). *The Dynamics of Social Practice: Everyday Life and How It Changes*. London: Sage Publishing Ltd.

Smith, S.L.J & Xiao, H. (2008). Culinary tourism supply chains: A preliminary examination. *Journal of Tourism Research*, 46(3), pp. 289–299.

Staiff, R. & Bushell, R. (2013). The rhetoric of Lao/French fusion: Beyond the representation of the western tourist experience of cuisine in the World Heritage City of Luang Prabang, Laos. *Journal of Heritage Tourism*, 8(2/3), pp. 133–144.

Su, C.-S. (2013). An importance performance analysis of dining attributes: A comparison of individual and packaged tourists in Taiwan. *Asia Pacific Journal of Tourism Research*, 18(6), pp. 573–597.

Sutton, D. (2001). *Remembrance of Repasts: An Anthropology of Food and Memory*. Oxford & New York: Berg.

Tan, C.B. (1979). Baba and Nyonya: A Study of the Ethnic Identity of the Chinese Peranakan in Malacca. Doctoral Thesis: Cornell University, New York.

Tan, C.B. (1988). *The Baba of Melaka*. Petaling Jaya: Pelanduk Publications.

Tan, C.B. (1993). *Chinese Baba-Nyonya Heritage in Malaysia and Singapore*. Kuala Lumpur: Penerbit Fajar Bakti.

Tan, C.B. (2007). Nyonya cuisine: Chinese, non-Chinese and the making of a famous cuisine in Southeast Asia. In Sidney C.H. Cheung and Tan Chee Beng *Food and Foodways in Asia* (pp. 171–182). Oxon: Routledge.

Teixera, V.A.V. & Ribeiro, N.F. (2013). The lamprey and the partridge: a multi-sited ethnography of food tourism as an agent of preservation and disfigurement in Central Portugal. *Journal of Heritage Tourism*, 8(2/3), pp. 193–212.

Teo, A. (2010). Malaysian's Peranakan struggle to keep culture alive. *News@AsiaOne*. http://news.asiaone.com

Teoh, K.M. (2015). Domesticating hybridity: Straits Chinese cultural heritage projects in Malaysia and Singapore. *Cross-Currents: East Asian History and Culture Review E-Journal*,17. http://cross-currents.berkeley.edu/e-journal/issue-17

Tikkanen, I. (2007). Maslow's hierarchy and food tourism in Finland: Five cases. *British Food Journal*, 109(9), pp. 721–734.

Timothy, D. & Ron, A.S. (2013). Understanding heritage cuisines and tourism: Identity, image, authenticity and change. *Journal of Heritage Tourism*, 8(2/3), pp. 99–104.

Tourism Malaysia (2017). Press Release: Malaysia's 2016 Tourist arrivals grow 4.0%. http://tourism.gov.my/media/view/malaysia-s-2016-tourist-arrivals-grow-4-0

Tsai, C-T.S. & Lu, P-H. (2012). Authentic dining experiences in ethnic theme restaurants. *International Journal of Hospitality Management*, 31, pp. 304–306.

Warde, A. (1997). *Consumption, Food and Taste: Culinary Antinomies and Commodity Culture*. London: Sage.

Wijaya, S., King, B., Nguyen, T-H. & Morrison, A. (2013). International visitor dining experiences: A conceptual framework. *Journal of Hospitality & Tourism Management*, 20, pp. 34–42.

Worden, N. (2001). 'Where it all Began': the representation of Malaysian heritage in Malacca. *International Journal of Heritage Studies*, 7(3), pp. 199–218.

Worden, N. (2003). National identity and heritage tourism in Malacca. *Indonesia and the Malay World*, 31(89), pp. 31–43.

Zainal, A., Zali, A.N. & Kassim, M.N. (2010). Malaysian gastronomy routes as a tourist destination. *Journal of Tourism, Hospitality & Culinary Arts*, 2(1), pp. 15–24.

17 Moving forward

Think 'dine and wine' with the Chinese tourists

Ian Phau and Christof Pforr

Introduction

As highlighted in many parts of this book, the Chinese tourist market is growing exponentially with approximately 140 million Chinese tourists travelling in 2016. These 'new rich' tourists are voraciously consuming new and novel experiences at tourist destinations all over the world. The appetite of Chinese tourists is not limited to unique travel experiences but extends to a demand for local food- and wine-related experiences. This creates an increasingly important niche segment for destinations in order to differentiate and diversify in a very competitive food and beverage market space. Given their unique culture and history, destination management organisations (DMOs) around the world are becoming keenly aware of the need to cater to the needs of these intrepid tourists. Chapters in this book cover a wide range of perspectives on the Chinese tourist market with regard to the food and wine industry. These studies encompass: (1) philosophical debates on whether to reorientate existing tourism strategies to focus on Chinese tourists; (2) trends for the outbound Chinese travel market; (3) economic assessments of Chinese tourists; (4) Chinese tourist behaviours; and (5) challenges in catering to this distinct tourist market. International perspectives, for instance from Australia, China, Italy and New Zealand, have also been included in order to broaden the scope of this book.

Undoubtedly, the immense potential of the Chinese tourist market, particularly with regards to food and wine tourism, is fully reflected throughout the chapters in this book. While the main focus of these studies was food and wine, the importance of the tourist destinations in which these products are consumed is also emphasised. With studies conducted in Australia, New Zealand, Taiwan, the USA and Italy, authors have examined interrelated areas and provoke new thoughts for future research in food and wine tourism. This chapter will reflect upon the key themes presented in this book. It also summarises the prominent opportunities, challenges and recommendations relevant to targeting Chinese tourists. The insight gleaned from these studies will undoubtedly benefit tourism and hospitality academics as well as practitioners in the area.

Chinese tourists in Western Australia

Western Australia (WA), next to other international cases, has been an exemplary focus of this book to explore opportunities and challenges associated with food and wine tourism in the context of the growing China outbound market. It has established itself as a prominent destination for tourists, both national and international, over recent decades. While WA only attracted 965,900 international visitors in 2016/17 (Tourism WA, 2017), unique natural and cultural resources offer an interesting backdrop to explore the food and wine tourism industry further. Due to its thriving agriculture and fishing industries, the state is well positioned to promote a plethora of food and wine products to potential tourists. Tourism Western Australia, the state's key DMO, considers Chinese travellers as a high-potential market segment and suggests nurturing and maintaining this lucrative market. Recent statistics highlight that the Chinese inbound tourism market is the fastest growing and second largest market segment for WA; the state's food and wine products have been identified as one of the key motivators that attract them to Australia. The fact that China is in geographical proximity to WA, that Chinese tourists spend a sizable amount of money on food and wine and value the variety of high quality edible food offered in WA, serves as a major justification for focusing efforts to cater and tailor products for the China outbound market (see Chapters 2, 4, 6, 7 and 8).

Chinese tourists do consider Western Australia as a relatively prestigious travel destination. In this regard, high quality service is necessary so as to align the various food and wine offerings to this existing image. Such expectations for prestige and high quality service have been found to result in conspicuous consumption which needs to be better understood by WA food and wine service providers. This prestige perception and gastronomic identity of the state is currently reflected in the 'Restaurant Australia' program, which is a marketing campaign run by Tourism Australia, the Australian national tourism authority. However, research suggests that WA hotels are still not currently geared for mainland Chinese guests' palates (see Chapter 9), particularly given the diversity of Chinese cuisines and high quality expectations. This may be a key hindrance in attracting more tourists in the upcoming years.

Chinese tourists in New Zealand

New Zealand has also been highlighted as an attractive destination for Chinese travellers, with the number of Chinese tourists to New Zealand increasing significantly over the last decade. Wineries have featured as a main attraction for these tourists, resulting in the growth in the country's wine export to China. Exports of New Zealand wine to China resulted in NZ$27.6m revenue at the end of June 2016 and market experts believe that there is still scope to expand the market further (New Zealand Winegrowers, 2017). It has been underscored that it is important to understand the experiential aspects of the wineries in order to further promote and advance the sector. Chinese tourists reportedly enjoy the landscape and scenery, a clean and unpolluted environment, and meeting friendly local people. Unfortunately, despite the apparent need for better understanding Chinese tourists, there has been limited academic and industry research on their motivations to visit New Zealand. Thus, greater attention needs to be paid to the winescape in which these tourists will experience the food and wine products of New Zealand and how this impacts on their subsequent attitudes and intentions toward the destination (see Chapter 15).

Moreover, research in this book (see Chapters 6, 7 and 10) suggests that product development, centred particularly on pairing food and wine in a tourism experience, will appeal to Chinese visitors, an area which deserves greater attention from both academic and industry researchers. The enjoyment of such culinary combinations does not only appeal to the more knowledgeable connoisseur market segments who have more time and money to spend on such experiences, but also to other groups of tourists for whom wine is still not a regular part of their lifestyles.

It was also highlighted Chapters 6, 7 and 9 that there exists a language barrier that has presented as an issue for many Chinese tourists. While the Australian service providers have begun to take steps to mitigate this communication gap, currently limited interpretation and translation services are offered at most wineries in New Zealand. This, coupled with the lack of information on the wineries and wines on the websites, greatly detract from the potential memorable experience for Chinese tourists. It is crucial that the New Zealand tourism industry tackles this issue in order to ensure that the various services offered take into account these unique needs of Chinese tourists.

'Home away from home' or 'Romans in Rome'

The studies in this book (see Chapters 2, 4, 6, 7 and 8) have demonstrated that most of the tourists in WA are interested in culinary tourism experiences beyond the basic necessities of eating and drinking. Tourists dedicated to food and wine, also known as 'gourmet travellers', account for almost 400,000 visitors to Western Australia each year (Tourism WA, 2017). These tourists consider the culinary experience a vital factor in choosing the travel destination. However, for Chinese travellers, food plays varying roles in their travel agenda.

The importance of food for Chinese tourists needs to be first contextualised to better understand their food preference while travelling. Food in China has traditionally been connected to ethnicity, place of birth, social status as well as cultural change and all calendar and family events. It is a key component of rituals and is offered to ancestors and gods. Therefore, the importance and relevance of food for the Chinese is multifaceted (see Chapters 2, 4, 5, 6, 7, 8 and 9).

As previously mentioned in this book, Chinese tourists generally want to have their own food while travelling but, at the same time, want to experience local foods too. It has emerged that tourists' preference for food is influenced by a number of factors, namely, physical, cultural, interpersonal, and status, prestige and past experiences. It is also noteworthy that trying local cuisine has become proof of the Chinese traveller's unique experiences abroad and is perceived to be 'fashionable and desirable'. Yet, while the desire to try local food is certainly prominent amongst younger Chinese travellers, the need to have Chinese food still remains. Tasting local food may satisfy the experiential needs of the visitors, but might not be enough to satiate their physiological needs so they tend to revert back to Chinese flavours to mitigate the unfamiliar taste of local food (see Chapters 4, 5, 7, 8, 9 and 13).

Research suggests that there are three types of Chinese tourists: observers, browsers and participators (see Chapter 6). On the one end of the spectrum, 'observers' have been found to constantly crave Chinese food while travelling; a predisposition resulting from a strong attachment to their original Chinese food culture. On the other end of the spectrum are 'participators' who have great gastronomic curiosity in local foods and believe the best way to appreciate local culture is through the participation in local dining experiences. Sitting in the middle of the continuum are the 'browsers' who are not fastidious about food selection while travelling and desire a balance between their own foods and local foods; for them, food is not a major evaluative criterion when assessing their satisfaction with their trip. Therefore, it is imperative that the service providers address this by providing 'Chinese-friendly' food offerings. This will help create memorable food experiences that will, in turn, enhance the attractiveness, satisfaction and revisit intention towards the destination (see Chapters 6 and 7).

Nevertheless, local food connects the tourists to the culture and heritage of the destination by symbolising local placeness and identity (see Chapter 16). Thus, local food remains an asset to be integrated further into the local tourism product development. However, the question arises as to whether international hotels are capable of providing the required foods and drinks to the Chinese tourist. The issue is not only relevant to the food ingredients, but also communicating the offerings in Chinese language. Research suggests that hoteliers should adapt their 'Asian-inspired' room service menus once they better understand their Chinese guest demographics and are able to gauge the culinary expectations of the sub-segments of this market (see Chapter 9).

Developing food tourism can contribute to job creation, economic development and sustainable tourism development. Destination marketing practitioners could, for example, develop dining activities that involve tourists partaking in cooking and tasting lessons, for them to gain greater insight and experiences into how food is grown and prepared as well as to explain to tourists the cultural significance of these local foods.

Further, in order to attract new tourists, food and wine service providers need to gain better insight into the changes in the Chinese tourism trends over the years. Chinese tourists spend a substantial amount of time searching for travel information over the Internet. Online forums, social networks and travel blogs/films have been cited as the key source for their information about a given destination. Chinese travellers consider online forums as reliable, realistic and recent platforms for their information. More traditional printed travel books, magazines and guides also hold an appeal for many enthusiastic Chinese travellers. However, with online group influence and peer pressure clearly impacting in particular younger 'face-conscious' tourists, the importance of online resources needs to be explored and understood further. Moreover, destination marketers should not neglect the prevalence of mobile phone applications in Chinese consumer culture. Applications in Chinese that engage, inspire and motivate the target tourist segments could be explored further to examine how this may aid and appeal to the Chinese tourist.

Dining, drinking and destination

Food and wine consumption has been acknowledged as a complex behaviour with diverse cultural, social, psychological and sensory acceptance factors affecting it. These factors play a critical role in tourists' preferences and choices for food and wine. Studies addressing this have identified three major factors that influence the decision-making process. First is the tourists' individual personality, which is shaped by socio-cultural, psychological and physiological factors. Second is the quality and variation of the food and wine being served; sensory attributes such as flavour, aroma, texture and appearance of the food and wine are important determinants of choice. The symbolic value of food and wine, which encompasses authenticity, prestige, culture, novelty and variety, plays a crucial role in the final choice. Finally, the cultural, social and economic context in which food and wine is consumed is critical as well. These three factors should not be considered in isolation, but rather, be approached holistically, as suggested in numerous studies in this book (see Chapters 4, 5, 7, 8, 9 and 13). Therefore, service providers need to ensure congruence among the tourist's personality, the food and wine itself as well as the consumption environment in order to enhance satisfaction and craft a competitive advantage.

'Wine in a new world'

Australia and New Zealand are two key destinations for wine-loving travellers and Chinese tourists represent a potentially large opportunity for wine producers and exporters. Wine has been considered as a means of building an economic future with China. However, there are several major challenges in entering and consolidating their presence in the vastly complex Chinese market. Given the substantial market opportunity for wine in China, overcoming barriers and risks is imperative to capitalise on the economic prospect that is being presented. The recently signed China Australia Free Trade Agreement (ChAFTA) is expected to reduce the barriers in upcoming years. Furthermore, the Western Australia-China Agribusiness Cooperation Conference 2014 enhanced relationships across the agriculture and food industries and brought together Chinese and Western Australian business and government leaders to facilitate business matching and to identify key trade and investment opportunities along the supply chain. Research suggests the positioning of Australian wine as 'fun', 'enjoyable' and 'healthy' would facilitate the expansion of Australian wine to China (see Chapters 7 and 10). Furthermore, wine producers might consider localised palate preferences in what they bring to the market in China; in other words, a 'customised' wine experience. Past empirical studies indicate that wine and food motives are still relatively contained when it comes to major pull factors for Chinese tourists. Recommendations have also been given into customised packaging and collaboration amongst smaller wine producers. International awards and success stories need to be packaged within the brand to develop a sound positioning strategy for a winery and wine region. Wine tourists noted the importance of complementary products in forming their attitude toward the winescape (see Chapter 14). It has been evident that wine tourists expect to participate in wine tasting as well as engaging in entertainment, visiting galleries as well as sampling of local produce. Researchers also suggest that the wine industry should recognise the value of co-investment that will allow producers to continue doing what they do best while the investors themselves secure the markets for quality wines. From an Italian perspective, studies in the area highlight four critical factors, namely the value of the wine tourism experience, the belief of having resources to afford the experience, the feeling of being familiar with the specific wine and tourism context, and the expected effects on the peer group, to interpret the limited but increasing interest into current wine and food offers in Italy among Chinese visitors (see Chapter 13).

Extant tourism literature suggests wine consumer segments based on the intensity of consumers' interest and knowledge regarding wine. While the groups named as wine lovers, wine connoisseurs, wine novices, curious tourists and wine interested have been widely used, the question remains unanswered whether these segments are applicable to Chinese tourists. It is necessary to examine how these groups behave differently while in their consumption decision. In addition, research might investigate whether the consumers share the experience with other consumers or if they bring the wine back home as souvenirs for their friends and families. Considering the prevalent competitiveness and complexity of consumer psychographics, it is imperative that wine marketers utilise the typology of the wine consumption situation. The consumers' wine knowledge, familiarity and normative influences need to be capitalised for better understanding the market and maintaining a sustainable business.

What next?

The studies published in this book will intrigue researchers to investigate further the link between food- and wine-related activities and experiences and the growing Chinese tourist market. It has been evident that the contextualisation of food and wine consumption is important in understanding Chinese travellers. One prominent issue is that many Chinese tourists still lack in-depth knowledge about a destination's food culture, which makes it difficult for them to appreciate the authenticity of its food and wine offerings. Therefore, destination marketers need to devise education programs that will help increase the domain-specific cultural capital of these tourists. In particular, destination marketers need to provide user-friendly, attractive and guided information about the destination's gastronomy to assist Chinese tourists to better understand the food culture. Social network and mobile application integration pose as opportunities to cater to the increasing number of younger Chinese tourists who prefer technologies for researching, booking and managing their travel plans.

The complexity introduced by the diverse psychographics of Chinese tourists has been acknowledged by various researchers. It is therefore imperative to identify the diverse Chinese tourist segments that have a particular interest or passion in the food and wine tourism industry. At the same time, it would be worth examining how these tourists segments can be targeted and serviced more effectively. Studies published in this book (see Chapters 2, 4, 6, 8, 10 and 12) call for strengthening strategic outlook, cross-sectoral understanding, organisational set-ups and umbrella branding approaches to further improve consistency as well as comprehensiveness in food and wine tourism value propositions. In fact, a wider conception of tourism destination management has been suggested through the studies.

The social implications and multifaceted cultural embeddedness of consumptive behaviour is a necessary focus for future studies. A mutual understanding amongst government tourism boards, tour operators and other industry stakeholders would be essential in crafting strategy in this area. The New Zealand Tourism Board has demonstrated a successful case of using social media for tourism promotion. Can tourism boards from other countries replicate the model? What kinds of cultural and communication considerations need to be factored into marketing communications to Chinese travellers? Will it be effective to leverage Chinese popular culture (e.g. TV dramas, music videos) for tourism promotion purposes? These are some of the questions that emerged across the chapters of this book and will undoubtedly feature in many studies to come.

References

New Zealand Winegrowers (2017), *Annual Report*. Retrieved from: https://www.nzwine.com/media/6600/nzw-annual-report-2017.pdf

Tourism WA (2017). *Visitor Statistics*. Retrieved from: https://www.tourism.wa.gov.au/Research-Reports/Latest_Visitor_Facts_and_Figures/Pages/Visitor-Statistics.aspx

Index

Note: References in *italics* are to figures, those in **bold** to tables.